WILLIAM F. MAAG LIBRARY
YOUNGSTOWN STATE UNIVERSITY

SEMICONDUCTORS AND SEMIMETALS
VOLUME 22
Lightwave Communications Technology
Part C
Semiconductor Injection Lasers, II
Light-Emitting Diodes

Semiconductors and Semimetals

A Treatise

Edited by R. K. WILLARDSON
CRYSCON TECHNOLOGIES, INC.
PHOENIX, ARIZONA

ALBERT C. BEER
BATTELLE COLUMBUS LABORATORIES
COLUMBUS, OHIO

SEMICONDUCTORS AND SEMIMETALS

VOLUME 22

Lightwave Communications Technology

Volume Editor

W. T. TSANG

AT&T BELL LABORATORIES
HOLMDEL, NEW JERSEY

Part C

Semiconductor Injection Lasers, II
Light-Emitting Diodes

1985

ACADEMIC PRESS, INC.
(Harcourt Brace Jovanovich, Publishers)

Orlando   San Diego   New York   London
Toronto   Montreal   Sydney   Tokyo

COPYRIGHT © 1985, BY BELL TELEPHONE LABORATORIES, INCORPORATED.
ALL RIGHTS RESERVED.
NO PART OF THIS PUBLICATION MAY BE REPRODUCED OR
TRANSMITTED IN ANY FORM OR BY ANY MEANS, ELECTRONIC
OR MECHANICAL, INCLUDING PHOTOCOPY, RECORDING, OR
ANY INFORMATION STORAGE AND RETRIEVAL SYSTEM, WITHOUT
PERMISSION IN WRITING FROM THE PUBLISHER.

ACADEMIC PRESS, INC.
Orlando, Florida 32887

United Kingdom Edition published by
ACADEMIC PRESS INC. (LONDON) LTD.
24–28 Oval Road, London NW1 7DX

LIBRARY OF CONGRESS CATALOG CARD NUMBER: 65-26048

ISBN 0–12–752152–6

PRINTED IN THE UNITED STATES OF AMERICA

85 86 87 88   9 8 7 6 5 4 3 2 1

Contents

LIST OF CONTRIBUTORS ix
TREATISE FOREWORD xi
FOREWORD xiii
PREFACE xvii

Chapter 1 Review of InGaAsP/InP Laser Structures and Comparison of Their Performance
R. J. Nelson and N. K. Dutta

I. Introduction 1
II. Heterojunction Laser Characteristics 4
III. Temperature Dependence of Threshold Current . . . 12
IV. Gain-Guided Laser Structures 25
V. Index-Guided Laser Structures 33
   References 57

Chapter 2 Mode-Stabilized Semiconductor Lasers for 0.7–0.8- and 1.1–1.6-μm Regions
N. Chinone and M. Nakamura

I. Introduction 61
II. Transverse-Mode Control 62
III. Longitudinal-Mode Control 72
IV. Conclusion 88
   References 89

Chapter 3 Semiconductor Lasers with Wavelengths Exceeding 2 μm
Yoshiji Horikoshi

I. Introduction 93
II. Semiconductor Materials 94
III. Crystal Growth and Laser Fabrication 103
IV. Laser-Diode Characteristics 120
   References 147

Chapter 4 The Functional Reliability of Semiconductor Lasers as Optical Transmitters

B. A. Dean and M. Dixon

I.	Introduction	153
II.	Experimental Conditions	156
III.	Life-Test Results for a 45-Mbit/sec, AlGaAs System	159
IV.	Functional Life Testing for Improved Design	170
V.	Functional Reliability of an AlGaAs Laser Transmitter for Space Communication	184
VI.	Summary	187
	References	189

Chapter 5 Light-Emitting-Diode Device Design

R. H. Saul, Tien Pei Lee, and C. A. Burrus

I.	Introduction	193
II.	LED Structures	194
III.	LED Power and Modulation Bandwidth	198
IV.	LED Device Characteristics	211
V.	Coupling of LEDs to Optical Fibers	221
VI.	Dual-Wavelength LEDs	232
VII.	LEDs Made by MBE and VPE	233
VIII.	Conclusions	234
	References	235

Chapter 6 Light-Emitting-Diode Reliability

C. L. Zipfel

I.	Introduction	239
II.	LED Life Testing	242
III.	GaAs/GaAlAs LEDs: Results of Accelerated Aging	243
IV.	Rapid Degradation Mechanisms in GaAs/GaAlAs LEDs	249
V.	Gradual Degradation in GaAs/GaAlAs LEDs	260
VI.	InP/InGaAsP LEDs: Results of Accelerated Aging	264
VII.	Degradation Mechanisms in InP/InGaAsP LEDs	268
VIII.	Conclusions	275
	References	276

Chapter 7 Light-Emitting-Diode-Based Multimode Lightwave Systems

Tien Pei Lee and Tingye Li

I.	Introduction	281
II.	Multimode-Fiber Properties	282
III.	System Considerations	286
IV.	System Experiments and Field Trials	292
V.	Conclusions	296
	References	297

Chapter 8 Semiconductor Laser Noise: Mode Partition Noise

Kinichiro Ogawa

I.	Introduction	299
II.	A Simple Model of Mode Partition Noise	300
III.	Analysis of Mode Partition Noise	302
IV.	Mode Partition Coefficient k Value	311
V.	Theoretical k Value with and without Interaction between Lasing Modes	315
VI.	Statistical Measurement of Mode Partition Noise	322
VII.	Effects of Optical Injection Locking on Mode Partition Noise	324
	Appendix I	328
	Appendix II	329
	References	330

INDEX	331
CONTENTS OF VOLUME 22	335
CONTENTS OF PREVIOUS VOLUMES	337

List of Contributors

Numbers in parentheses indicate the pages on which the authors' contributions begin.

C. A. BURRUS, *AT&T Bell Laboratories, Crawford Hill Laboratory, Holmdel, New Jersey 07733* (193)

N. CHINONE, *Central Research Laboratory, Hitachi Ltd., Kokubunji, Tokyo 185, Japan* (61)

B. A. DEAN, *AT&T Bell Laboratories, Allentown, Pennsylvania 18103* (153)

M. DIXON, *AT&T Bell Laboratories, Allentown, Pennsylvania 18103* (153)

N. K. DUTTA, *AT&T Bell Laboratories, Murray Hill, New Jersey 07974* (1)

YOSHIJI HORIKOSHI, *Musashino Electrical Communication Laboratory, Nippon Telegraph and Telephone Public Corporation, Musashino-shi, Tokyo 180, Japan* (93)

TIEN PEI LEE,* *AT&T Bell Laboratories, Crawford Hill Laboratory, Holmdel, New Jersey 07733* (193, 281)

TINGYE LI, *AT&T Bell Laboratories, Crawford Hill Laboratory, Holmdel, New Jersey 07733* (281)

M. NAKAMURA, *Central Research Laboratory, Hitachi Ltd., Kokubunji, Tokyo 185, Japan* (61)

R. J. NELSON,† *AT&T Bell Laboratories, Murray Hill, New Jersey 07974* (1)

KINICHIRO OGAWA, *AT&T Bell Laboratories, Holmdel, New Jersey 07733* (299)

R. H. SAUL, *AT&T Bell Laboratories, Murray Hill, New Jersey 07974* (193)

C. L. ZIPFEL, *AT&T Bell Laboratories, Murray Hill, New Jersey 07974* (239)

---

* Present address: Bell Communications Research, Murray Hill, New Jersey 07974.
† Present address: Lytel, Inc., Somerville, New Jersey 08876.

Treatise Foreword

This treatise continues the format established in the books of Volume 21, in which a subject of outstanding interest and one possessing ever-increasing practical applications is treated in a multivolume work organized by a guest editor of international repute. The present series, which consists of five volumes (designated as Volume 22, Parts A through E) deals with an area that is experiencing a technological revolution and is destined to have a far-reaching impact in the near future—not only in the communications and data-processing fields, but also in numerous ancillary areas involving, for example, control systems, interconnects that maintain individual system isolation, and freedom from noise emanating from stray electromagnetic fields.

That the excitement engendered by the rapid pace of developments in lightwave communications technology is universal is borne out by the large number of contributions to this series by authors from abroad. It is indeed fortunate that W. T. Tsang, who is most highly knowledgeable in this field and has made so many personal contributions, has been able to take the time to put together a work of the extent and excellence of the present series. The treatise editors are also greatly indebted to Dr. Patel and the other colleagues of Dr. Tsang at AT&T Bell Laboratories, without whose understanding and encouragement this group of books would not have been possible.

R. K. WILLARDSON
ALBERT C. BEER

Foreword

Lightwave technology is breaking down barriers in communications in a manner similar to the way barriers in computing came down thanks to semiconductor integrated circuit technology. Increased packing densities of components on integrated circuit chips made possible a phenomenal amount of information processing capacity at continually decreasing cost. The impact of lightwave technology on communications is quite similar. We are reaching a point where an exponentially increasing transmission capacity is resulting in our capability to provide vast amounts of information to the most distant reaches of the world at a nominal cost. This revolution in information transmission capacity is engendered by the rapid developments in lightwave communications.

Along with the very large transmission capacity predicted in the late fifties when the laser was invented have come a number of additional advantages. Of these advantages, I single out those arising from the nonmetallic nature of the transmission medium. These fall under the broad category of what may be called an immunity from unanticipated electromagnetic coupling. The following rank as very important benefits: freedom from electromagnetic interference, absence of ground loops, relative freedom from eavesdropping (i.e., secure links), and potential for resistance to the electromagnetic pulse problems that plague many conventional information transmission systems utilizing metallic conductors as well as satellite and radio technology. Each of these benefits arises naturally from the medium through which the light is propagated and is, therefore, paced by the progress in optical fibers.

However, what we take for granted today was not so obvious for many decades following the first practicable use of light for communications by Alexander Graham Bell in 1880. The use of heliographs in ancient Greece, Egypt, and elsewhere and the smoke signaling by various American Indian tribes notwithstanding, Bell's experiments on the use of sunlight for transmitting spoken sounds over a distance of a few hundred meters was undoubtedly the first step toward practical optical communications, since it represents a quantum jump in the increase in the bandwidth used for information transmission. The excitement he felt is keenly expressed in his words:

> I have heard articulate speech produced by sunlight. I have heard a ray of sun laugh and cough and sing. I have been able to hear a shadow, and I have even perceived by ear the passing of a cloud across the sun's disk.

The results of his experiments were presented at a meeting of the American Association of Scientific Persons in Boston, Massachusetts. But the generally favorable reaction to Bell's photophone in the popular press was tempered with some skepticism. The following paragraph is taken from an article that appeared on the editorial pages of the August 30, 1880, issue of the *New York Times,* which reported on Bell's results.

> What the telephone accomplishes with the help of a wire the photophone accomplishes with the aid of a sunbeam. Professor Bell described his invention with so much clearness that every member of the American Association must have understood it. The ordinary man, however, may find a little difficulty in comprehending how sunbeams are to be used. Does Professor Bell intend to connect Boston and Cambridge, for example, with a line of sunbeams hung on telegraph posts, and, if so, of what diameter are the sunbeams to be, and how is he to obtain them of the required size? . . .

Bell reported optical communication through free atmosphere, but the reported, unintentionally, seemed to have foreseen the time when optical-fiber cables would be strung from pole to pole or buried underground.

A unique set of circumstances and a host of advances resulting from extensive interdisciplinary efforts have fueled the revolution in lightwave communications and the acceptance of this new technology. The tremendous progress in lightwave communications is a result of necessity as well as of the response of the scientists and engineers to the formidable challenges. The large bandwidth possible with lightwave communications is a direct result of the very high carrier frequency of electromagnetic radiation in the optical region. This advantage was recognized at least as early as the late fifties and early sixties. Yet almost fifteen years elapsed before lightwave communications technology became economically viable. Two primary components of the communications technology paced this development: the light source and the transmission medium. A third component, the receiver, is also important but was not the pacing one in the early years of development of lightwave systems.

The laser was invented in 1958, and within a very few years laser action was demonstrated in a variety of solids, liquids, and gases. The semiconductor injection laser, the workhorse of contemporary optical communications, was invented in 1962, but its evolution to a practical transmitter in a lightwave system took another eight years. In 1970 Hayashi and Panish (and, independently, Alferov in the Soviet Union) demonstrated the first continuous wave (cw) room-temperature-operated semiconductor laser. The potentials of small size, high reliability, low cost, long life, and ability to modulate the light output of the semiconductor laser at very high rates by merely modulating the drive current were recognized early in the game. With the demonstration of the cw room-temperature operation the race was on to exploit all these advantages.

Again, while laser light propagation through the atmosphere was considered in the mid-sixties, everyone recognized the limitations due to unpredictable and adverse weather conditions. To avoid these limitations, propagation in large hollow pipes was also studied, but again practical difficulties arose. It was the development of optical fiber technology to reduce transmission losses to acceptable levels that has led to the practical implementation of lightwave communications. While light transmission through very small-diameter fibers was demonstrated in the early fifties, it was a combination of theoretical advances by Kao and inventive experimentation by Maurer in the late sixties that resulted in the realization of 20-dB/km fiber. Additional fuel was thus provided to speed up the revolution.

Today, new records are continually being set for the longest and the highest-capacity lightwave communications system. Yet these records are thousands of times below the fundamental bandwidth limits set by the carrier frequency of optical radiation on the rate of information transmission. Furthermore, from very fundamental considerations of light-transmitting materials, there is no reason why the currently achieved lowest losses for optical fibers, in the region of 0.1 dB/km at 1.55 μm, will not be considered too high in the future. It is not inconceivable that fiber losses as low as 10^{-4} dB/km may someday be achieved. It does not take a great deal of imagination to realize the impact of such development.

This is where we are. What future developments will pace the exploitation of lightwave communications? The five-volume minitreatise on lightwave communications technology aims both to recapitulate the existing developments and to highlight new science that will form the underpinnings of the next generation of technology. We know a lot about how to transmit information using optical means, but we know less than enough about how to switch, manipulate, and process information in the optical domain. To take full advantage of all the promise of lightwave communications, we have to be able to push the optical bits through the entire communications system with the electronic-to-optical and optical-to-electronic interfaces only at the two ends of the lightwave communications system. To achieve this, we will need practical and efficient ways of switching, storing, and processing optical information. This is a must before lightwave communications is able to touch every single subscriber of the present telephone and other forms of communications technology.

We have come a long way since Bell's experiments of 1880, but there is a lot more distance ahead. That is what the field of lightwave communications is all about — more challenges, more excitement, more fun for those who are the actors, and a greater opportunity for society to derive maximum benefit from the almost exponentially increasing information capacity of lightwave systems.

AT&T Bell Laboratories C. K. N. PATEL
October 9, 1984

Preface

When American Indians transmitted messages by means of smoke signals they were exploiting concepts at the heart of modern optical communications. The intermittent puffs of smoke they released from a mountaintop were a digital signal; indeed, the signal was binary, since it encoded information in the form of the presence or absence of puffs of smoke. Light was the information carrier; air was the transmission medium; the human eye was the photodetector. The duplication of the signal at a second mountaintop for the transmission to a third served as signal reamplification, as in today's electronic repeater. Man had devised and used optical communications even long before the historic event involving the "photophone" used over a hundred years ago (1880) by Alexander Graham Bell to transmit a telephone signal over a distance of two hundred meters by using a beam of sunlight as the carrier. It was not until 1977, however, that the first commercial optical communications system was installed. Involved in the perfection of this new technology are the invention and development of a reliable and compact near-infrared optical source that can be modulated by the information-bearing signal, a low-loss transmission medium that is capable of guiding the optical energy along it, and a sensitive photodetector that can recover the modulation error free to re-treat the information transmitted.

The invention and experimental demonstration of a laser in 1958 immediately brought about new interest and extensive research in optical communications. However, the prospect of practical optical communications brightened only when three major technologies matured. The first technology involved the demonstration of laser operation by injecting current through a semiconductor device in 1962 and the achievement of continuous operation for over one million hours in 1977. The second technology involved the attainment of a 20-dB/km doped silica fiber in 1970, the realization that pure silica has the lowest optical loss of any likely medium, the discovery in 1973 that suitably heat-treated, boron-doped silica could have a refractive index less than that of pure silica, and the recent achievement of an ultralow loss of 0.157 dB/km with Ge-doped silica-based fibers. The third technology is the development of low-noise photodetectors in the 1970s, which made possible ultrahigh-sensitivity photoreceivers. It is the simultaneous achievement of reliable semiconductor current-injection lasers, low loss in optical fibers, and low-noise photodetectors that thrusts lightwave communications technology into reality and overtakes the conventional transmission systems employing electrical means.

Since optical-fiber communications encompasses simultaneously several other technologies, which include the systems area of telecommunications and glass and semiconductor optoelectronics technologies, a tremendous amount of research has been conducted during the past two decades. We shall attempt to summarize the accumulated knowledge in the present series of volumes of "Semiconductors and Semimetals" subtitled "Lightwave Communications Technology." The series consists of seven volumes. Because of the subject matter, the first five volumes concern semiconductor optoelectronics technology and, therefore, will be covered in "Semiconductors and Semimetals." The last two volumes, one on optical-fiber technology and the other on transmission systems, will be covered in the treatise "Optical Fiber Communications," edited by Tingye Li and W. T. Tsang.

Volume 22, Part A, devoted entirely to semiconductor growth technology, deals in detail with the various epitaxial growth techniques and materials defect characterization of III–V compound semiconductors. These include liquid-phase epitaxy, molecular beam epitaxy, atmospheric-pressure and low-pressure metallo-organic chemical vapor deposition, and halide and chloride transport vapor-phase deposition. Each technique is covered in a separate chapter. A chapter is also devoted to the treatment of materials defects in semiconductors.

In Volume 22, Parts B and C, the preparation, characterization, properties, and applications of semiconductor current-injection lasers and light-emitting diodes covering the spectral range of 0.7 to 1.6 μm and above 2 μm are reviewed. Specifically, Volume 22, Part B, contains chapters on dynamic properties and subpicosecond-pulse mode locking, high-speed current modulation, and spectral properties of semiconductor lasers as well as dynamic single-frequency distributed feedback lasers and cleaved-coupled-cavity semiconductor lasers. Volume 22, Part C, consists of chapters on semiconductor lasers and light-emitting diodes. The chapters on semiconductor lasers consist of a review of laser structures and a comparison of their performances, schemes of transverse mode stabilization, functional reliability of semiconductor lasers as optical transmitters, and semiconductor lasers with wavelengths above 2 μm. The treatment of light-emitting diodes is covered in three separate chapters on light-emitting diode device design, its reliability, and its use as an optical source in lightwave transmission systems. Volume 22, Parts B and C, should be considered as an integral treatment of semiconductor lasers and light-emitting diodes rather than as two separate volumes.

Volume 22, Part D, is devoted exclusively to photodetector technology. It includes detailed treatments of the physics of avalanche photodiodes; avalanche photodiodes based on silicon, germanium, and III–V compound semiconductors; and phototransistors. A separate chapter discusses the sensitivity of avalanche photodetector receivers for high-bit-rate long-wavelength optical communications systems.

Volume 22, Part E, is devoted to the area of integrated optoelectronics and other emerging applications of semiconductor devices. Detailed treatments of the principles and characteristics of integrable active and passive optical devices and the performance of integrated electronic and photonic devices are given. A chapter on the application of semiconductor lasers as optical amplifiers in lightwave transmission systems is also included as an example of the important new applications of semiconductor lasers.

Because of the subject matter (although important to the overall treatment of the entire lightwave communications technology), the last two volumes will appear in a different treatise. The volume on optical-fiber technology contains chapters on the design and fabrication, optical characterization, and nonlinear optics in optical fibers. The final volume is on lightwave transmission systems. This includes chapters on lightwave systems fundamentals, optical transmitter and receiver design theories, and frequency and phase modulation of semiconductor lasers in coherent optical transmission systems.

Thus, the series of seven volumes treats the entire technology in depth. Every author is from an organization that is engaged in the research and development of lightwave communications technology and systems.

As a guest editor, I am indebted to R. K. Willardson and A. C. Beer for having given me this valuable opportunity to put such an important and exploding technology in "Semiconductors and Semimetals." I am also indebted to all the contributors and their employers who have made this series possible. I wish to express my appreciation to AT&T Bell Laboratories for providing the facilities and environment necessary for such an endeavor and to C. K. N. Patel for preparing the Foreword.

CHAPTER 1

Review of InGaAsP/InP Laser Structures and Comparison of Their Performance

R. J. Nelson* and N. K. Dutta

AT&T BELL LABORATORIES
MURRAY HILL, NEW JERSEY

I.	INTRODUCTION	1
	1. Materials for the 1.3–1.6-μm-Wavelength Region	2
	2. Band-Gap–Composition Relation for InGaAsP	3
II.	HETEROJUNCTION LASER CHARACTERISTICS	4
	3. Heterojunctions.	5
	4. Stimulated Emission and Gain	9
	5. Light–Current Characteristics	12
III.	TEMPERATURE DEPENDENCE OF THRESHOLD CURRENT ...	12
	6. Experimental Temperature Dependence of Threshold ..	14
	7. Auger Recombination: Theory	16
	8. Auger Recombination: Experiment	23
IV.	GAIN-GUIDED LASER STRUCTURES	25
	9. Current Spreading	26
	10. Gain Guiding	27
	11. Threshold Current	29
	12. Maximum Operating Temperature	30
	13. Nonlinearities in Light Output	31
	14. Modulation Characteristics.	32
V.	INDEX-GUIDED LASER STRUCTURES	33
	15. Waveguiding.	36
	16. Threshold Current	43
	17. Maximum Operating Temperature	51
	18. Spectral Characteristics	53
	19. Far Field: Fiber Coupling	53
	20. Modulation Characteristics.	55
	21. High-Output-Power Devices	56
	REFERENCES	57

I. Introduction

Numerous lightwave communications systems have been installed throughout the world, with most of the systems installed to date employing

---

* Present address: Lytel Inc., Somerville, New Jersey.

short-wavelength ($\lambda = 0.82-0.88$ μm) GaAlAs/GaAs lasers as light sources. However, longer-wavelength sources have been developed and are now being used in many new installations to take advantage of the lower loss and lower dispersion at $\lambda = 1.3-1.55$ μm in silica fibers. Using $\lambda = 1.3$-μm sources, repeater spacings of 65 km at bit rates as high as 432 Mbits/sec have been achieved in systems experiments (Cheung *et al.*, 1983). Wavelength-division-multiplexed operation employing buried heterostructure lasers emitting at 1.273 and 1.330 μm has also been demonstrated (Kaiser *et al.*, 1983). The use of single-frequency lasers emitting at the 1.55-μm fiber attenuation minimum has allowed the demonstration of a system operating at 420 Mbits/sec over a repeaterless span of 119 km of single-mode fiber (Tsang *et al.*, 1983). This represents a major advance in the past few years over the <10-km spacing and ~90-Mbit/sec systems in commercial use with GaAlAs lasers. These advances have been made possible by the rapid evolution of silica fibers and by lasers fabricated from the InGaAsP/InP-material system.

In this chapter, InGaAsP/InP laser structures and their performance characteristics are reviewed. No attempt is made to identify an optimum structure, since the choice of a laser structure for a given application is dependent not only on the requirements for that application but also on cost and/or manufacturing issues as well. Various InGaAsP laser structures are described, including both gain-guided and refractive-index-guided devices. The gain-guided laser that was extensively used at short wavelengths has proven to be much more difficult to optimize at longer wavelengths for fundamental reasons, as discussed later. Real-refractive-index-guided lasers, which were extensively developed in the GaAlAs system largely by Japanese companies in the mid-1970s, have become the predominant device structures at long wavelengths. These index-guided lasers, which require a more sophisticated growth and fabrication technology than gain-guided devices, and their operating characteristics are described in detail. Differences in the operating characteristics of short- and long-wavelength lasers are emphasized, including the higher temperature sensitivity of threshold than GaAlAs lasers.

1. Materials for the 1.3–1.6-μm-Wavelength Region

Several semiconductor compounds have exhibited efficient light emission in the near-infrared region (see Fig. 1). Specifically, II–VI-compounds materials have been used at wavelengths of >2 and <0.8 μm; III–V compounds are now used in the $\lambda = 0.8-1.65$-μm range. Although several III–V materials emit in the $\lambda = 1.3-1.55$-μm region, the InGaAsP-material system is the most highly developed.

1. InGaAsP/InP LASER STRUCTURES AND PERFORMANCE

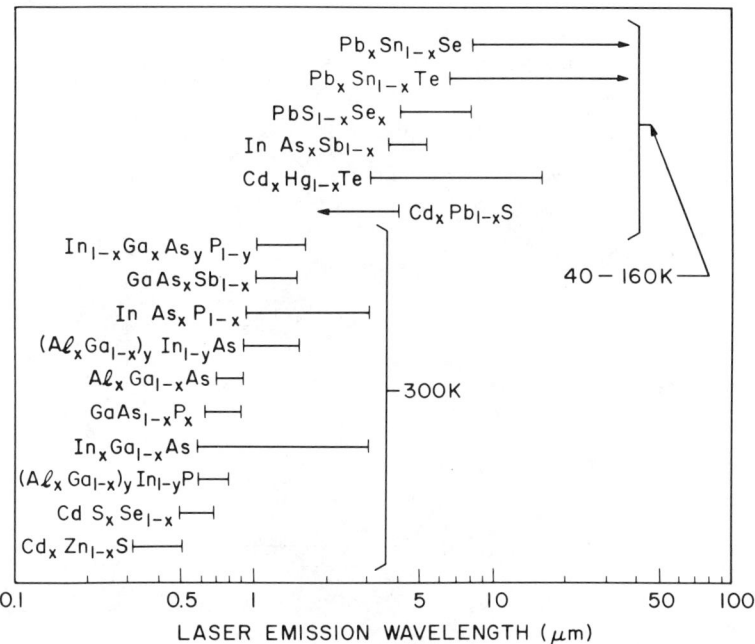

FIG. 1. Emission wavelengths of selected III–V and IV–VI compounds at 300 K. Parameters x and y refer to $In_{1-x}Ga_xAs_yP_{1-y}$. [From Casey and Panish (1978).]

2. BAND-GAP–COMPOSITION RELATION FOR InGaAsP

Figure 2 shows the band-gap–composition relationship for the InGaAsP-material system (Wright et al., 1977). The band gaps of the four binary compounds are shown at the corners of Fig. 2. Isolattice constant lines are dashed, and the heavy solid line represents the compositions that are lattice matched to InP substrates. Compositions of constant energy are represented by the solid lines. The band structure of InGaAsP is direct outside the shaded region. In this direct band-gap region, the minimum in the conduction band occurs at the same point in the Brillouin zone as the maximum in the valence band, so that radiative emission occurs directly as a result of a momentum-conserving electron–hole recombination event. In the shaded indirect band-gap region, the conduction-band minimum does not occur at the same point in the Brillouin zone as the valence-band maximum, so that a phonon must be involved in a recombination event between electrons and hole. The lifetime associated with this radiative process in the indirect region is several orders of magnitude longer than the lifetime associated with nonradiative recombination involving impurities. As a result, the internal radiative re-

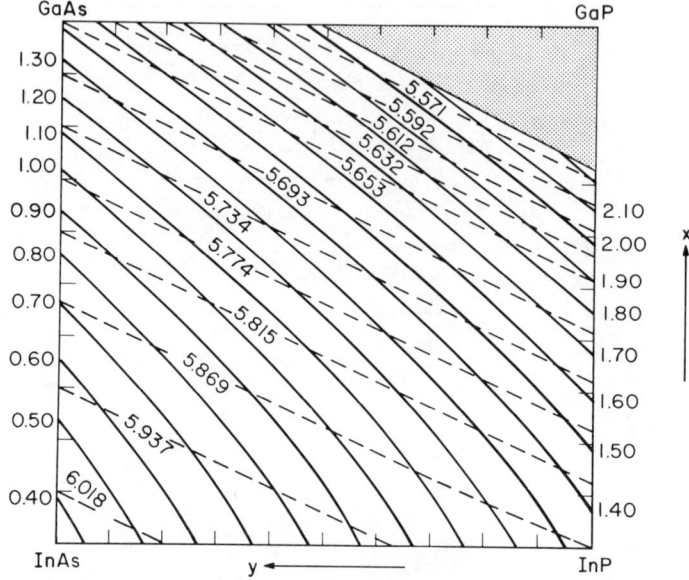

FIG. 2. Composition diagram for $In_{1-x}Ga_xAs_yP_{1-y}$ on which isolattice constant lines are shown by the dashed lines and iso-band-gap lines are shown by the solid lines. Numbers at ends of curves are photon energies in electron volts. Numbers on curves are wavelengths in micrometers. [From Wright et al. (1977). © North-Holland Physics Publishing, Amsterdam, 1977.]

combination efficiency of indirect material is often less than a few percent. In the direct band-gap region, by contrast, the radiative lifetime is of the same order of magnitude as that associated with nonradiative recombination through traps, so that the radiative efficiency can approach 100% (Nelson and Sobers, 1978b). The emission wavelength is related to the band-gap energy by $\lambda = 1.24$ μm eV$/E_g$. Compositions lattice matched to InP can be made to emit over the range 1.65–0.93 μm (Fig. 3). The solid line shows the fit to the data points, given by (Nahory et al., 1978),

$$E_g(y) = 1.35 - 0.72y + 0.12y^2 \text{ eV}. \tag{1}$$

The commonly used light-emitting region in 1.3-μm lasers is $In_{0.74}Ga_{0.26}As_{0.6}P_{0.4}$, whereas for 1.55-μm devices, it is $In_{0.6}Ga_{0.4}As_{0.9}P_{0.1}$.

II. Heterojunction Laser Characteristics

InGaAsP lasers typically contain two heterojunctions that form boundaries around the light-emitting active region to provide the proper refractive index step to form a dielectric waveguide and also to provide a potential barrier that confines the injected carriers to a small volume. The high carrier

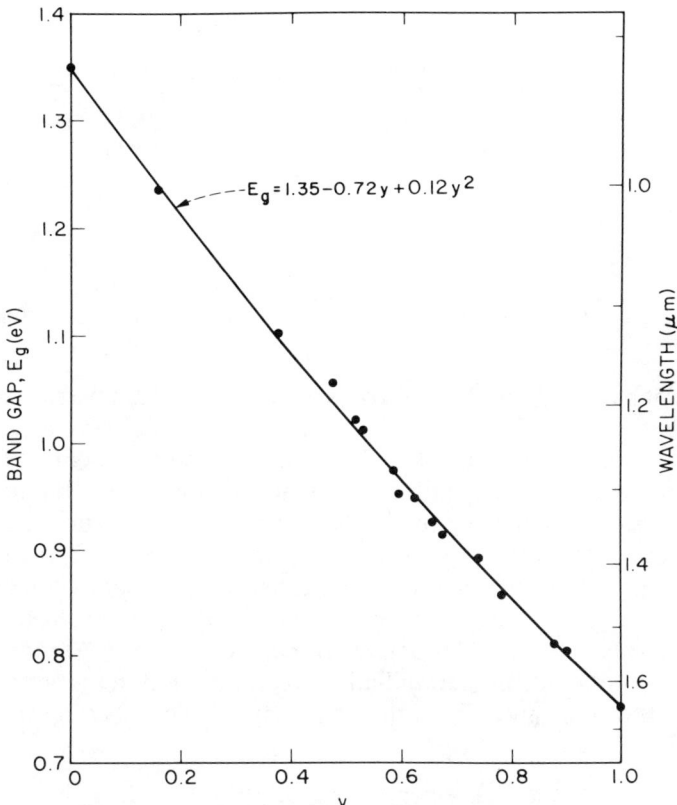

FIG. 3. Band-gap–composition diagram for $In_{1-x}Ga_xAs_yP_{1-y}$ lattice matched to InP. [From Nahory et al. (1978).]

concentrations needed for laser action can therefore be produced at low current densities.

3. Heterojunctions

Figure 4 shows a standard double-heterostructure laser that consists of two different types of heterojunctions (Casey and Panish, 1978). An anisotype junction is composed of two materials with different energy gaps and different doping types. The injection efficiency of majority carriers from the high-gap layer into the low-gap layer in this type of junction is enhanced by a factor of $\exp(\Delta E_g/kT)$ (where ΔE_g is the heterobarrier height) over that of a p–n homojunction with the same doping levels. The double heterojunction is formed when an isotype junction is placed within a diffusion length of the injecting junction. An isotype junction, which is formed by two materials of

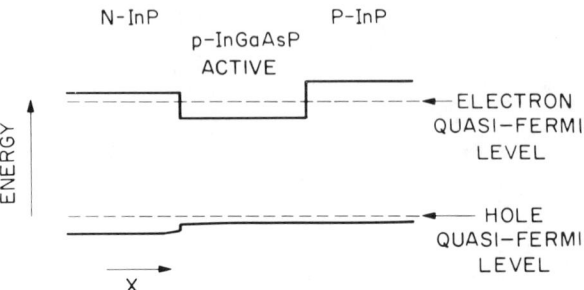

Fig. 4. Band-gap diagram for a forward-biased InP/InGaAsP double-heterostructure.

different band gap but the same doping type, provides a barrier that serves to localize the injected carriers and therefore produces a high carrier density for a given injection current. The effectiveness of the confinement provided by the isotype junction depends on the heterobarrier height, the minority-carrier diffusion length, and the drift field in the high-gap region, along with the thickness of the high-gap layer (Casey and Panish, 1978). Carriers in the high-energy tail of the electron or hole distribution that have energies larger than the heterobarrier height are not confined by the heterobarrier. These carriers may leak out of the active layer and constitute a current component that does not contribute to radiative recombination in the active layer. The carrier transport away from the active layer is described by the continuity equation.

$$dN/dt = g - r + D_n \nabla^2 N + \mu_n \nabla \cdot (N\mathbf{E}), \qquad (2)$$

where N is the local electron concentration, g the generation rate, D_n is the diffusion coefficient for electrons, μ_n is the electron mobility, $\mathbf{E}$ the electric field, $r = N/\tau_n$ is the recombination rate, and τ_n is the electron lifetime. If the field is sufficiently small, the transport of carriers through the cladding layers is by diffusion. If, for the p cladding layer, it is assumed that the minority carrier concentration goes to zero at the contact layer that is a distance w from the active layer, then the electron diffusion current is given by $qD_n \, dN/dx$ evaluated at $x = 0$:

$$i_n = -qD_n\{N/[L_n \tanh(w/L_n)]\}, \qquad (3)$$

where $L_n = (D_n \tau_n)^{1/2}$ is the electron diffusion length. The minority-carrier concentration N in Eq. (3) is the value in the cladding layer at the interface with the active layer where the continuity of the quasi-Fermi level across the interface is assumed. The hole diffusion current through the n cladding layer is often neglected because of the low hole mobility and the absence of an effective sink for holes within a few diffusion lengths of the active layer. A calculation of the diffusive leakage current (Dutta and Nelson, 1982b) at threshold over the heterobarrier as a function of active-layer composition is shown in Fig. 5 for a diffusion length of 5 and 1 μm in the p and n cladding layers, respectively, and doping levels of 3×10^{17} cm^{-3} in both the 2-μm-thick p-InP layer and the n cladding layer. The carrier density at threshold is approximated by $1.6 \times 10^{18}(T/300)^{3/2}$ cm^{-3}. The calculated diffusive leakage current at 300 K represents less than 2% of the observed threshold current density (~ 1000 A cm^{-2}) for wavelengths longer than 1.2 μm.

In the presence of a strong electric field in the cladding layers, which may

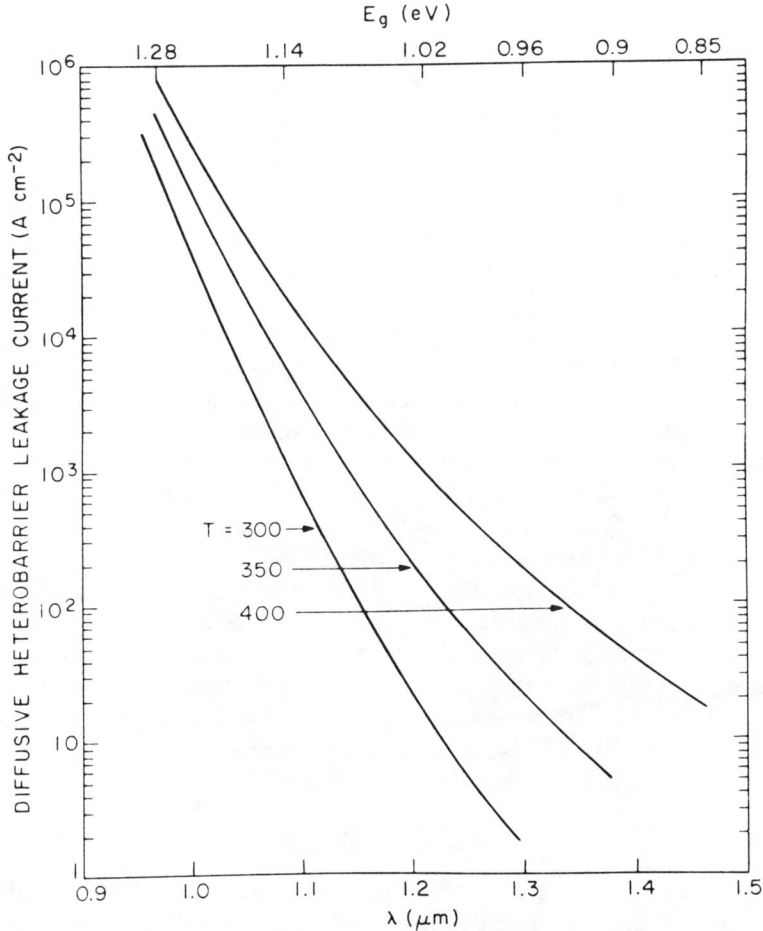

FIG. 5. Diffusive leakage current plotted as a function of active-layer emission wavelength and band gap for InGaAsP/InP lasers. [From Dutta and Nelson (1982c). © IEEE 1982.]

arise as a result of low doping in one of those layers, the leakage of carriers may be dominated by a drift component (Rode, 1974; Anthony and Schumaker, 1980). In that case, the total leakage current is approximated (when the minority-carrier current is less than the majority carrier current; i.e., $E =$ const by (Dutta, 1981),

$$i_n = qD_nN\frac{(z - m_1)e^{m_2w} + (m_2 - z)e^{m_1w}}{e^{m_2w} - e^{m_1w}},\qquad(4)$$

where $z = qE/kT$ and

$$m_{2,1} = \tfrac{1}{2}z \pm (1/L_n^2 + \tfrac{1}{4}z^2)^{1/2}.$$

For $E = 0$, Eq. (4) reduces to Eq. (3). For large diffusion lengths and cladding-layer widths or high fields, Eq. (4) reduces to $i_n = qN_p\mu E$, the expression for drift transport alone. Drift current is not expected to be important for p-cladding-layer dopings greater than 4×10^{17} cm^{-3} (Dutta et al., 1981). This is consistent with experiments that have shown the leakage current to be < 20% of the total current at threshold for devices with p-InP layers doped to 2×10^{17} cm^{-3} (Chen et al., 1983).

A stringent requirement on the heterojunctions in lasers is that the nonradiative recombination rate at the interface be sufficiently small so as not to significantly affect the threshold current. If the interface is characterized by a interfacial recombination velocity s, the recombination rate at the interface is then given by $R = 2sN/d$, where N is the injected carrier density and d the width of the active region surrounded by the two interfaces. It is assumed that the effect of carrier diffusion can be neglected; i.e., $d \ll L$, where L is the carrier diffusion length. The interfacial nonradiative current density is then given by $J = 2sNe/d$, where e is the electron charge. For high-quality GaAlAs/GaAs, heterojunctions, the interfacial recombination velocity is 450 cm/ sec^{-1} (Nelson and Sobers, 1978a). The interfacial recombination velocity has not been determined for InGaAsP/InP interfaces using extensive measurements on layers of different thickness as was done for GaAlAs. However, an estimate for s determined from the efficiency data of Wada et al. (1979) for light-emitting diodes of various active-layer thicknesses is shown in Fig. 6. In the absence of other nonradiative recombination, the recombination rate in the light-emitting-diode (LED) active layer is given by

$$J/ed = BN^2 + (2sN/d),\qquad(5)$$

where B is the radiative recombination coefficient. The relative radiative output power, which can be determined directly from Eq. (5), is plotted in Fig. 6 for $s = 400, 1000,$ and 5000 cm sec^{-1}. It can be concluded that $s \leq 1000$ cm sec^{-1} for InGaAsP/InP interfaces of the quality described by Wada et al. (1979). The contribution of the interfacial recombination veloc-

1. InGaAsP/InP LASER STRUCTURES AND PERFORMANCE

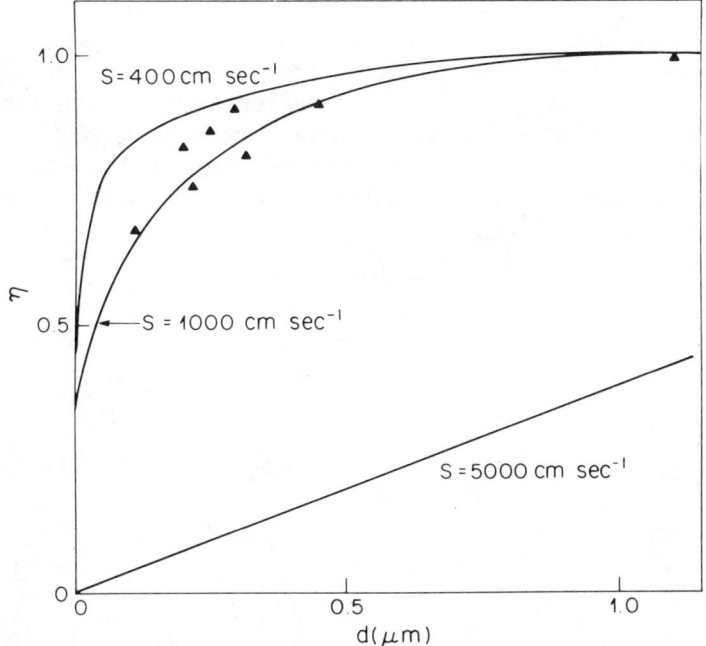

FIG. 6. Relative output power of $\lambda = 1.3$-μm InGaAsP/InP at constant current plotted as a function of active-layer thickness.

ity to the threshold current in a laser device is $<5\%$ of the threshold value for $s = 1000$ cm sec^{-1}.

4. STIMULATED EMISSION AND GAIN

Lasing threshold is reached when the total cavity losses are overcome by the optical gain caused by the injected carriers. The relationship between the active-layer gain g_{th} at threshold and the various losses is given by (Casey and Panish, 1978)

$$\Gamma g_{th} = \alpha_{cl}(1 - \Gamma) + (1/2L) \ln(1/R_1 R_2) + \Gamma \alpha_{fc}, \qquad (6)$$

where Γ is the fraction of the optical mode energy propagating in the active layer, α_{cl} the absorption coefficient of the cladding layers at the emission wavelength, α_{fc} the free-carrier absorption in the active layer, L the cavity length, and R_1 and R_2 the reflectivities of the two mirror facets. The optical absorption or gain is given by (Dutta, 1980),

$$\alpha(E) = \frac{e^2 h}{2\varepsilon_0 m_0^2 cnE} \int_{-\infty}^{\infty} \rho_c(E') \rho_v(E'') $$
$$|M(E', E'')|^2 (f(E' - E) - f(E')) \, dE', \qquad (7)$$

where m_0 is the free electron mass, e the electron charge, E the photon energy, ε_0 the permittivity of free space, n the refractive index at energy E, and ρ_c and ρ_v the densities of state in the conduction and valence bands, respectively; $f(E')$ is the probability that the state at energy E' is occupied by an electron, $E'' = E' - E$, and M is the effective matrix element between the valence-band state at energy E'' and the conduction-band state at energy E'. The injected carrier density N in the active layer is related to the current density J by the equation

$$J/ed = (N/\tau_r) + (N/\tau_{nr}), \tag{8}$$

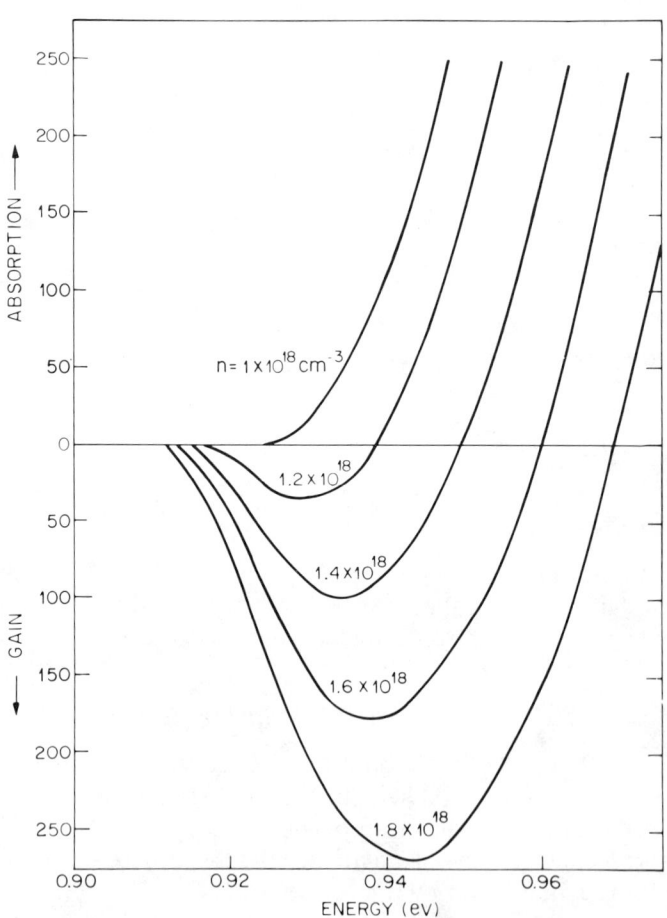

FIG. 7. Calculated spectral dependence of absorption or gain at several injected-electron densities. Fixed parameters are $T = 297$ K, $N_A = N_D = 2 \times 10^{17}$ cm^{-3}, and $m_c/m_0 = 0.059$. [From Dutta (1980).]

where e is the electron charge, τ_{nr} the nonradiative lifetime, τ_r the radiative lifetime, and d the width of the active layer. As the injected carrier density increases, the spontaneous emission rate increases, followed by a saturation of the absorption and the observation of optical gain at sufficiently high injection levels. The calculated spectral dependence of absorption or gain is shown in Fig. 7 at various injected electron densities for 1.3-μm InGaAsP material. Note that gain appears at lower energy as the injection level is increased as a result of a band-gap shrinkage. The peak gain, on the other hand, increases in energy as N increases.

The relationship between the peak gain and the normalized current density is shown for two different effective mass ratios in Fig. 8. For $g > 25$ cm^{-1}, the gain g may be expressed as $g = 0.057(J_{nom} - 2400)$ cm^{-1} for $m_c/m_0 = 0.059$. The threshold current density can be expressed by using Eqs. (6)–(8) in the following form:

$$J_{th} = 2400\frac{d}{\eta} + 17.54\frac{d}{\eta\Gamma}\left[\alpha + \frac{1}{2L}\ln(R_1 R_2)\right]\frac{d}{\eta\Gamma}, \qquad (9)$$

where η is the internal quantum efficiency. The calculated variation with temperature of the gain–current density relationship is shown in Fig. 9.

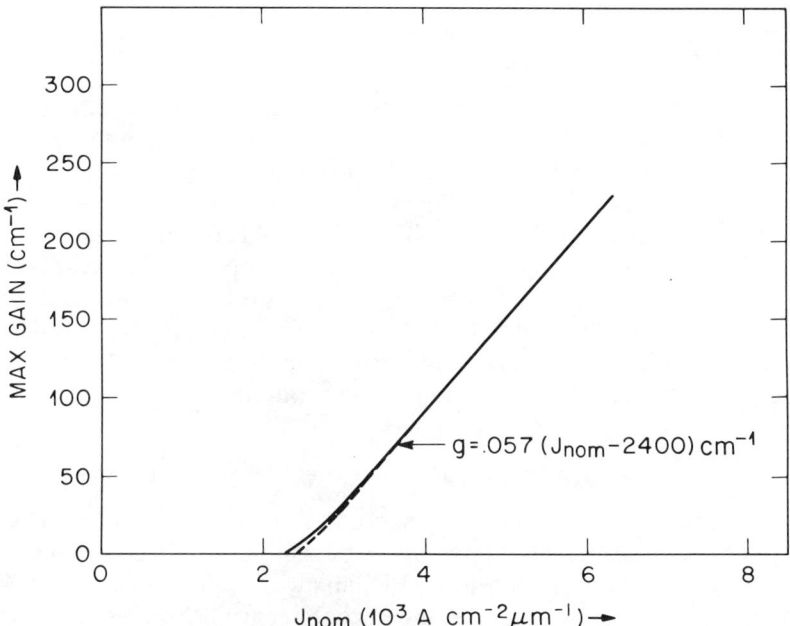

FIG. 8. Calculated peak gain as a function of normalized current density. Fixed parameters are the same as given in Fig. 7. [From Dutta (1980).]

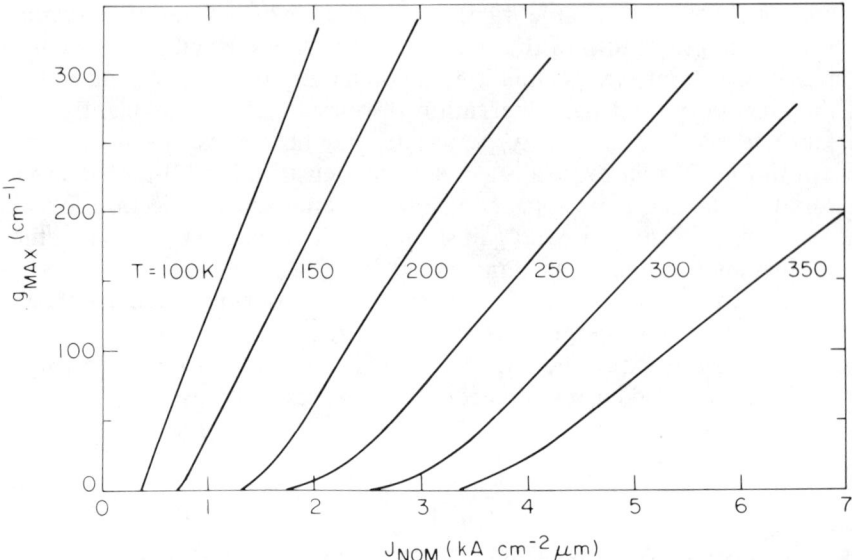

FIG. 9. Calculated peak gain as a function of normalized current density at various temperatures for $In_{0.72}Ga_{0.28}As_{0.6}P_{0.4}$ material. $N_A = N_D = 10^{17}$ cm^{-3}. [From Dutta and Nelson (1982a). © IEEE 1982.]

5. Light–Current Characteristics

A light–current characteristic is shown in Fig. 10 for a broad-area laser with a cavity length of 250 μm and a width of 380 μm. The threshold current is 800 mA, corresponding to a threshold current density of 840 A cm^{-2}. The external differential quantum efficiency above threshold for this device is 0.22 mW mA^{-1} per facet. The broad-area threshold current density is found to be a function of active layer thickness, with the optimum thickness being 0.15 ± 0.05 μm (Greene and Henshall, 1979). Similar results are found for 1.55-μm lasers (Horikoshi and Furukawa, 1979). Table I gives a comparison of the lowest broad-area threshold current densities reported for 1.3-μm lasers grown by the alternative growth techniques.

III. Temperature Dependence of Threshold Current

One major difference between the operating characteristics of GaAlAs and InGaAsP lasers is the higher temperature sensitivity of threshold observed for InGaAsP devices (Horikoshi and Furukawa, 1979; Dutta and Nelson, 1980). A related phenomena is the greater sublinearity observed in the emission characteristic of InGaAsP LEDs of the surface-emitting type than for similar GaAlAs diodes (Uji et al., 1981; Dutta and Nelson, 1982b). The

1. InGaAsP/InP LASER STRUCTURES AND PERFORMANCE

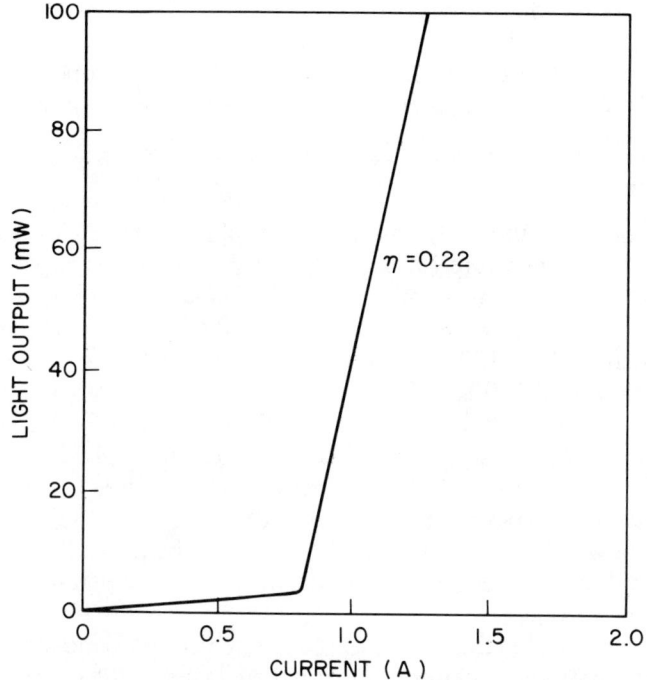

FIG. 10. Light versus current characteristic for a broad-area laser emitting at $\lambda = 1.3$ μm.

high-temperature dependence of threshold imposes severe limitations for high-temperature operation, resulting in a concentration on lasers with low-threshold currents for many applications. The ability to make high-temperature-stress aging experiments is also limited by the high-temperature sensitivity of threshold.

TABLE I

BROAD-AREA THRESHOLD CURRENT DENSITIES[a]

Growth technique	J_{th} (A cm^{-2})	Reference
LPE	670	Nelson (1979)
	730	Hsieh and Shen (1978)
CVD	1500	Olsen et al. (1979)
MOCVD	800 (430[b])	Razeghi et al. (1983)
MBE	1800	Tsang et al. (1982)

[a] For cavity lengths of ~380 μm.
[b] For a cavity length of 1550 μm.

Part III is devoted to a discussion of experimental results on the temperature dependence of threshold of InGaAsP double-heterostructure (DH) lasers and calculations of the magnitude, as well as the temperature dependence of mechanisms suggested to explain the experimental results.

6. Experimental Temperature Dependence of Threshold

It is found experimentally that the threshold current J_{th} of a DH laser varies with the temperature T as $J_{th} = J_0 \exp(T/T_0)$, where the parameter T_0 describes the temperature sensitivity. For an InGaAsP laser emitting at 1.3 μm, the quantity T_0 is ~60–70 K for $T > 250$ K and $T_0 \simeq 110$ K for $T <$ 250 K, whereas $T_0 \simeq 150$–180 K for GaAlAs DH lasers over the entire temperature range 100 K $< T <$ 350 K. Figure 11 shows the measured threshold current plotted as a function of temperature for InGaAsP DH lasers emitting at 1.3 and 1.55 μm. The T_0 values for $\lambda = 1.55$-μm lasers are typically 50–60 K for $T > 250$ K. Values for T_0, as large as 120 K have been observed for some types of InGaAsP lasers (Tamari et al., 1982), although it has been shown that these higher T_0 values can be a result of significant current flow through parallel shunt paths with lower temperature sensitivity (Dutta et al., 1981).

It has been suggested that the observed temperature dependence of threshold for InGaAsP lasers cannot be explained in terms of the temperature dependence of the radiative recombination component (Dutta and Nelson, 1982d) of the current. The calculated T_0 value is ~200 K for radiative recombination alone, assuming parabolic bands. Several possible mechanisms have been suggested to explain the low T_0 values for InGaAsP lasers, including carrier leakage over the heterobarrier (Yano et al., 1981); intervalence-band absorption (Adams et al., 1980); and Auger recombination (Thompson and Henshall, 1980; Dutta and Nelson, 1980).

Carrier leakage is known to be responsible for the higher temperature sensitivity of some $Ga_{1-x}Al_xAs$ DH lasers. The amount of carrier leakage increases rapidly with decreasing heterobarrier energy height ΔE_g; J_{th} increases and T_0 decreases rapidly if the aluminum fraction x in the ternary cladding layers is less than 0.25, i.e., for $\Delta E_g < 0.3$ eV (Casey and Panish, 1978). The leakage current is also found to increase with decreasing p-ternary carrier concentration because of increased drift field (Anthony and Schumaker, 1980; Dutta, 1981). However, for properly doped InGaAsP DH lasers with emission wavelengths longer than ~1.2 μm, there are considerable experimental data and calculations suggesting that carrier leakage by diffusion or drift mechanisms are not responsible for the low T_0 values. In the discussion of Fig. 5 it was shown that the calculated carrier leakage by diffusion in InGaAsP lasers is insignificant at room temperature for $\lambda >$

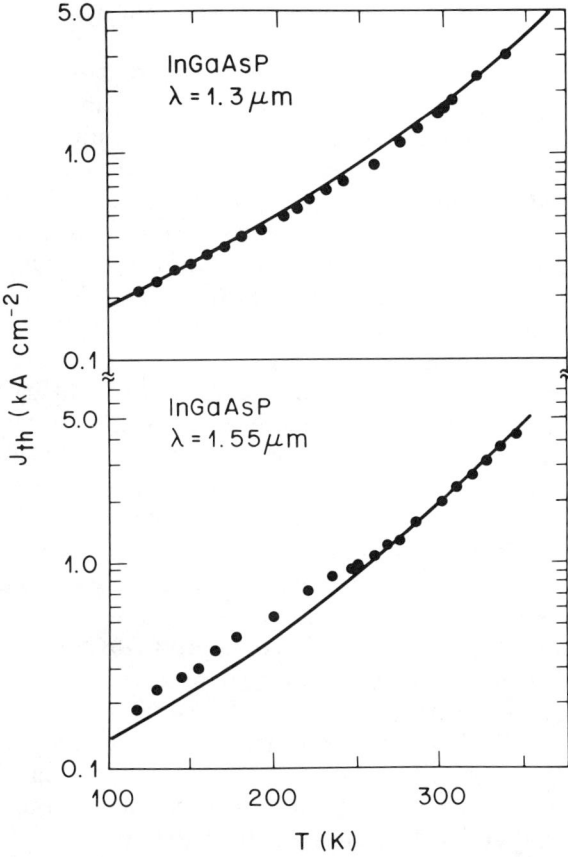

FIG. 11. Threshold current versus temperature for $\lambda = 1.3$- and $\lambda = 1.55$-μm InGaAsP/InP lasers. [From Dutta and Nelson (1981).]

1.2 μm. These calculated results suggest that diffusive leakage is not a major factor for $\lambda > 1.2$ μm. For p-cladding-layer dopings greater than 4×10^{17} cm^{-3}, the drift leakage is negligible, as discussed in Section 3. As expected, low T_0 values (35 K) have been observed for p-cladding dopings less than 10^{17} cm^{-3}, where drift leakage is important (Henshall and Thompson, 1982). More recently, it has been shown that for devices with p-InP doping levels of 2×10^{17} cm^{-3}, the leakage current at threshold is $\sim 20\%$ of the total current (Chen et al., 1983). These results suggest that conventional drift or diffusion leakage current is not the primary mechanism responsible for the low T_0 of InGaAsP lasers with p-cladding doping above $\sim 4 \times 10^{17}$ cm^{-3}. However, a fraction of the hot carriers produced by the Auger process, as discussed later, may escape from the active layer and constitute a type of leakage current. This may be the explanation for the experimental results of Yano et al. (1982), which are consistent with some type of carrier leakage.

Inter-valence-band absorption has been suggested as a possible mechanism for low T_0; however, the absorption needed is an order of magnitude too small from the calculated result at 1.3 μm (Sugimura, 1981). Experimental measurements of the intravalence-band absorption support this conclusion (Henry et al., 1983b). In addition, measurements on well-behaved lasers with no parallel shunt-path leakage have shown that for 1.3-μm lasers, the maximum increase in the absorption inferred from the change in the external differential quantum efficiency may account for less than 15% of the threshold increase observed (Dutta and Nelson, 1982c). It should be pointed out, however, that 1.55-μm lasers have exhibited sharper decreases in the external efficiency than have 1.3-μm lasers (Asada et al., 1980). This may be an indication that the longer wavelength lasers are affected more severely by intravalence-band absorption than are the shorter wavelength lasers, although the data of Henry et al. (1983b) suggest that even at 1.55 μm, the measured absorption is not strong enough to have a major effect on the temperature dependence of threshold current.

A nonradiative thermally activated trap has been suggested as a possible alternative reason for the low T_0 values (Horikoshi and Furukawa, 1979; Nahory et al., 1978). However, a trap of appropriate density, capture cross section, and activation energy has not been observed experimentally. It has also been suggested that the low T_0 values could be explained by carrier temperatures higher than those of the lattice and a function of the injection level. Although experimental evidence was initially presented that showed evidence of carrier heating (Shah et al., 1981), more recent results by several groups showed that carrier heating is not major factor in 1.3-μm devices (Henry et al., 1983a; Su et al., 1982).

Considerable evidence has been compiled that suggests that Auger recombination contributes to, if not fully accounts for, the low T_0 values in 1.3-μm lasers. Although uncertainties both in calculations and in experimental results prevent one from concluding with 100% certainty that Auger processes account for all the nonradiative recombination, the theoretical modeling and experimental results described here all suggest that the major mechanism is in fact Auger recombination.

7. Auger Recombination: Theory

The band-to-band Auger processes in direct-gap semiconductors are shown in Fig. 12a, whereas the phonon-assisted processes are shown in Fig. 12b. The CCCH recombination mechanism, which involves two electrons and a hole, is dominant in n-type material (In the notation of recombination mechanisms, C stands for the conduction band, H for heavy holes, L for light

1. InGaAsP/InP LASER STRUCTURES AND PERFORMANCE

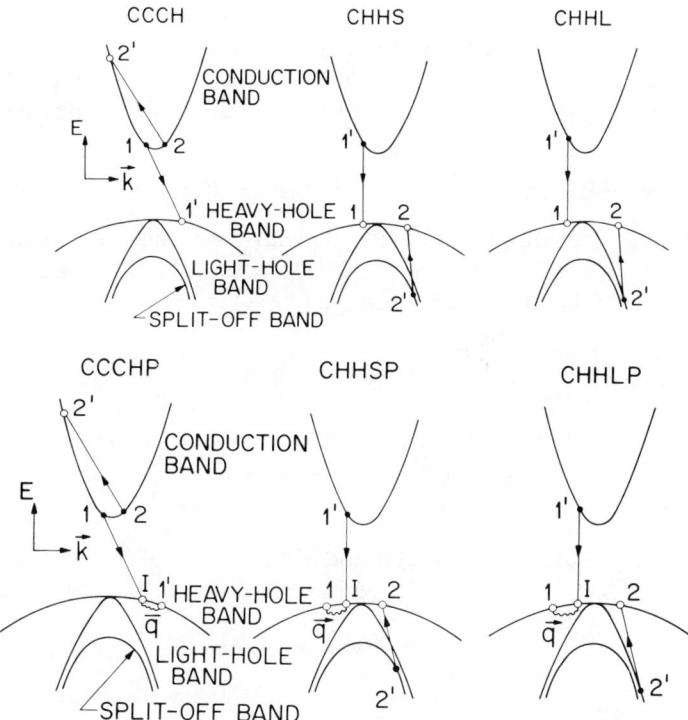

FIG. 12. (a) Band-to-band Auger processes in direct band-gap material. (b) Phonon-assisted Auger processes. [From Dutta and Nelson (1982d).]

holes, and S for the spin split-off band.) Beattie and Landsberg (1959) considered this process to explain the experimental results of Tauc (1959) on InSb. The CHHS and the CHHL mechanisms are dominant in p-type material. Under the high-injection conditions present in lasers, all three band-to-band mechanisms must be taken into consideration.

The transition probabilities for the CCCH, CHHS, and CHHL can be calculated taking into account nonparabolic bands, Fermi statistics, and screening effects, which are important at the injection levels in laser devices. The transition rate for the CCCH Auger process can be expressed (Dutta and Nelson, 1982d) in the following form, where Bloch functions are used and Umklapp processes have been neglected:

$$R = \frac{2\pi}{\hbar} \left(\frac{1}{8\pi^3}\right)^4 \iiiint |M_{if}|^2 P(1, 1', 2, 2') \, \delta(\mathbf{k}_1 + \mathbf{k}_2 - \mathbf{k}_{1'} - \mathbf{k}_{2'}) \\
\times \delta(E_i - E_f) \, d^3\mathbf{k}_1 \, d^3\mathbf{k}_2 \, d^3\mathbf{k}_{1'} \, d^3\mathbf{k}_{2'}. \qquad (10)$$

where R denotes the total Auger rate per unit volume and $P(1, 1', 2, 2')$ the product of the occupancy probability of the various states. The momentum conservation arises from the assumption of the states being the product of a plane wave and a Bloch function. The matrix element M_{if} is the interaction potential

$$V(\mathbf{r}_1 - \mathbf{r}_2) = (e^2/\varepsilon|\mathbf{r}_1 - \mathbf{r}_2|)\exp[-\lambda|\mathbf{r}_1 - \mathbf{r}_2|], \quad (11)$$

where λ is the screening factor, ε the dielectric constant, e the electronic charge, and $\mathbf{r}_1, \mathbf{r}_2$ are the position of the interacting electrons. The quantity M_{if}^2 as derived by Beattie and Landsberg (1959) is

$$|M_{if}|^2 = \left(\frac{4\pi e^2}{\varepsilon}\right)^2 \left\{\left|\frac{F(1, 1')F(2, 2')}{\lambda^2 + g^2}\right|^2 + \left|\frac{F(1, 2')F(2, 1')}{\lambda^2 + h^2}\right|^2 \right.$$
$$\left. + \left|\frac{F(1, 1')F(2, 2')}{\lambda^2 + g^2} - \frac{F(1, 2')F(2, 1')}{\lambda^2 + h^2}\right|\right\}, \quad (12)$$

where $g^2 = |\mathbf{k}_1 - \mathbf{k}_1'|^2 = |\mathbf{k}_2' - \mathbf{k}_2|^2$ and $h^2 = |\mathbf{k}_1 - \mathbf{k}_2|^2 = |\mathbf{k}_1' - \mathbf{k}_2|^2$, and $F(i, j)$ denotes the overlap integral of the Bloch functions. It has been shown (Dutta and Nelson, 1982d) that the recombination rate may be expressed as

$$R = (4e^4 m_{c0} \alpha_{ch} \beta / \pi^3 \hbar^3 \varepsilon^2 E_g) I, \quad (13)$$

where

$$I = \int_0^\infty dZ_1 \int_{A(Z_1)}^\infty dZ_2 \int_{-1}^1 dt' \, Z_1^2 Z_2^2 |J_0| f_v(1')[1 - f_c(2')] F(\mathbf{k}_1', \mathbf{k}_2') \quad (14)$$

and $t' = \cos\theta'$, where θ' is the angle between $\mathbf{Z}_1$ and $\mathbf{Z}_2$,

$$\mathbf{z}_1 = \mathbf{k}_1 + [\mathbf{k}_2'/(1 + 2\mu)], \quad \mathbf{z}_2 = \mathbf{k}_2', \quad (15a)$$

$$|J_0|^2 = 2a_c Z_2^2 - Z_1^2(1 + 2\mu) - (4m_{c0} E_g/\hbar^2), \quad (15b)$$

$$a_c = \mu_c - [\mu/(1 + 2\mu)], \quad (16)$$

$$A^2(Z_1) = \frac{1}{2a_c}\left[(1 + 2\mu)Z_1^2 + \frac{4m_{c0} E_g}{\hbar^2}\right], \quad (17)$$

$$F(\mathbf{k}_1', \mathbf{k}_2') = \frac{f_B^2}{b(1 - f_B^2)}\ln\left[\frac{e^b + f_B}{1 + f_B e^b}\right], \quad (18)$$

$$f_B = \exp\left[\frac{E_{Fc}}{k_B T} - \frac{\hbar^2}{8m_{c0} k_B T}(|J_0|^2 + |h|^2)\right], \quad (19)$$

$$b = \frac{\hbar^2}{8m_{c0} k_B T^2}|h||J_0|. \quad (20)$$

The recombination rate for the CHHS process has been shown (Sugimura, 1981) to be given by

$$R = \frac{4e^4 m_{c0} \alpha_{ch} \alpha_{sh}}{\pi^3 \hbar^3 \mu \varepsilon^2 E_g^2} \left\{ 1 - \exp\left[\frac{E_{Fv} - E_{Fc} - E_g}{k_B T}\right] \right\}$$

$$\times \int_0^\infty dZ_1 \int_{A_1(Z_1)}^\infty dZ_2 \int_{-1}^1 dt' |j_1| L f_c$$

$$\times \left[Z_1 - \frac{Z_2}{2/\mu + 1} \right] F \left[Z_1 - \frac{Z_2}{2/\mu + 1}, Z_2 \right], \qquad (21)$$

where

$$|j_1|^2 = \frac{2}{\mu} a_s Z_2^2 - \frac{2}{\mu}\left(1 + \frac{\mu}{2}\right) Z_1^2 - \frac{4 m_{c0}}{\hbar^2 \mu}(E_g - \Delta),$$

$$a_s = \mu_s - 1 + 1/(1 + \mu/2), \qquad \mu_s = m_{c0}/m_s,$$

$$L = \frac{(|h|^2 + |j_1|^2)^2}{(4\lambda^2 + |h|^2 + |j_i|^2)^2},$$

$$A_1(Z_1) = (1/a_s)(1 + 2/\mu) Z_1^2 + (2 m_{c0}/a_s \hbar^2)(E_g - \Delta).$$

The function F is the same as defined in Eq. (18), with E_{Fc} replaced by E_{Fv} in the equation for f_B. The quantity α_{sh} arises from the overlap integral of the modulation parts of the Bloch function between the heavy-hole band and the spin split-off band. The split-off band mass m_s at the threshold energy E is given by

$$E_T = \frac{2 m_v + m_{c0}}{2 m_v + m_{c0} - m_s}(E_g - \Delta). \qquad (22)$$

The recombination rate for the CHHL process is given (Dutta and Nelson, 1982d) by

$$R = \frac{4e^2 m_{c0} \alpha_{ch} \alpha_{lh}}{\pi^3 \hbar^3 \mu \varepsilon^2 E_g^2} \left\{ 1 - \exp\left[\frac{E_{Fv} - E_{Fc} - E_g}{k_B T}\right] \right\}$$

$$\times \int_0^\infty dZ_1 \int_{A_2/Z_1}^\infty dZ_2 \int_{-1}^1 dt' Z_1^2 Z_2^2 |j_2| L f_c(\mathbf{k}_1') F(\mathbf{k}_1', \mathbf{k}_2'), \qquad (23)$$

where, as before, the quantity α_{lh} arises from the overlap of the modulating part of the Bloch function between the heavy-hole and the light-hole band. The various quantities in Eq. (23) are given by

$$\mathbf{Z}_1 = \mathbf{k}_1 + \frac{\mathbf{k}_2'}{2/\mu + 1}, \qquad \mathbf{Z}_2 = \mathbf{k}_2', \qquad (24)$$

TABLE II

EXPRESSIONS FOR BAND-STRUCTURE PARAMETERS OF $In_{1-x}Ga_xAs_yP_{1-y}$, LATTICE-MATCHED TO InP[a]

Energy gap at zero doping	$E_g(eV) = 1.35 - 0.72y + 0.12y^2$
Heavy-hole mass	$m_v/m_0 = (1-y)[0.79x + 0.45(1-x)] + y(0.45x + 0.4(1-x))$
Light-hole mass	$m_{l0}/m_0 = (1-y)[0.14x + 0.12(1-x)] + y[0.082x + 0.026(1-x)]$
Dielectric constant	$\varepsilon = (1-y)[8.4x + 9.6(1-x)] + y[13.1x + 12.2(1-x)]$
Spin–orbit coupling	$\Delta(eV) = 0.11 + 0.31y - 0.09y^2$
Conduction-band mass	$m_c/m_0 = 0.080 - 0.039y$

[a] $x = 0.4526y/(1-0.031y)$.

$$|j_2|^2 = \frac{2}{\mu}a_l Z_2^2 - \frac{2}{\mu}\left(1 + \frac{\mu}{2}\right)Z_1^2 - \frac{4m_{c0}}{\hbar^2\mu}E_g, \quad (25)$$

$$a_l = \mu_l - 1 + 1/(1 + \mu/2), \quad (26)$$

$$\mu_l = m_{c0}/m_l, \quad (27)$$

$$E_T = \frac{2m_v + m_{c0}}{2m_v + m_{c0} - m_l}F_g, \quad (28)$$

where m_l denotes the light-hole mass at the threshold energy given by Eq. (28):

$$A_2^2(Z_1) = \frac{1}{a_l}\left(1 + \frac{2}{\mu}\right)Z_1^2 + \frac{2m_{c0}}{a_l\hbar^2}E_g. \quad (29)$$

The quantity $F(\mathbf{k}_1', \mathbf{k}_2')$ is the same as in Eq. (18) with E_{Fc} replaced by E_{Fv} in the equation for F_B. The quantity L is the same as in Eq. (21).

By using the band-structure parameters of InGaAsP as given in Table II, the band-to-band Auger rates can be related to a nonradiative lifetime τ_A by the relation $R = N/\tau_A$ where N is the injected carrier density and R is the total Auger rate. Figure 13 shows the calculated Auger lifetime at threshold for band-to-band processes. Note that the Auger lifetime decreases rapidly with increasing temperature. Figure 14 shows the calculated Auger lifetimes for both band-to-band and phonon-assisted processes as a function of composition for InGaAsP. By using the Auger rates described here, the threshold current density at 300 K for an InGaAsP laser with $\lambda = 1.3$ μm and an active layer thickness of 0.2 μm is 2.87 kA cm^{-2}. This is significantly larger than the commonly observed threshold current density of ~1.5 kA cm^{-2} for this type of laser. Because of the uncertainties in many of the band-structure parameters and the approximations made in calculating the Auger rate, this disagreement is not surprising. For example, a 20% increase in nonparabolicity lowers the calculated J_{th} to 1.8 kA cm^{-2}. The calculation also produces a T_0

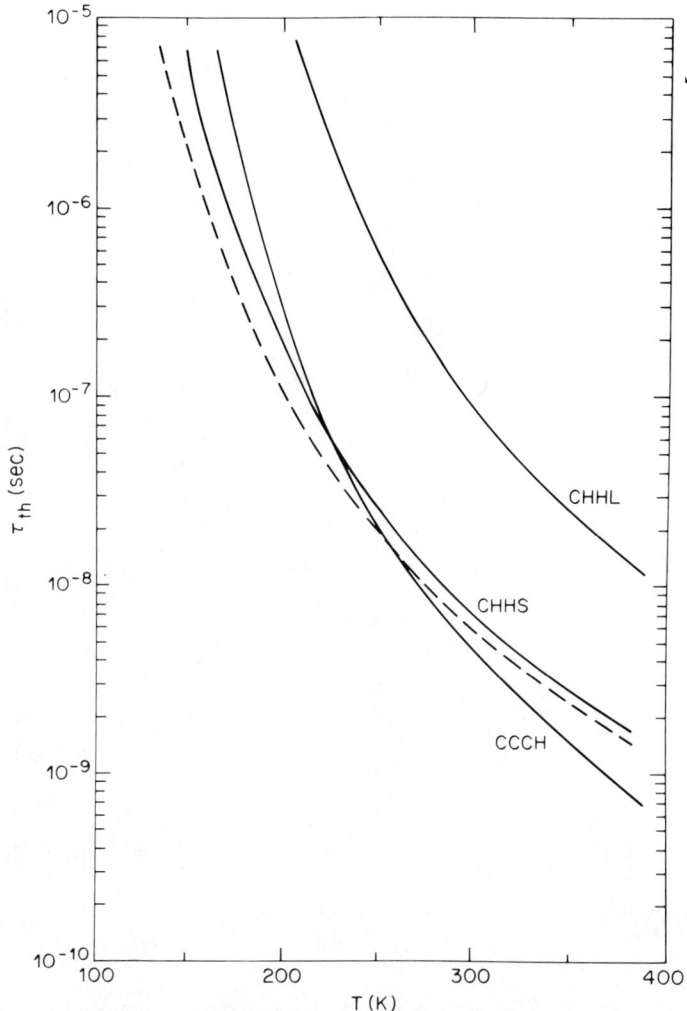

FIG. 13. Calculated band-to-band Auger lifetimes at threshold for $\lambda = 1.3$-μm InGaAsP/InP lasers. [From Dutta and Nelson (1982d).]

value of 49 K for $\lambda = 1.3$-μm lasers, compared with the usually observed value of ~70–75 K. A somewhat smaller value for the Auger recombination coefficient defined (at high injection levels) by $C = R/N^3$ would produce better agreement with experiment. The previous calculations give an Auger coefficient of ~1×10^{-28} cm^6 sec^{-1}. Figure 15 shows the relationship between the Auger coefficient and the predicted T_0 value for two different assumed values of the carrier concentration at threshold, with the assumption that the temperature dependence of the Auger coefficient is represented

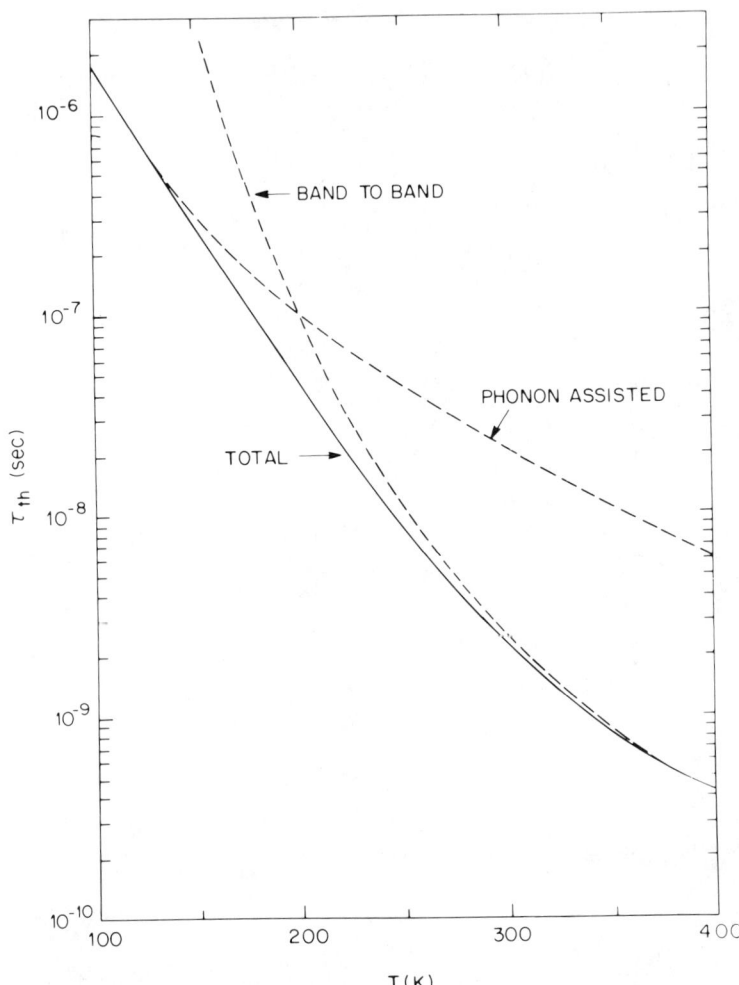

FIG. 14. Band-to-band and phonon-assisted Auger lifetimes at threshold for $\lambda = 1.3\text{-}\mu\text{m}$ InGaAsP lasers. [From Dutta and Nelson (1982d).]

by the ratio k between the Auger coefficient at 350 K to the value at 300 K. The smaller value of k used in Fig. 15 is that calculated earlier for the phonon-assisted processes, whereas the larger value is that calculated for the band-to-band processes.

Several authors (Sugimura, 1981; Chiu et al., 1982) have reported Auger calculations that use a calculation similar to that described earlier. All of these calculations result in Auger coefficients that lead to calculated threshold current densities larger than those experimentally observed. However,

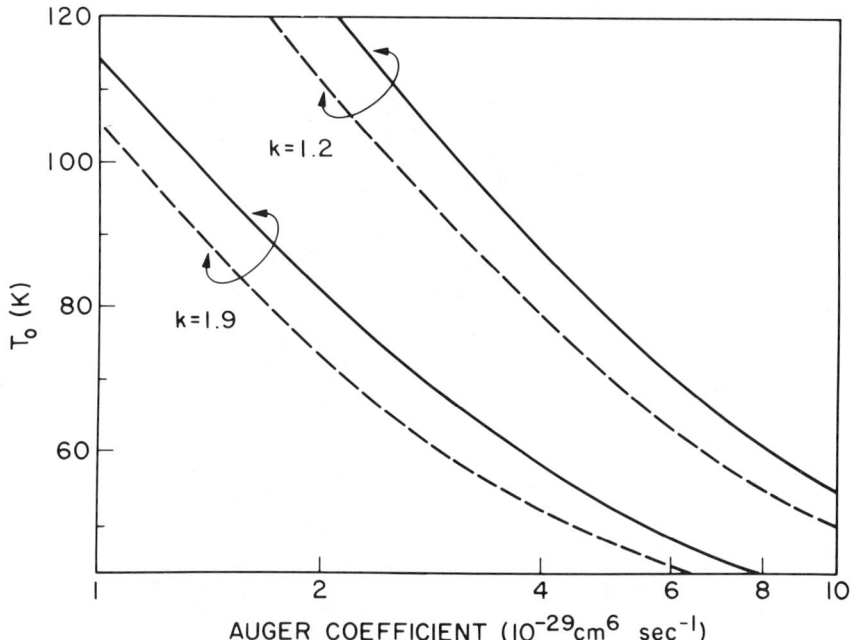

FIG. 15. Calculated T_0 versus Auger coefficient for two different values of the carrier density at (———, 2×10^{18} cm^{-3}; - - -, 2.5×10^{18} cm^{-3}) and two different values of the temperature dependence of the Auger coefficient.

Haug (1983) has shown that the dominant Auger process for InGaAsP material that emits at 1.3 μm is a phonon-assisted CCCH process if the band curves of Chelikowsky and Cohen (1976) for InP are used, but the energy gap is assumed to be that of InGaAsP with $\lambda = 1.3$ μm. Haug determined a value for the total Auger coefficient of 2.5×10^{-29} cm^6 sec^{-1}. By using this value, one can then calculate the T_0 value for $\lambda = 1.3$-μm lasers to be 100–110 K [using the calculated temperature dependence of the phonon-assisted processes (Dutta and Nelson, 1982d)], which is somewhat higher than that observed experimentally. The uncertainty concerning band-structure parameters and the carrier concentration at threshold for InGaAsP lasers is such that it is unreasonable to expect better agreement between calculation and experiment.

8. AUGER RECOMBINATION: EXPERIMENT

Considerable experimental evidence now exists to suggest that Auger recombination is important in determining the temperature sensitivity of InGaAsP lasers. Direct confirmation of the importance of Auger recombination in this material has been provided by time-decay measurements (Fig.

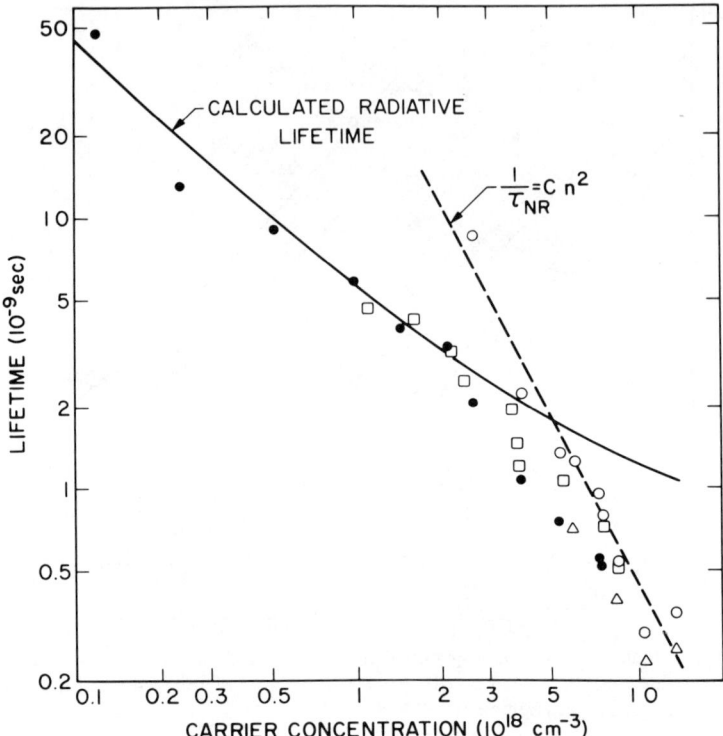

FIG. 16. Experimentally determined values for photoinjected-carrier lifetime as a function of injected-carrier density in InGaAsP ($\lambda = 1.3$ μm) at 300 K. Calculated radiative lifetime is shown by the solid curve; derived nonradiative lifetime is shown by dashed curve. [From Sermage et al. (1983).]

16) of optically excited material up to carrier densities of nearly 10^{19} cm^{-3}. The Auger coefficient determined from these experiments is $C = (2.3 \pm 1) \times 10^{-29}$ cm^6 sec^{-1}. The Auger coefficient given by the data of Fig. 16 predicts a T_0 value of either 66–77 K if the temperature dependence of the band-to-band process is assumed or 104–114 K if phonon-assisted processes are used. Further experimental confirmation of Auger recombination in InGaAsP has been shown by the workers listed in Table III, with the majority of the Auger coefficients near 4×10^{-29} cm^6 sec^{-1}. The T_0 values predicted from Fig. 15 for this Auger coefficient are 80–90 K for phonon-assisted processes and 50–60 K for band-to-band processes. These predicted values agree quite well with experimental T_0 data, given the experimental uncertainties. It can be concluded therefore that Auger recombination is responsible for the low T_0 values of InGaAsP lasers. Other III–V compounds

TABLE III

MEASURED AUGER COEFFICIENTS

C (cm^6 sec^{-1}) (total)	Comment	Reference
5×10^{-29}	$\lambda = 1.3\ \mu$m[a]	Mozer et al. (1982)
4×10^{-29}	$\lambda = 1.58\ \mu$m	Asada and Suematsu (1982)
$2.3 \pm 1 \times 10^{-29}$	$\lambda = 1.3\ \mu$m, optically pumped	Sermage et al. (1983)
1×10^{-29}	$\lambda = 1.3\ \mu$m, doping dependence of τ	Henry et al. (1981)
$<3 \times 10^{-29}$	$\lambda = 1.3\ \mu$m	Su et al. (1982)
3×10^{-29}	Fit to data of Su et al.	Thompson (1983)
$3 - 8 \times 10^{-29}$	$\lambda = 1.3\ \mu$m LEDs	Uji et al. (1983)

[a] CCHS process only.

that emit at 1.3 or 1.55 μm have band structures similar to InGaAsP, leading also to low predicted T_0 values (Nelson and Dutta, 1983).

IV. Gain-Guided Laser Structures

The optical mode and the injected carriers in semiconductor lasers are confined in the direction perpendicular to the $p-n$ junction by the heterostructure layers. Broad-area lasers as described in Section 5 emit over the entire chip width of 2–300 μm. Most applications, however, require that the width of the light source in the junction plane be comparable to the width in the perpendicular direction. This requires that special attention be given to the laser design and fabrication techniques to provide optical and carrier confinement in the direction along the junction plane over widths less than 5–10 μm. Laser designs with this type of confinement are refered to as stripe-geometry lasers (Casey and Panish, 1978).

The two general classifications of stripe-geometry lasers are gain-guided and real-index-guided lasers. Real-index-guided lasers, which are described in Part V, contain a dielectric step in the active-layer plane to confine the optical mode using either additional heterostructures or a layer-thickness change. Because of the fabrication difficulties involved in making narrow (<3-μm) dielectric waveguides, early devices were of the gain-guided design in which injected carriers were localized by restricting the contact area (Fig. 17a) to a rectangular region running the length of the laser cavity (typically, ~250 μm) with a width of 3–20 μm. Since carriers are injected into the active layer only in the vicinity of the stripe, optical gain can exist only in this area with the remainder of the chip exhibiting optical loss. For narrow stripe widths, the lowest-order optical mode will be preferred. However, optical power can be deflected out of the lowest-order mode by nonuniformities in

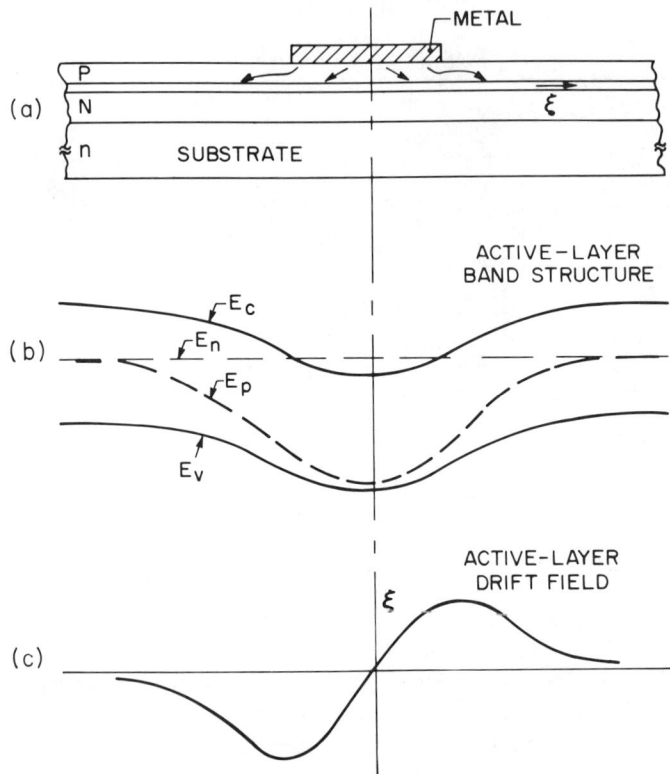

FIG. 17. (a) Schematic cross section of stripe-geometry laser showing current spreading and field in active layer. (b) Active-layer band structure under forward bias. (c) Active-layer drift field. [From Joyce (1982).]

the optical characteristics of the material caused by variations in material composition or layer thickness as well as by the negative index step produced by the injected carriers. Although careful design and fabrication of this type of laser have resulted in useful laser devices in the shorter-wavelength GaAlAs-material system (Dixon *et al.*, 1976; Dixon, 1980), significant problems inherent to the InGaAsP system make the gain-guided laser fabricated from InGaAsP more difficult to optimize. In addition, the InGaAsP gain-guided device is useful over a narrower temperature range than the GaAlAs lasers, as described later.

9. CURRENT SPREADING

Although the contact area or the region of current flow is restricted in a gain-guided laser, the injected-carrier distribution can be considerably wider as a result of ohmic spreading in the cladding layers in addition to spreading caused by diffusion and drift of carriers in the active layer (Joyce, 1982). A lateral voltage drop in the p-InP cladding layer (which has a higher sheet resistivity than the n cladding and the n substrate) results in the current-crowding effect shown Fig. 17. Under the commonly used assumption that the lateral voltage drop in the high-conductivity n layer is zero, the electron Fermi level E_n in the n-layer is taken to be a constant. If one also makes the assumption that the electron Fermi level is continuous across the interface into the active layer, then as Joyce (1982) has pointed out, the lateral electron current density is therefore given by

$$\mathbf{J}_n = \mu_n N \nabla E_n = 0, \tag{30}$$

where μ_n is the electron mobility. The lateral hole current density can be shown (Joyce, 1982) to be given by

$$\mathbf{J}_p = -qD_e \nabla E_p, \tag{31}$$

where the effective diffusion coefficient is given by

$$qD_e/kT = 2\mu_p[1 + \tfrac{1}{2}A_1(u+v) + A_2(u^2+v^2) + \cdots], \tag{32}$$

with $u + v = p(N_c^{-1} + N_v^{-1})$, $u^2 + v^2 = p^2(N_c^{-2} + N_v^{-2})$, etc. The coefficients for parabolic bands are $A_1 = \tfrac{1}{2} 2^{1/2}$, $A_2 = 3/16 - 3^{1/2}/9$, etc., and where N_c and N_v are the conduction- and valence-band densities of states. The effective diffusion coefficient therefore takes both the drift field and the electron distribution into account. Joyce (1980) has also shown that the effect of ohmic current spreading in the p-cladding layer and the hole drift and diffusion can also be combined in to a single diffusion equation.

10. Gain Guiding

As a result of the current confinement discussed in Section 9, the injected carriers have a lateral profile similar to that of the schematic diagram in Fig. 18a. At high carrier densities, optical gain is present along the central region of the stripe, as shown in Fig. 18b. The injected carriers produce a reduction of the refractive index as a result of free-carrier interactions (proportional to the injected-carrier concentration) and by means of band-to-band processes that relate (via the Kramers–Kroenig relations) the refractive index to the dispersion of the gain (absorption) spectra (Casey and Panish, 1978). The carrier-induced dielectric constant changes produce an antiguiding effect that tends to spread the optical wave (Fig. 18c) instead of confine it. In addition, diffraction effects that are related to the beam dimensions also lead to loss of optical power from the mode.

The complex dielectric constant $\varepsilon = \varepsilon_1 + i\varepsilon_2$ can be shown (Casey and Panish, 1978) to be related to the real refractive index and the gain by the

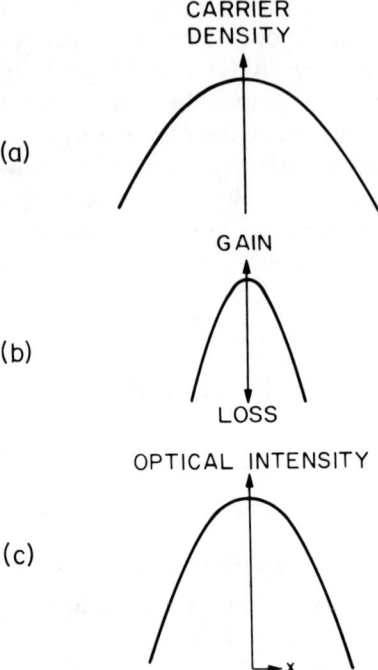

FIG. 18. Schematic representation of (a) carrier-density profile, (b) gain profile, and (c) optical intensity profile in gain-guided laser.

following equations:

$$\varepsilon_1/\varepsilon_0 = n^2 + \lambda_0 g/4\pi, \tag{33}$$

$$\varepsilon_2/\varepsilon_0 = n\lambda_0 g/2\pi, \tag{34}$$

where ε_0 is the permittivity of free space, n the real refractive index, g the gain coefficient, and λ_0 the free-space wavelength. The ratio $R = -\Delta\varepsilon_1/\Delta\varepsilon_2$ between the change in the real and the imaginary parts of the dielectric constant produced by the injected carriers is critical in determining the propagation characteristics of the optical mode. Figure 19 shows the sensitivity of the mode width s to the value of R for several contact stripe widths (G. P. Agrawal, unpublished, 1983). This variation of s in turn affects the threshold current as shown in Fig. 20, where I_{th} is plotted as a function of stripe width for several values of R. For larger values of R, i.e., for more effective carrier-induced antiguiding, the threshold current becomes larger for fixed stripe width w.

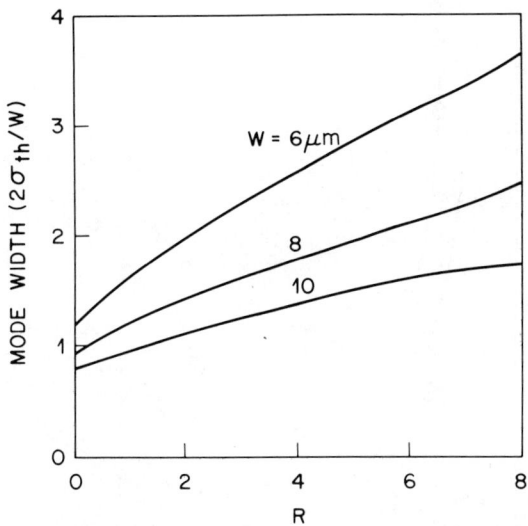

FIG. 19. Normalized optical mode width versus the index and guiding parameter R and three values of the contact-stripe width. [From G. P. Agrawal, unpublished (1983).]

11. THRESHOLD CURRENT

Experimentally, it has been found that the threshold current of InGaAsP gain-guided lasers (Kawaguchi et al., 1981a) increases more rapidly with decreasing stripe width than that of comparable GaAlAs lasers. The observed differences between the GaAlAs and InGaAsP devices could be accounted for by a larger R value for InGaAsP (Kawaguchi et al., 1981b) than for GaAlAs. Indeed, Turley (1982) has reported index changes to threshold approximately 4 times larger for InGaAsP devices than for GaAlAs devices. Other authors have reported somewhat smaller values for the index change for InGaAsP lasers (Stubkjaer et al., 1980); however, calculations of the variation of threshold current with stripe width do not agree with experiment for any value of R, as shown in Fig. 20.

An additional mechanism contributing to the threshold-current sensitivity to stripe width in InGaAsP lasers is the Auger effect. The threshold carrier concentration at the center of the stripe is expected to be higher for narrow-stripe devices than for wider-stripe devices because of increased losses from diffraction and the decreased overlap between the optical mode and that portion of the active layer that exhibits gain. The presence of significant Auger recombination in InGaAsP, along with the negligible Auger rate in GaAlAs, therefore leads to a more rapid increase of I_{th} for narrow stripe widths in InGaAsP than for GaAlAs lasers. Figure 21 shows the good agree-

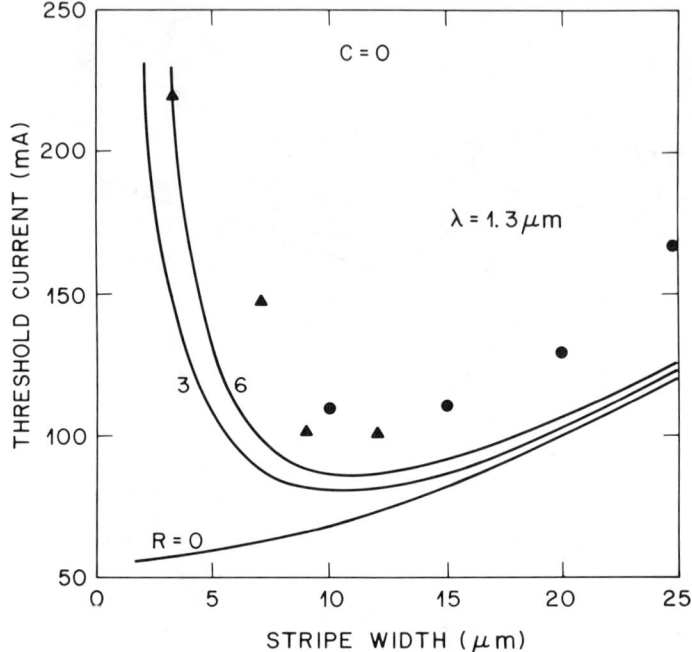

FIG. 20. Calculated variation of the threshold current with stripe width for several values of the parameter R.

ment between experimental data and calculated curves for $R = 3$ and an Auger coefficient of 3×10^{-29} cm^6 sec^{-1} for $\lambda = 1.3$ μm and $C = 9 \times 10^{-29}$ cm^6 sec^{-1} for $\lambda = 1.55$ μm.

12. Maximum Operating Temperature

Gain-guided InGaAsP lasers have not been as useful as similar GaAlAs structures because of the combination of higher threshold currents and the higher temperature sensitivity (Part III). The resulting thermal runaway of InGaAsP lasers can easily be calculated from the temperature dependence of threshold described in Part III and the temperature rise of the device, which is given by

$$\Delta T = R_{th}(IV_j + IR_s), \tag{35}$$

where R_{th} is the thermal resistance, I the operating current, V_j the junction voltage, and R_s is the series resistance. Figure 22 illustrates the calculated thermal runaway for InGaAsP lasers with pulsed threshold currents of 25 and 100 mA at 20°C. The drive current at 5-mW output (assuming $\eta = 0.2$ mW mA^{-1}, $R_{th} = 30$ K W^{-1}, and $T_0 = 60$ K) is shown as a function

1. InGaAsP/InP LASER STRUCTURES AND PERFORMANCE

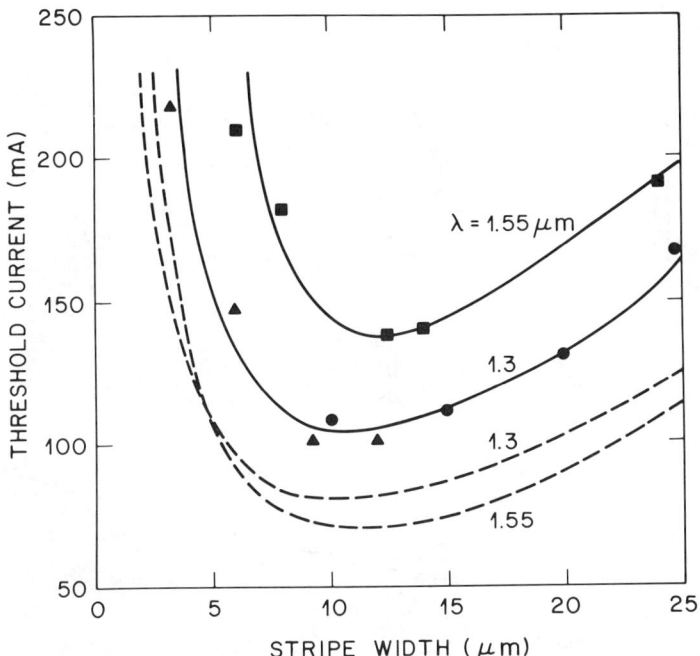

FIG. 21. Observed variation of threshold current with stripe width for $\lambda = 1.3$- and 1.55-μm InGaAsP/InP. Solid calculated curves include both index antiguiding and Auger recombination (see text). Dashed curves include only index antiguiding. [From Agrawal and Dutta (1983).]

of temperature. The 100-mA threshold is characteristic of state-of-the-art gain-guided lasers. The lower value is characteristic of state-of-the-art index-guided lasers, as discussed next in Section 13. Since it is often desirable to hold drive currents below 200–250 mA, it can be seen that the useful temperature range for gain-guided lasers is limited to ~50°C. Shown for comparison in Fig. 22 is the calculated temperature dependence of the drive current for a state-of-the-art GaAlAs/GaAs gain-guided laser. The temperature range over which the short-wavelength gain-guided laser is useful is much larger than either of the InGaAsP lasers in Fig. 22.

13. Nonlinearities in Light Output

It is well known that gain-guided lasers exhibit kinks in the $L-I$ characteristic (inset, Fig. 23) as a result of lateral mode instabilities. It was shown for GaAlAs lasers that the power at the kink can be increased by decreasing the width of the contact stripe (Dixon *et al.*, 1976). This stabilizes the optical mode by providing a gain profile that is narrower than the optical mode.

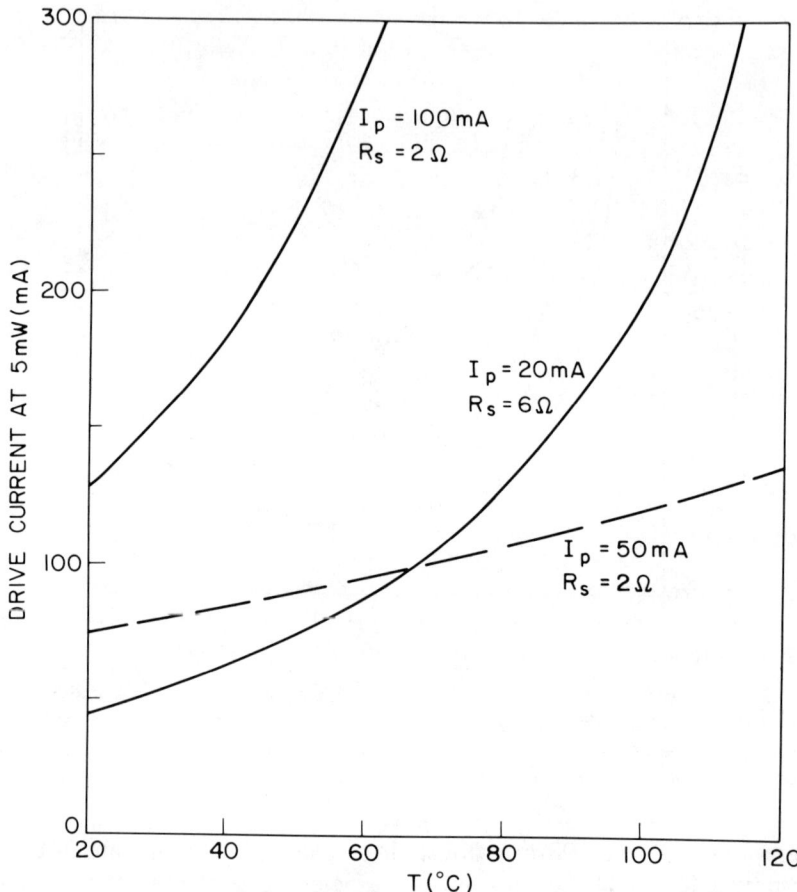

FIG. 22. Calculated cw drive current for 5-mW operation as a function of temperature for InGaAsP/InP and GaAlAs lasers. ———, $\lambda = 1.3$ μm; - - -, $\lambda = 0.8$ μm.

Although the kink power is also observed to increase with narrowing stripe width for InGaAsP lasers (Fig. 23), the rapid increase of threshold current with decreasing stripe width, coupled with the thermal runaway discussed previously in Section 12 make gain-guided InGaAsP lasers difficult to optimize and useful only over a narrow temperature range.

14. Modulation Characteristics

In addition to low kink power, a major problem with the wide-stripe gain-guided laser, which exhibits low threshold current, is that self-pulsations are often observed in the light output at frequencies of 150–300 MHz

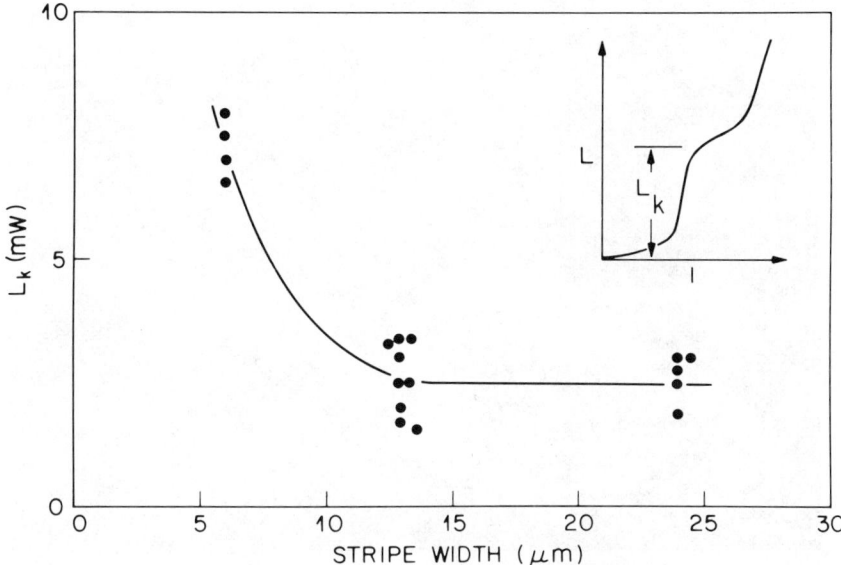

FIG. 23. Dependence of kink power P_k on stripe width for $\lambda = 1.3$-μm gain-guided lasers. Inset shows a representative light output versus current characteristic illustrating a kink. [From Kawaguchi *et al.* (1981b) © IEEE 1981.]

(Nelson and Dutta, 1980) as in Fig. 24. Self-pulsations have been shown to degrade system performance, as described by Iwashita and Nakagawa (1983).

V. Index-Guided Laser Structures

Laser structures that employ a variation in the real-refractive index along the junction plane to form an optical waveguide are referred to as index-guided lasers (Casey and Panish, 1978). The characteristics of the dielectric waveguide are carefully chosen to produce a stable zero-order lateral mode. The dielectric step producing the waveguide must be sufficiently large ($\gtrsim 0.02$) to overcome carrier-induced antiguiding and thermal effects (Section 4) but small enough ($\lesssim 0.1$) to allow the waveguide dimensions to be large enough for reproducible fabrication. The mode stability provided by the waveguide of properly designed index-guided lasers makes these lasers free from the $L-I$ nonlinearities often exhibited by gain-guided lasers. In addition, the high-threshold currents for narrow stripe widths described in Section 4 for gain-guided devices are not observed in index-guided lasers having index steps larger than those induced by the injected carriers. As a result, InGaAsP index-guided lasers are generally superior to gain-guided

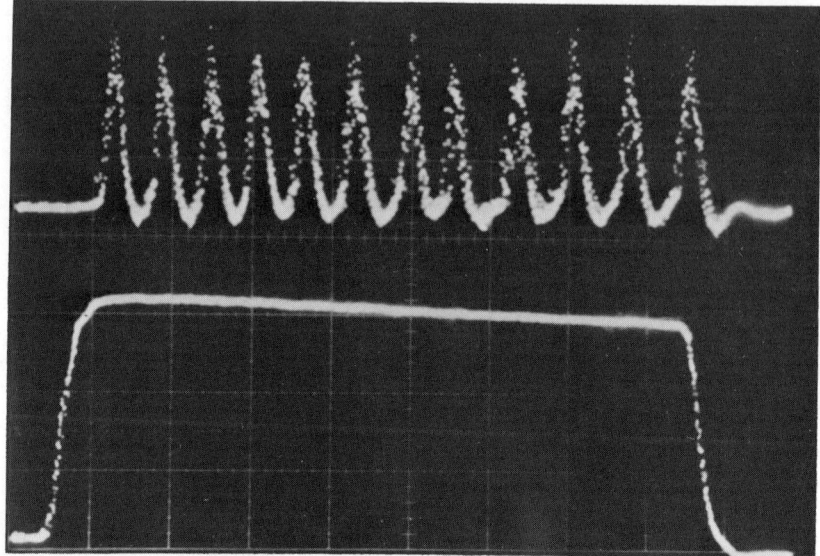

FIG. 24. Drive-current pulse (top) and light emission showing self-pulsations for an InGaAsP/InP ($\lambda = 1.3$ μm) gain-guided laser. Vertical scale: 50 mV; horizontal scale: 5 nsec. [From Nelson and Dutta (1980).]

lasers for most applications. However, the optimization of index-guided lasers, including both performance and manufacturability criteria, requires great attention to issues of device design and control of critical parameters during growth and fabrication processes.

The fabrication of high-performance index-guided lasers requires state-of-the-art techniques in both materials growth and wafer processing. Since the cutoff value for the width of a passive waveguide is given by $w = m\lambda/(4\,\Delta n^{1/2})$, where m is the mode number and Δn the refractive index step, it can be seen that a single-mode waveguide with a lateral index step larger than 0.01 must have a width less than 2.5 μm. Although this width is well within the current state of the art of lithographic and etching techniques, producing such a small index step from the materials commonly used in heterostructure lasers is difficult because the index differences among these materials are of the order of 0.3. The device designer has the choice of working with much narrower widths (~ 1 μm) or with wider but specially designed waveguides consisting of several adjacent material layers that are sufficiently thin (~ 2000 Å) to produce effective lateral index steps, ~ 0.01. These specially designed waveguides (described later) require carefully controlled nonplanar growth or fabrication techniques, since deviations of less than a few hundred angstroms in layer thickness or placement can produce a waveguide with nonoptimal characteristics. Other factors requiring careful attention in order to produce satisfactory performance are the techniques for both current and carrier confinement. Threshold current, linearity, high-frequency response, reflection sensitivity, and reliability are all dependent on the way in which current and carrier confinement are achieved.

InGaAsP index-guided lasers can be divided into two general classifications. Structures employing lateral heterojunctions to provide for carrier confinement are of the buried-heterostructure type. As a result of the lateral carrier confinement, the threshold currents of buried-heterostructure lasers are lower than those of other index-guided lasers. Figure 25a illustrates a buried-heterostructure laser with a planar InGaAsP active layer surrounded by higher band-gap InP. Because of the large index step (0.3) between InGaAsP ($\lambda = 1.3 \mu$m) and InP, active layer widths of $\sim 1.5 \mu$m are required for single-mode operation. An alternative type of buried-heterostructure with a nonplanar active layer is shown in Fig. 25b. Table IV compares the general performance of these structures. The actual operating characteristics of these devices are descirbed later.

In the second type of index-guided structure, a lateral waveguide is provided, although lateral carrier confinement is not. One device of this type is

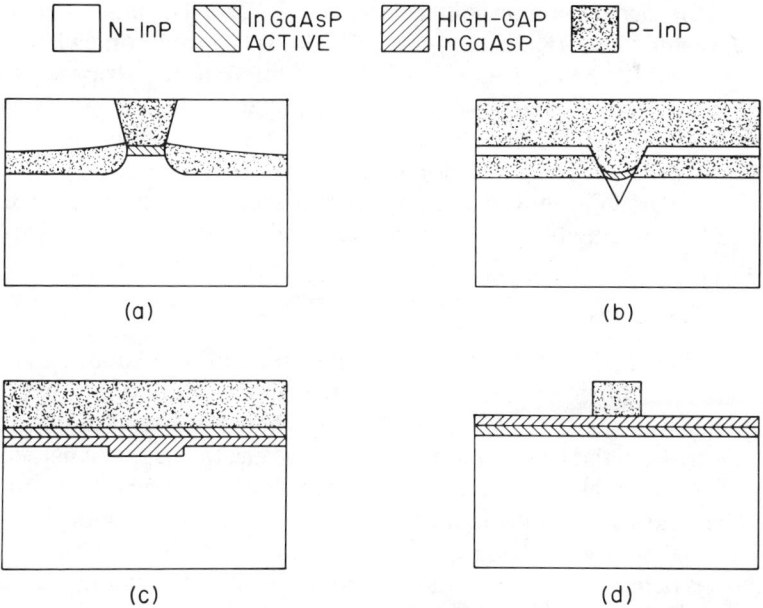

FIG. 25. Representative real-refractive index-guided lasers: (a) planar active-layer BH, (b) nonplanar active BH, (c) continuous-channel substrate laser, and (d) continuous active layer with adjacent waveguide layer.

TABLE IV

COMPARISON OF LASER STRUCTURE PERFORMANCE

Type	Threshold current[a] (mA)	Maximum operating temperature[b] (°C)	Reflection sensitivity
Planar active BH	10–20	100–110	Low
Nonplanar active BH	10–20	70–80	Low
Continuous active index guided	50–70	50–80	High

[a] 23 C.
[b] 5 mW per facet.

shown in Fig. 25c. A continuous planar active layer is grown above a nonplanar passive layer of an index intermediate between that of the InP cladding layers and that of the InGaAsP active layer. With passive guide layer thicknesses of ~4000 Å in the channel and 2000 Å outside the channel, a single-mode waveguide can be formed for channel widths of 2.5 μm, although it is found to be difficult to optimize this structure without the additional mode discrimination provided by gain guiding (Ueno et al., 1982). Another example of an index-guided device without lateral heterojunctions for carrier confinement is shown in Fig. 25d. The index step in this ridge waveguide laser (Kaminow et al., 1983a) is typically somewhat larger than that used in the device of Fig. 25c, and the required stripe width for single lateral-mode operation is therefore smaller for the ridge waveguide device. Table IV shows the characteristics of the continuous active-layer structures to be generally inferior to those of the BH designs. In the remainder of Part IV, the waveguiding criteria are discussed in more detail, followed by a detailed comparison of the performance of InGaAsP index-guided lasers.

15. WAVEGUIDING

The characteristics of the dielectric waveguides of semiconductor lasers are often determined by using the effective index approximation. For a rectangular waveguide as shown in Fig. 26, where the lateral dimension is much larger than the transverse dimension, the electromagnetic field distribution $F(x, y, z)$ is assumed to be separable into the functions $f(x)$, $g(y)$, and $h(z)$. The field distribution is first established in the x direction by using conventional slab waveguide theory. Then the field distribution in the y direction is determined also from slab waveguide theory, although an effective index of refraction is used for the central portion of the lateral waveguide in this case. This effective index is determined from the propagation constant of the modal distribution in the y direction, as described next.

1. InGaAsP/InP LASER STRUCTURES AND PERFORMANCE

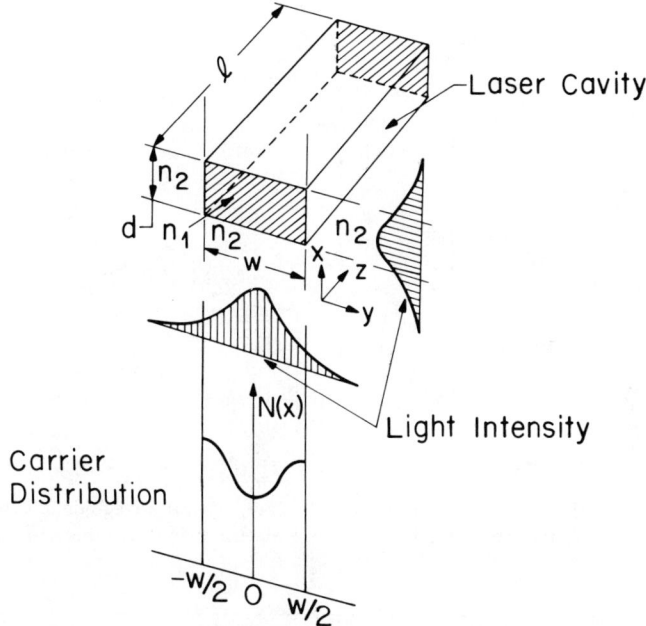

FIG. 26. Schematic diagram of (a) active-layer waveguide in a planar active-layer BH. Active layer of refractive index n_1 is surrounded by lower index (n_2) InP material, (b) representative light distribution in the planes parallel y and perpendicular x to the junction, and (c) representative carrier distribution in active layer above threshold showing spatial hole burning.

Conventional slab waveguide theory applied to a waveguide of the form shown in Fig. 27 gives (Thompson, 1980) a field distribution.

$$E_X = \begin{cases} A_1 \exp[P(D/2 - |X|)], & |X| > D/2, \\ A_2 \cos QX + \beta_2 \sin QX, & |X| < D/2, \end{cases} \quad (36)$$

in the center region expressed in terms of a normalized position X defined as $X = x\,(\delta\varepsilon)^{1/2}\,\beta_0$ with $D = d\,(\delta\varepsilon)^{1/2}\,\beta_0$, $\beta_0 = 2\pi/\lambda$, $P = p^2/(\delta\varepsilon\,\beta_0^2)$, and $p = (\beta_z^2 - \varepsilon_2\beta_0^2)^{1/2}$. The field distribution may be found by solving the secular equation

$$QD = 2\cos^{-1} Q + m\pi, \quad (37)$$

$$P^2 + Q^2 = 1, \quad (38)$$

where m is an integer for Q in terms of the normalized thickness D. The effective index ε_{eff} is then found by determining the propagation coefficient in the z direction from the equation

$$\beta_z^2 = (\varepsilon_1 - \delta\varepsilon\,Q^2)\beta_0^2 = (\varepsilon_2 + \delta\varepsilon\,P^2)\beta_0^2, \quad (39)$$

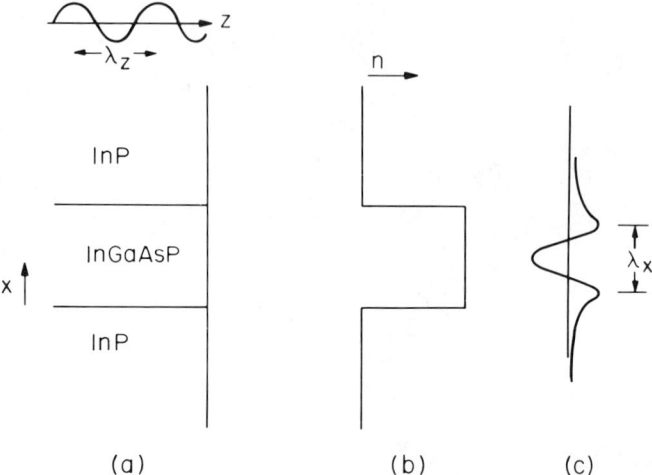

FIG. 27. Schematic cross section of (a) the three-layer optical waveguide, (b) the refractive index as a function of the x coordinate, and (c) the electric field distribution of one mode.

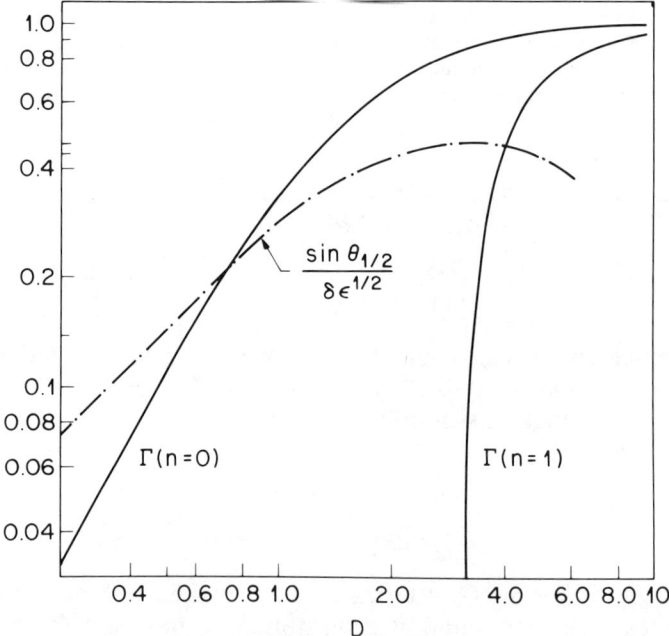

FIG. 28. Calculated confinement factors for two lowest order modes versus normalized cavity width D. Also shown is the variation of $\sin\theta_{1/2}/\delta\varepsilon^{1/2}$ versus D for the lowest order mode, where $\theta_{1/2}$ is the far field half-width.

which gives the longitudinal propagation coefficient β_z and

$$\varepsilon_{\text{eff}} = \varepsilon_1 - \delta\varepsilon\, Q^2 = \varepsilon_2 + \delta\varepsilon\, P^2, \tag{40}$$

The optical intensity distribution, which can be directly measured, is obtained by squaring the field distribution. A characteristic of the waveguide of interest for laser design is the ratio of the integral of the portion of the distribution located in the central layer to the total integral of the intensity. This confinement factor Γ is given by

$$\Gamma = \int_D E_X^2\, dX \Big/ \int_{-\infty}^{\infty} E_X^2\, dX, \tag{41}$$

For a symmetrical waveguide ($n_1 = n_3$), the dependence of the confinement factor on the normalized thickness D is shown in Fig. 28 for the two lowest order modes. Below $D = 3.14$, only the lowest order mode will propagate. The emission (far-field) pattern from the waveguide into air is characterized by the angle $\Theta_{1/2}$, which is the full width of the emission pattern at the half-intensity points. The dependence of $\Theta_{1/2}$ on D is shown in Fig. 28. A comparison of the calculated far-field angle in the direction perpendicular to the active layer with experimental data is shown in Fig. 29.

The effective index for InGaAsP/InP lasers emitting at 1.3 and 1.55 μm is shown as a function of active layer thickness in Fig. 30. By using the effective indices plotted in Fig. 30, the effective lateral index step can be easily found as a function of active-layer thickness for planar active-layer BH lasers where the active layer is surrounded by InP. Since typical active-layer thicknesses are in the 0.1–0.2-μm range, the lateral index step for planar active-layer BH lasers is in the range 0.06–0.1 for 1.3-μm emitters and 0.03–0.11 for 1.55-μm emitters. The passive waveguide lateral dimension must therefore be less than 1.3 μm for the cutoff of higher order modes for $\lambda = 1.3$-μm BH lasers and less than 2 μm for 1.55-μm BH lasers.

The effective index approximation is also commonly used to characterize the waveguides of channeled-substrate continuous active-layer structures such as shown in Fig. 25c. Effective indices are first found by using standard slab-waveguide theory for the four-layer waveguides in the channel and outside the channel region. The lateral index step is then found by using these effective indices in a three-layer symmetric slab model. The effective lateral index step for the channeled substrate planar laser emitting at 1.3 μm is shown as a function of groove depth in Fig. 31. For the active-layer thickness of 0.2 μm used in Fig. 31, it can be seen that lateral index steps of 0.009–0.013 are possible for sufficiently shallow groove depths such that the passive layer will planarize during growth to provide a planar active layer. Somewhat larger or smaller index steps can be achieved for various active-layer thicknesses.

To provide stable single lateral-mode operation, the lateral waveguide

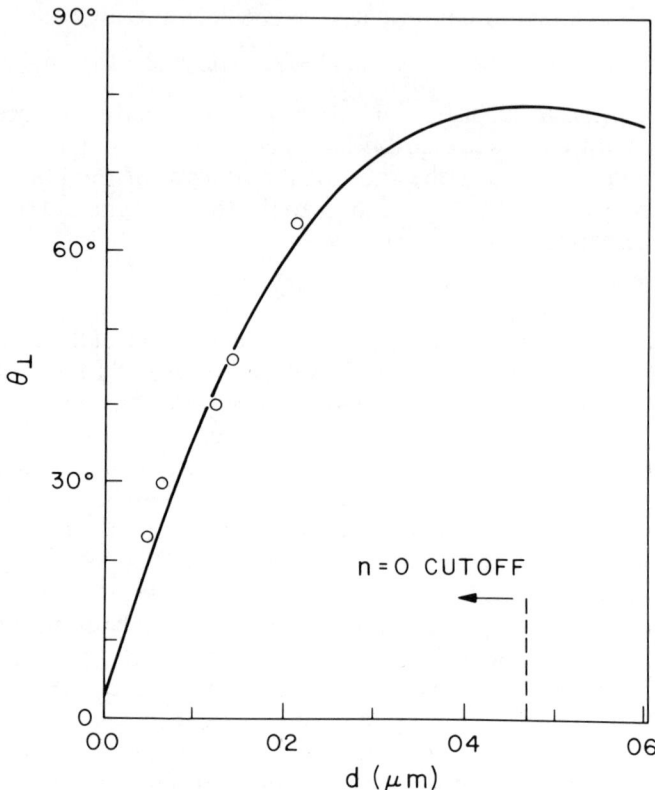

FIG. 29. Far-field width in the direction perpendicular to the junction for 1.3-μm lasers. [From Itaya *et al.* (1979).]

must have an index step sufficiently large such that thermal- or carrier-induced index changes will not dominate the mode guiding. In addition, the width of the lateral guide must be sufficiently small for the laser to operate in the lowest order mode only as described earlier. The waveguide width at which cutoff of the first-order mode occurs for 1.3-μm lasers is plotted as a function of lateral index step in Fig. 32. Experimental data are also plotted in Fig. 32, showing the maximum width for which stable single-mode operation has been obtained (to > 5 mW) for several types of index-guided lasers.

Gain guiding from narrow contact stripes (2.5 μm) must be employed with the weak dielectric guide of the continuous active-layer devices to provide stable lateral-mode operation. A critical submicron realignment must therefore be employed to properly align the contact stripe (Ueno *et al.*, 1982). Any yield advantage over stronger dielectric guide BH lasers provided by the wider guide width is likely overcome by yield loss from this critical

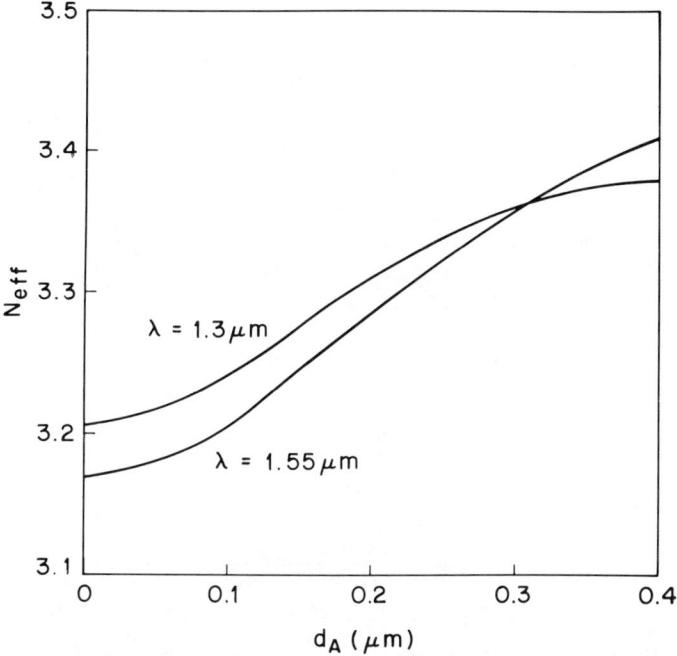

FIG. 30. Calculated effective index versus active-layer thickness for $\lambda = 1.3$- and 1.55-μm InGaAsP/InP lasers.

realignment. It is often observed that stable single-mode operation is observed for stripe widths somewhat larger than those predicted by the cutoff condition for larger Δn. For example, the cutoff condition for planar active-layer BH lasers is approximately 1 μm for 0.1-μm-thick active layers, whereas experimentally it is found that devices with somewhat larger widths will operate in the lowest order mode, since total cutoff is not necessary in a laser waveguide to prevent higher-order-mode operation. A gain discrimination between the two lowest order modes of approximately 10% may be sufficient to allow stable zero-order-mode operation in the absence of spatial hold burning. The effect of spatial hole burning may be neglected for $w < L$, where L is the injected carrier diffusion length. However, for $w > L$, spatial hole burning becomes a dominant factor above threshold so that the cutoff condition must be adhered to for these wider cavity devices. Commonly used values of the diffusion length are between 1 and 2.5 μm.

Calculated results that take spatial hole burning into account for planar active-layer BH lasers are shown in Fig. 33 for two different diffusion lengths. Note that the experimental data show that active widths less than 1.8 μm are needed for devices that are kink free to more than 8 mW output. The effect of

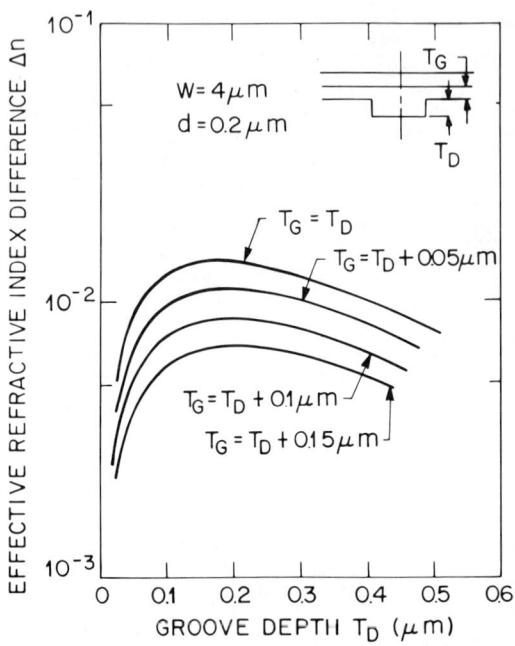

FIG. 31. Effective lateral refractive index step for 1.3-μm PCW lasers as a function of channel depth for several different guide layer thicknesses. The cladding layers are InP and 1.1-μm InGaAsP, respectively. [From Ueno et al. (1981).]

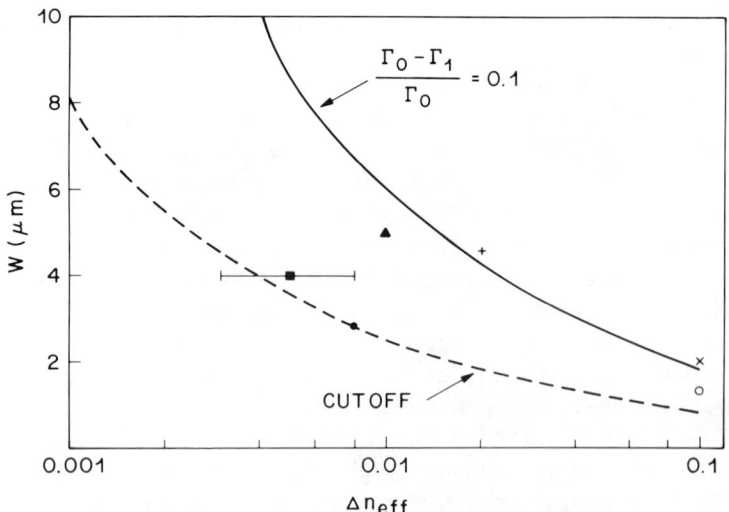

FIG. 32. Calculated cavity width versus lateral index step for cutoff of $n = 1$ mode (shown by dashed curve) and for 10% discrimination in confinement factors (shown by solid curve).

1. InGaAsP/InP LASER STRUCTURES AND PERFORMANCE

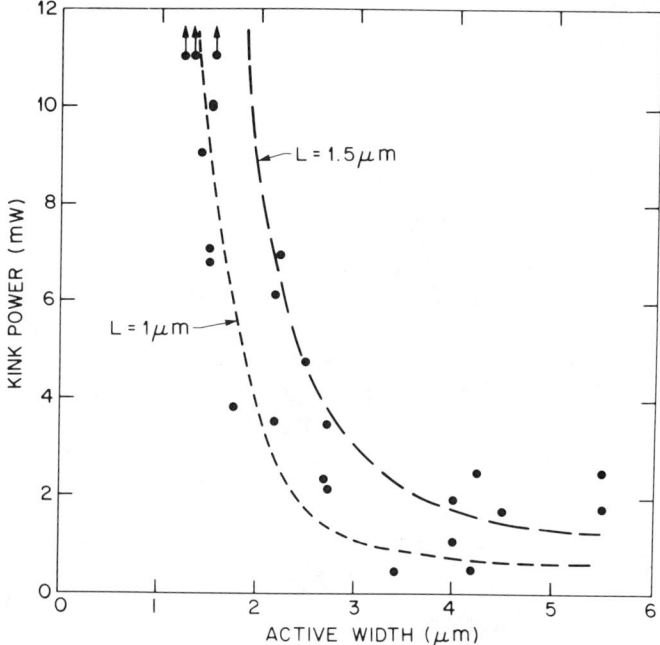

FIG. 33. Calculated output power for $n = 0$ mode at threshold for $n = 1$ mode versus stripe width for planar active-layer BH laser. The calculation assumes an active-layer thickness of 0.2 μm. [From Dutta et al. (1984a). © 1984 IEEE.]

this spatial hole burning on the far field and the $L-I$ characteristic of a BH laser is shown in Fig. 34. Note the onset of higher order lateral mode operation at an operating current of ~ 45 mA. A pronounced "kink" is observed in the $L-I$ characteristic for this device. More often, however, the onset of higher order mode operation is accompanied by a $L-I$ kink that can be observed only in the dL/dI characteristic (Fig. 35). Figure 36 shows the active-layer dimensions needed for stable operation of nonplanar active-layer BH lasers. Active-layer widths as large as 3 μm support kink-free operation to ~ 8 mW for an active-layer center thickness of 0.15 μm. Careful control of fabrication techniques, coupled with sensitive screening techniques such as dL/dI measurements, can be employed to produce devices with stable lateral-mode operation.

16. THRESHOLD CURRENT

The threshold currents for narrow-stripe real-index-guided lasers are nearly 10 times lower than those for narrow-stripe gain-guided lasers. Buried-heterostructure lasers containing lateral heterojunctions for carrier confinement have the lowest threshold currents (Table IV). However, even

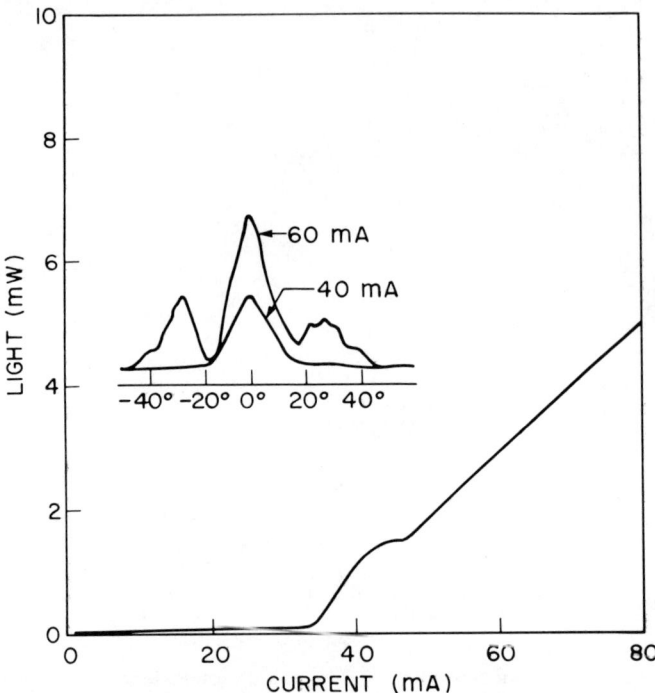

FIG. 34. Light versus current characteristic and far-field pattern parallel to the junction (inset) of planar active-layer BH laser showing effect of spatial hole burning and onset of higher lateral mode operation. [From Dutta et al. (1984a). © 1984 IEEE.]

real-index-guided lasers with a continuous active layer (Fig. 25c,d) have threshold currents a factor of 2 lower than the best gain-guided lasers. Auger recombination and carrier-induced index antiguiding as discussed in Part IV cause an increase in the threshold current of gain-guided lasers as the stripe width is narrowed. The built-in waveguide of the index-guided laser, if it is sufficiently large, overcomes the carrier-induced index antiguiding providing for a better overlap between the optical mode and the gain profile. Threshold currents below 15 mA have been achieved for 1.3-μm BH lasers of both the planar and nonplanar active-layer types (Table V). Threshold currents for continuous active-layer lasers with built-in waveguides as low as 50 mA have been reported for stable single lateral-mode lasers. The lowest reported threshold currents for several different types of index-guided lasers in each of three general classifications are given in Table V. Also given in Table V are the T_0 values reported for each structure along with the highest operating temperature reported for 5 mW output. Further discussion of the latter is given in Section 17.

1. InGaAsP/InP LASER STRUCTURES AND PERFORMANCE

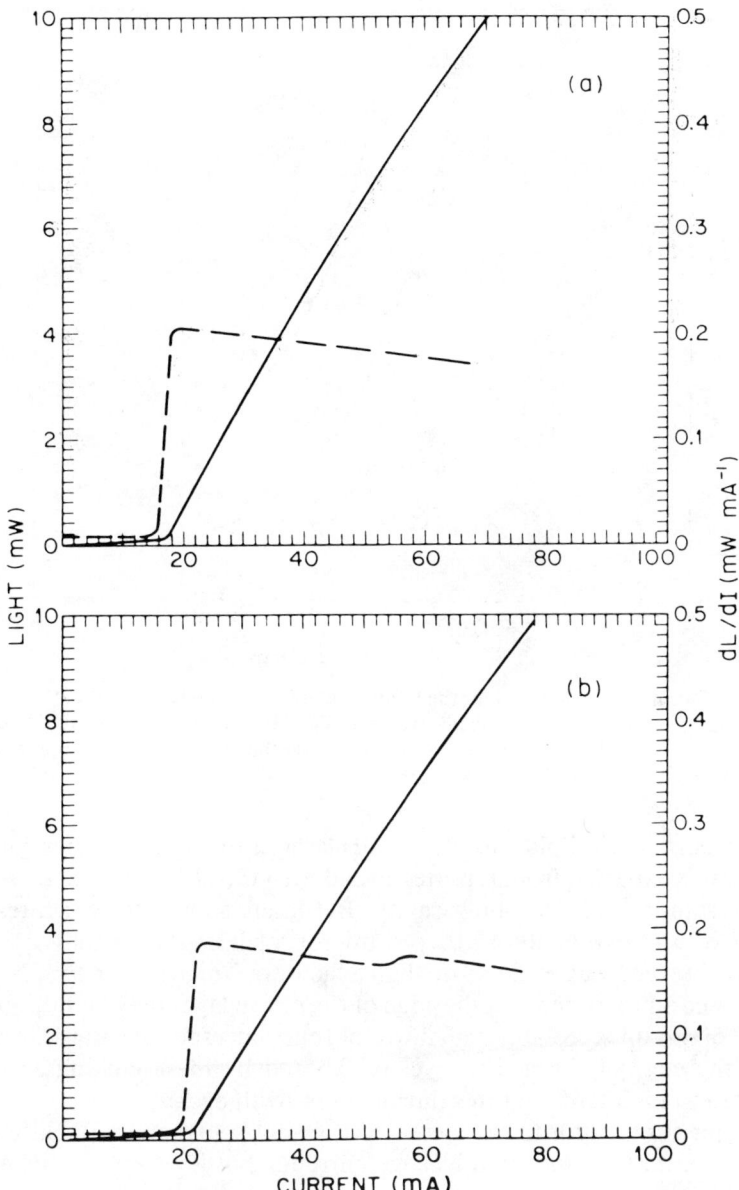

FIG. 35. L–I and dL/dI–I characteristics for (a) a BH laser operating in the lowest order mode, and (b) a BH laser exhibiting higher order mode operation above 55 mA. [From Dutta *et al.* (1984c).]

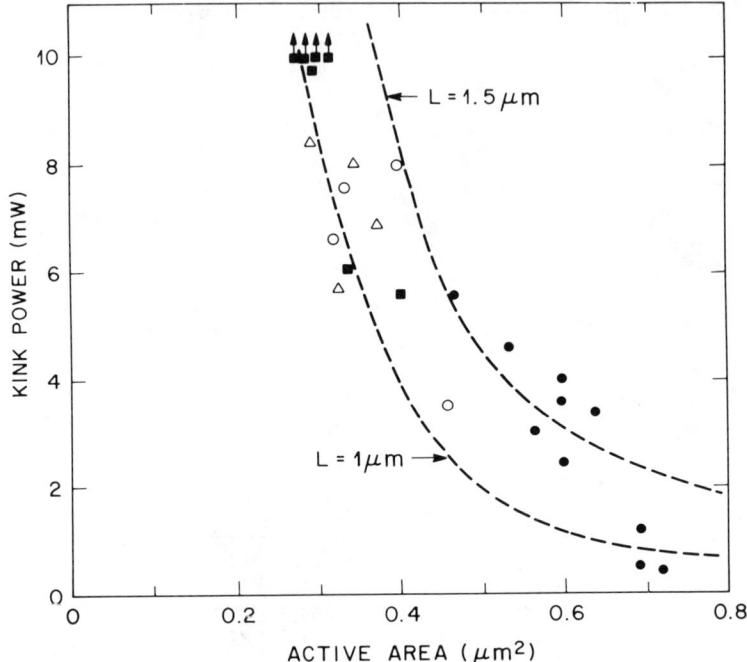

FIG. 36. Calculated light output for the fundamental ($n = 0$) mode at threshold for the first higher order ($n = 1$) mode for nonplanar active-layer BH laser versus active-layer cross-sectional area $A = \frac{2}{3}d/W$, where d, and W are the active-layer thickness and width, respectively, of the device structure of Fig 25b. [From Dutta et al. (1984c).]

The reported threshold currents for BH lasers are generally higher than the value that is expected from reported broad-area threshold current densities. For the standard 250-μm-long cavity, BH lasers should have a threshold current of approximately 4 mA per micron of lateral active-layer width. Typical observed values are more than a factor of 2 higher than this. Nonradiative recombination along the edge of the active layer may be responsible for part of this unexpectably high threshold current, especially in BH lasers in which the edge of the active layer is exposed by mesa etching and also subjected to high temperatures during a regrowth cycle.

A major contributor to the increased threshold currents in BH lasers has been shown to be shunt-path leakage currents that flow around the active region and through the reverse-biased junctions commonly used for current confinement (Nelson et al., 1981; Wright et al., 1982). This has proved to be a much more difficult problem in the InGaAsP/InP materials system than in the GaAlAs/GaAs system. The reason for this may be a combination of the specific bulk material and junction characteristics of InGaAsP/InP material in addition to the problems brought about by any residual thermal damage

1. InGaAsP/InP LASER STRUCTURES AND PERFORMANCE

TABLE V

REAL-INDEX-GUIDED LASER PERFORMANCE CHARACTERISTICS

Structure	I_{th}^a (mA)	T_0 (K)	T_{max} (°C)	Reference
Graphite				
$\lambda = 1.3\ \mu m$				
Planar active BH				
Double channel	11	106	110	Nelson et al. (1984)
etched mesa	10	70		Hirao et al. (1980)
Non-planar-active BH				
V-groove	10	55	100	Nelson et al. (1984
buried crescent	20	78	60	Oomura et al. (1981)
Continuous active				
inverted ridge	80	62	80	Turley et al. (1981)
ridge waveguide	34	60	50	Kaminow et al. (1983a)
self-aligned	120			Nishi et al. (1979)
planoconvex	60	70	70	Ueno et al. (1982)
$\lambda = 1.55\ \mu m$				
Planar active BH				
double channel	19	70	70	Nelson et al. (1984)
Nonplanar active BH				
channeled substrate	30	55	70	Nelson et al. (1984)
buried crescent	60	87	65	Murrell et al. (1982)
Continuous active				
multiclad	90	60	60	Imai et al. (1981)
buried rib	50	54		Yuasa et al. (1983)
ridge waveguide	40	55	<50	Kaminow et al. (1983b)

[a] At 23°C.
[b] $L = 5$ mW.

to materials subjected to high-temperature cycles during a regrowth step for the fabrication of a planar active-layer BH lasers.

Figure 37 shows the combinations of reverse- and forward-biased *pn* junctions used for current confinement in various BH laser designs along with the resistive, diode, transistor, and silicon controlled rectifier (SCR) elements in the equivalent circuits. In order to optimize a particular device structure, the layer thicknesses, placements, and doping levels need to be correctly specified. The etched-mesa BH laser shown in Fig. 37a is dominated by diode leakage when the *p*-type regrown layer at the side of the active layer has a high sheet conductivity. Reducing the sheet conductivity of this layer can lead to transistor-type leakage through the thin low-doped *p* layer, which can act as a base of an *NPN* transistor. A thyristor-like (*PNPN*) leakage path is also possible at high currents.

Figure 37b shows a channel-substrate BH laser along with its equivalent circuits. The major leakage path for this structure is through the high-gain

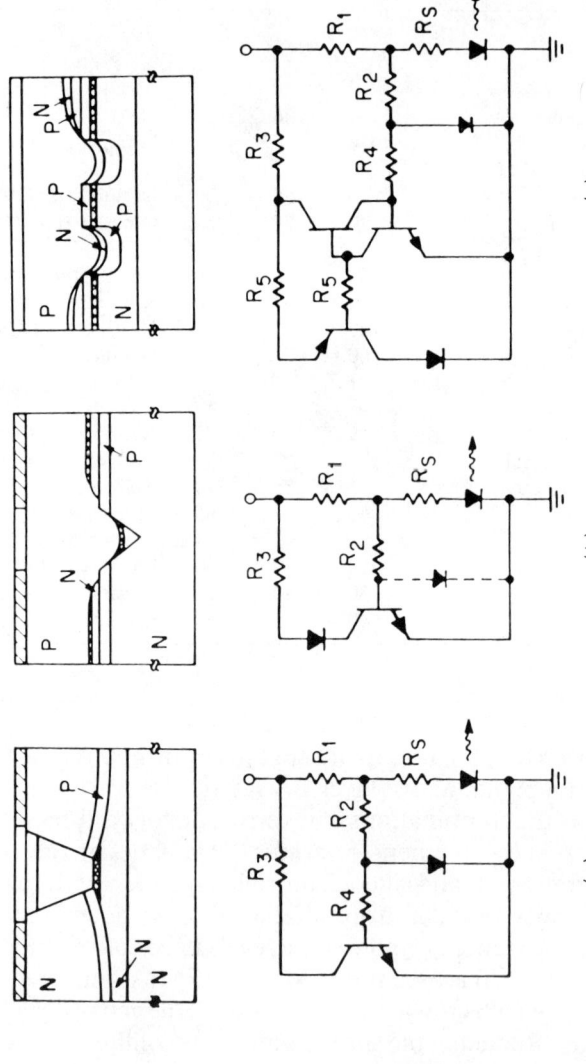

Fig. 37. Equivalent circuit diagrams for several BH designs. (a) EMBH, (b) CSBH, (c) DPCBH. [From Dutta et al. (1984b). © 1984 IEEE.]

1. InGaAsP/InP LASER STRUCTURES AND PERFORMANCE

NPN transistor composed of the *N*-substrate, the *P*-buffer layer and the *N*-regrown layer.

Figure 37c shows a double-channel planar BH laser and its equivalent circuit. Since the transistor leakage path is through a low gain *PNP* structure, the transistor leakage is low. Diode and thyristor leakage in this device is also low because the double channel limits the area of the leakage path to a few microns on either side of the active layer.

Electrical derivative characterization of BH lasers is an excellent technique to determine whether leakage currents are present in laser devices (Nelson *et al.*, 1980; 1981; Wright *et al.*, 1982). Both the $I\, dV/dI - I$ and the $dV/dI - I$ characteristics are useful in this regard. If the laser is treated using the standard model consisting of an ideal $p-n$ junction in series with a resistance R_s, the current–voltage characteristic would be $I = I_0[\exp \eta(V - IR_s) - 1]$, where $\eta = q/nkT$, and the other parameters have their usual meaning. In this model, $I\, dV/dI = 1/\eta + IR_s$ in the subthreshold region and $I\, dV/dI = IR_s$ above threshold when the junction voltage has saturated.

The $I\, dV/dI - I$ characteristic of an ideal laser device is shown in Fig. 38a.

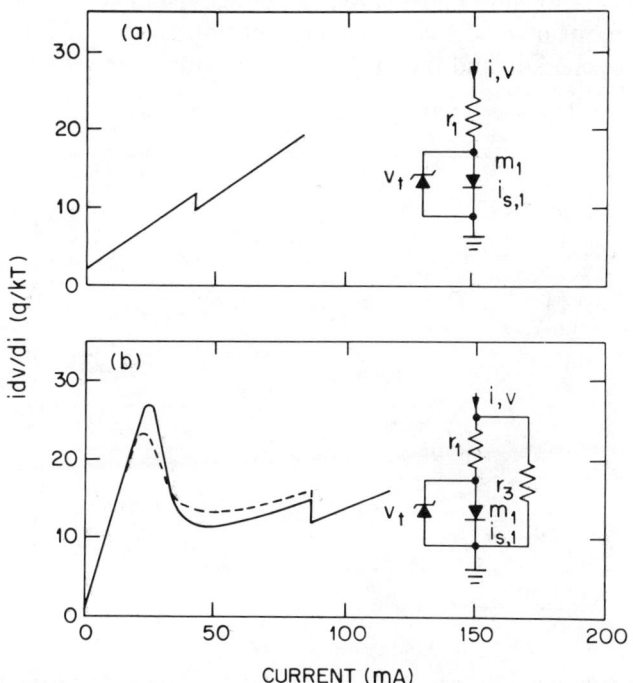

FIG. 38. Calculated $i\, dv/di$ characteristics for (a) an idealized device and (b) a device with a resistive parallel shunt path. ———, $m_1 = 2$; - - -, $m_1 = 3$. [From Wright *et al.* (1982).]

In a device with a resistive shunt-path leakage, the slope of the $I\, dV/dI - I$ curve at low currents is nearly equal to R_{sh}, the resistance of the leakage path as shown in Fig. 38b. At a sufficiently high drive current, the slope of the curve becomes that of the parallel combination of R_{sh} and the series resistance of the path through the active layer. The current that flows through the leakage path contributes to the high-threshold current of the device. Resistive leakage paths are often a result of localized growth imperfections in which a blocking junction is not properly placed.

Leakage currents flowing through confining junctions can often be detected by a decreased $I\, dV/dI$ step at threshold caused by the lack of saturation by the confining junctions at threshold. An alternative way to detect this type of leakage path is through the examination of the $dV/dI - I$ characteristic. For an ideal laser diode, the dV/dI curve exhibits an abrupt step at threshold followed by a constant dV/dI value owing to the saturation of the laser junction voltage. In the presence of leakage through the confining junctions, which do not saturate at threshold, the dV/dI curve will continue to decrease above threshold, as shown in Fig. 39. The presence of this leakage current is typically accompanied by nonlinear $L-I$ characteristics, as shown in Fig. 39. Also shown in Fig. 39 are the characteristics of a nearly ideal BH laser that exhibits no evidence of leakage current. Note that the linearity of the $L-I$ characteristic and the excellent saturation of the $dV/dI - I$ curve

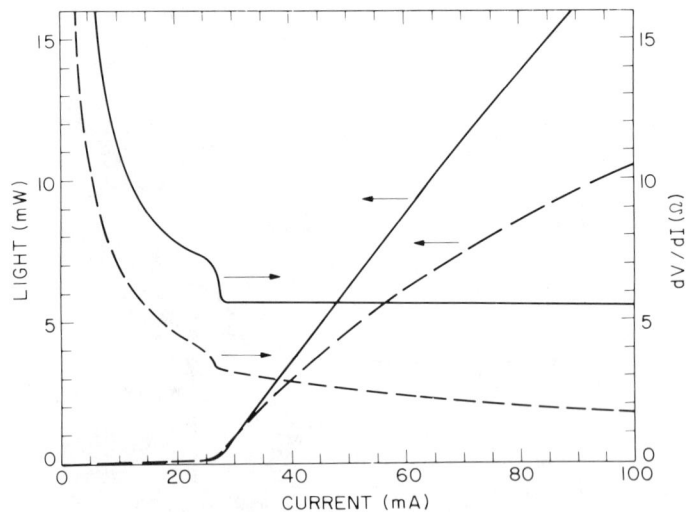

Fig. 39. Experimental $L-I$ and $dv/dI-I$ characteristics for a device showing near-ideal saturation at threshold (solid curves) and for a device showing poor saturation in the dV/dI and rollover of the $L-I$ curve due to leakage.

above threshold changes in the dV/dI characteristics of aged lasers have proved to be useful in identifying the importance of changes in leakage currents on laser reliability (Oomura *et al.*, 1983).

17. MAXIMUM OPERATING TEMPERATURE

Because of their low-threshold currents, real-index-guided lasers have higher operating temperatures than do gain-guided devices, as discussed in Section 12. Among the various types of index-guided lasers, the planar active-layer BH lasers have been operated cw to 110°C at 5 mW output per facet (Fig. 40), whereas the continuous active-layer devices have operated to only 80°C because of their somewhat higher operating current. Nonplanar active-layer BH lasers are reported to have greater temperature sensitivity (lower T_0) than do planar active-layer devices, apparently as a result of the leakage of hot carriers from the narrowest region of the active layer. As shown in Fig. 41, devices of this type will operate to ~ 80°C. Table V summarizes the maximum operating temperature for the 5-mW operation reported for various index-guided-laser structures. InGaAsP lasers emitting at $\lambda = 1.55\ \mu$m have slightly higher threshold currents than do $\lambda = 1.3$-μm devices, possibly because of a larger Auger recombination rate. However, low-threshold devices emitting at $\lambda = 1.55\ \mu$m of both the planar active-layer design (Fig. 42) and the nonplanar active-layer type (Fig. 43) have been fabricated that operate to 5 mW output at temperatures to 60 and 50°C, respectively.

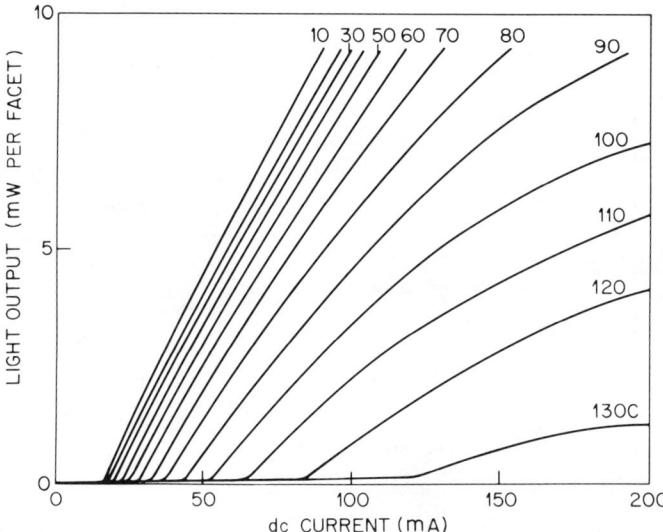

FIG. 40. $L-I$ characteristics for a planar active-layer BH laser ($\lambda = 1.3\ \mu$m) from 10 to 130°C.

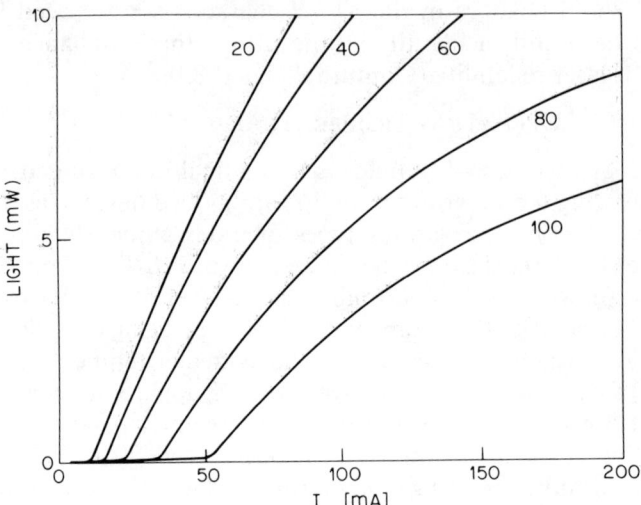

FIG. 41. $L-I$ and $dL/dI-I$ characteristics for a nonplanar active-layer BH laser ($\lambda = 1.3$ μm) from 20 to 100°C.

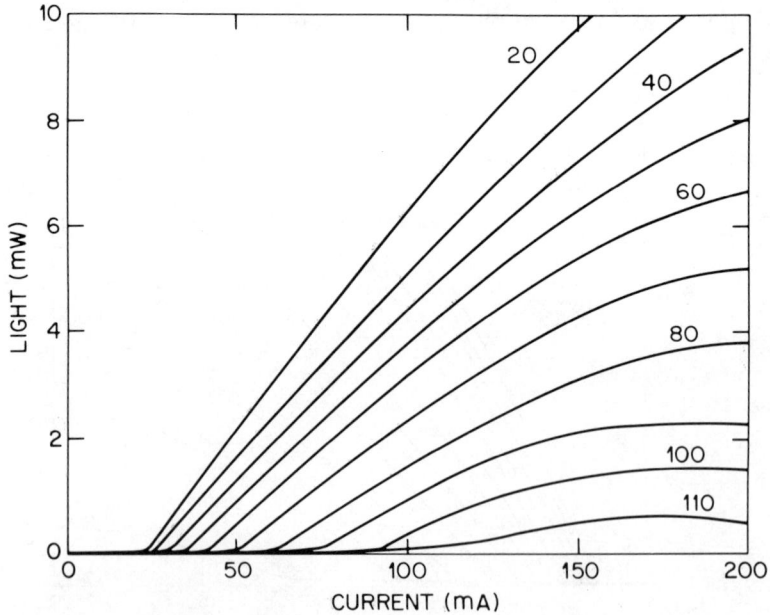

FIG. 42. $L-I$ $dL/dI-I$ and $dV/dI-I$ characteristics for a planar active-layer BH laser ($\lambda = 1.55$ μm) from 20 to 110°C.

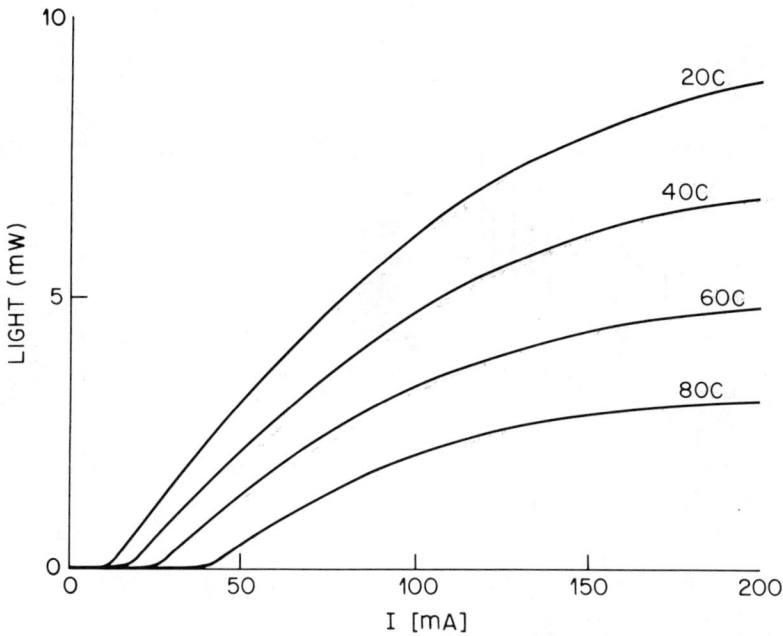

FIG. 43. $L-I$ characteristics of a nonplanar active-layer BH laser ($\lambda = 1.55$ μm).

18. Spectral Characteristics

InGaAsP/InP lasers exhibit spectral characteristics that are similar to those of GaAlAs/GaAs lasers. Figure 44 shows cw spectral data to 8 mW per facet for a BH laser emitting at 1.3 μm. Just above threshold, several longitudinal modes dominate the spectrum. At higher output powers, the device is observed to operate in several longitudinal modes. Single longitudinal mode operation is observed in some devices under cw operation; however, as with GaAlAs/GaAs lasers, the spectra broaden under high-speed modulation (Kishino et al., 1982). Typical spectral half-widths are approximately 10–20 nm. The lasing peak for a 1.3-μm laser is found to vary with temperature at 0.31 meV °C^{-1} or 4.2 Å °C^{-1} (Nelson et al., 1981). For comparison, the lasing peak of AlGaAs/GaAs lasers varies at 0.45 meV °C^{-1}.

19. Far Field: Fiber Coupling

The successful utilization of semiconductor lasers in single-mode optical fiber communication systems requires stable fundamental lateral mode operation. As described earlier, the built-in waveguide must be sufficiently narrow in dimension to provide this type of operation. The efficiency of fiber coupling depends upon several laser characteristics in addition to the mode

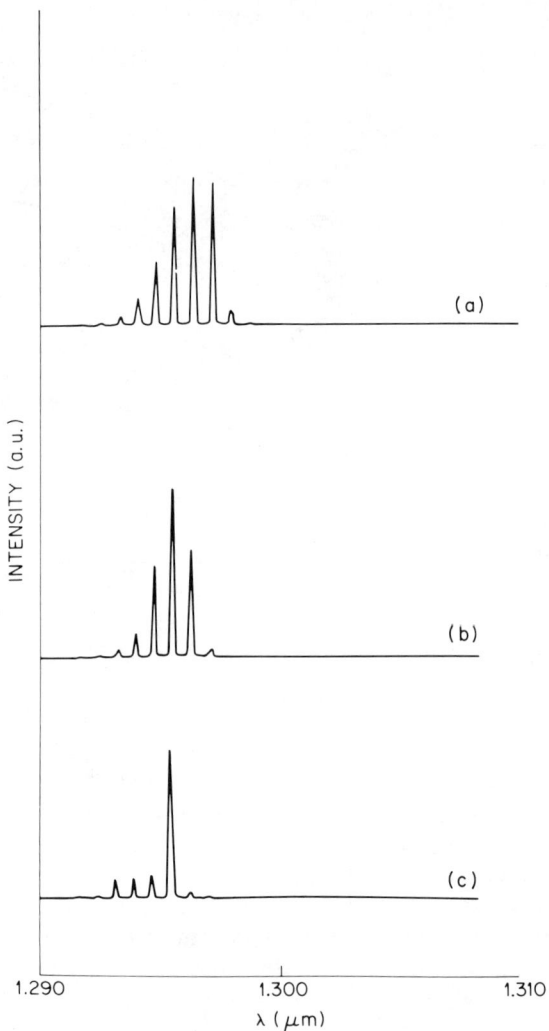

FIG. 44. Optical spectra at three different output power levels for a BH laser emitting at 1.3 μm. Power: (a) 8 mW, (b) 3 mW, (c) 1 mW.

stability. These include the optical beam divergence and the degree of astigmatism of the optical beam. Since index-guided lasers do not exhibit the astigmatism associated with gain-guided lasers, the output of the former devices can be more effectively coupled into single-mode fibers. Although cylindrical optics may be used to optimize the fiber coupling of lasers that have nonidentical beam divergences in the directions parallel and perpendicular to the junction plane, it is more typically found that spherical optics are used, in which case the highest coupling efficiencies are expected to be for devices with equal beam divergences in the two directions. As discussed in Section 15, the beam divergence in the direction perpendicular to the junction plane is determined by the active-layer thickness in the case of InP-cladding layers. Somewhat narrower far-field patterns are observed for devices with additional waveguide layers, as would be expected from dielectric waveguide theory. In the direction parallel to the active layer, the beam divergence is determined by the lateral index step and the waveguide dimensions, as shown in Fig. 28.

20. MODULATION CHARACTERISTICS

Although laser diodes are capable of being modulated at frequencies in excess of 2 GHz, the performance of a device in an optical fiber system depends in detail upon the sensitivity of the device to reflection of light from optical elements of the system, upon the rise and fall time of the diode emission, and upon features of the driver and receiver circuitry and design.

Structure-dependent differences in system performance are known to exist. For example, devices with reverse-biased current-confining junctions are known to have somewhat lower modulation bandwidths (~ 1 GHz) than ridge-guide lasers (~ 2 GHz) (Linke, 1984). The BH lasers (Fig. 25a,b), however, are known to be less sensitive to reflection than the continuous

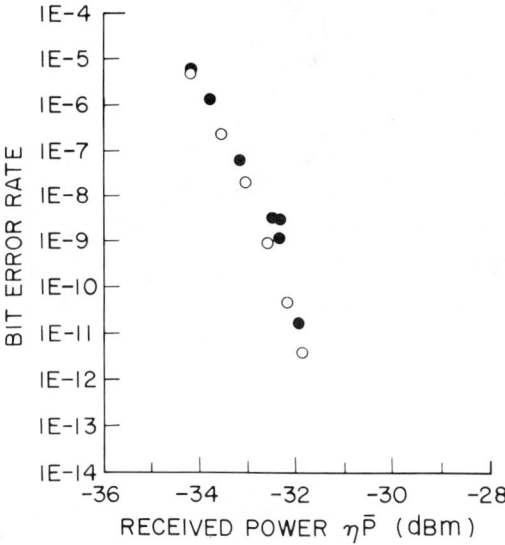

FIG. 45. Bit error rate for a 1.3-μm BH laser without and with 4% reflection (Cheung and Sandahl, 1983). O, no reflection, ●, maximum reflection.

active-layer laser (Fig. 25d) (Cheung and Sandahl, 1983). As shown in Fig. 45, BH lasers operate in a 430 mb sec^{-1} single-mode fiber system with an insignificant power penalty (~0.1 dB) for a 10^{-9}-bit error rate, even for a 12% reflection from a connector 2.6 m from the laser. The ridge-guide device (Fig. 25d), however, cannot be operated below a 10^{-4}-bit error rate for a 4% reflection. Although this reflection sensitivity may be overcome by careful elimination of reflections by the use of index-matching fluids and other techniques, it does represent a problem for the systems engineer.

21. High-Output-Power Devices

The output power of InGaAsP/InP lasers is found to be higher than that for equivalent GaAlAs/GaAs designs. The catastrophic mirror damage that limits the output power from GaAlAs/GaAs BH lasers to only a few milli-

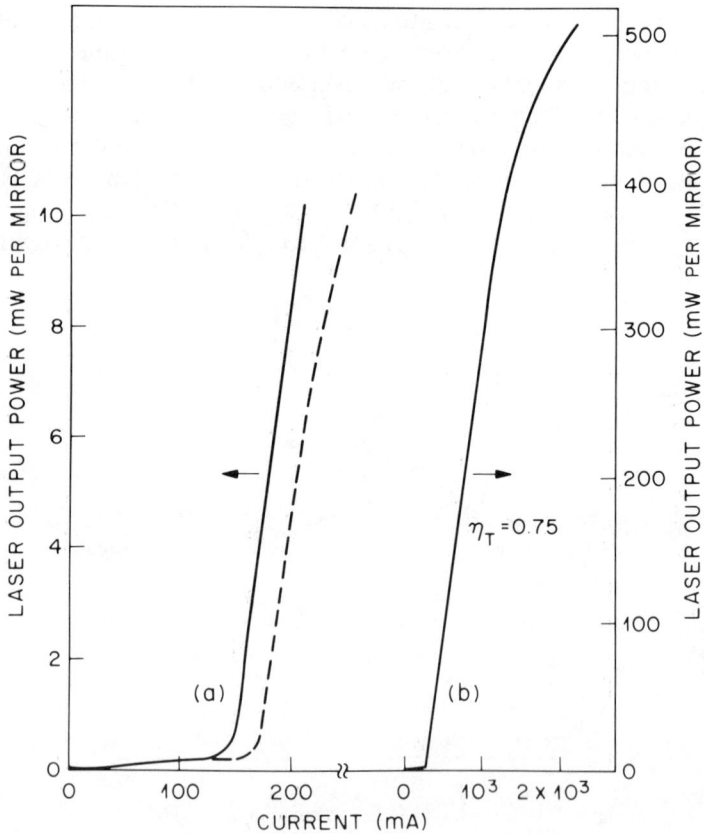

FIG. 46. $L-I$ characteristic of a pulsed-strip BH laser. ———, Pulsed; - - -, cw. [From Nelson et al., 1980.]

watts depends critically upon the surface recombination velocity of the active layer at the cleaved mirror facet. As shown by Henry et al. (1979), the absorption of the optical power by the active layer near the facet (where the injected-carrier density is lower than in the rest of the active layer) causes melting of the GaAlAs, which destroys the laser facet. Since the surface recombination velocity of the InGaAsP/InP material system is two orders of magnitude lower than that for GaAs, the catastrophic damage level is high.

This estimate is verified by the experimental observations that output powers as high as 40 mW can be achieved for 1.3-μm BH lasers. Devices with optical waveguides, such as strip BH lasers, are found to produce powers over 500 mW pulsed as shown in Fig. 46.

References

Adams, A. R., Asada, M., Suematsu, Y., and Arai, S. (1980). *Jpn. J. Appl. Phys.* **19**, L621.
Agrawal, G. P. (1983). Unpublished work.
Agrawal, G. P., and Dutta, N. K. (1983). *Electron. Lett.* **19**, 974.
Agrawal, G. P., Joyce, W. B., Dixon, R. W., and Lax, M. (1982). Unpublished.
Anthony, P. J., and Schumaker, N. E. (1980). *IEEE Electron Device Lett.* **EDL-1**, 58.
Asada, M., Adams, A. R., Stubkjar, K. E., Suematsu, Y., Itaya, Y., and Arai, S. (1980). *IEEE J. Quantum Electron.* **QE-16**, 611.
Beattie, A. R., and Landsberg, P. T. (1959). *Proc. R. Soc. London Ser. B* **249**, 16.
Casey, H. C., and Panish, M. B. (1978). "Heterostructure Lasers," Part A. Academic Press, New York.
Chelikowsky, J. R., and Cohen, M. L. (1976). *Phys. Rev. B* **14**, 556.
Chen, T. R., Margalit, S., Koren, V., Yu, K. L., Chiu, L. C., Hasson, A., and Yariu, A. (1983). *Appl. Phys. Lett.* **42**, 1000.
Cheung, N. K., and Sandahl, C. R. (1983). *Top. Meet. Op. Fiber Commun., New Orleans, 1983*, Paper TUJI.
Cheung, S. S., Gardner, W. B., McGrath, C. J., and Kaiser, P. (1983). *Top. Meet. Op. Fiber Commun., New Orleans, 1983*, Paper PD No. 8.
Chiu, L. C., Chen, P. C., and Yariv, A. (1982). *IEEE J. Quantum Electron.* **QE-18**, 938.
Dixon, R. W. (1976). *Bell Syst. Tech. J.* **55**, 973.
Dixon, R. W. (1980). *Bell Syst. Tech. J.* **59**, 669.
Dixon, R. W., Nash, F. R., Hartman, R. L., and Hepplewhite, R. T. (1976). *Appl. Phys. Lett.* **29**, 372.
Dutta, N. K. (1980). *J. Appl. Phys.* **51**, 6095.
Dutta, N. K. (1981). *J. Appl. Phys.* **52**, 70.
Dutta, N. K., and Nelson, R. J. (1980). *Conf. Ser.—Inst. Phys.* **56**, 193.
Dutta, N. K., and Nelson, R. J. (1981). *Appl. Phys. Lett.* **38**, 407.
Dutta, N. K., and Nelson, R. J. (1982a). *IEEE J. Quantum Electron.* **QE-18**, 4.
Dutta, N. K., and Nelson, R. J. (1982b). *IEEE J. Quantum Electron.* **QE-18**, 375.
Dutta, N. K., and Nelson, R. J. (1982c). *IEEE J. Quantum Electron.* **QE-18**, 871.
Dutta, N. K., and Nelson, R. J. (1982d). *J. Appl. Phys.* **53**, 74.
Dutta, N. K., Nelson, R. J., and Barnes, P. A. (1981). *Electron. Lett.* **16**, 653.
Dutta, N. K., Nelson, R. J., Wright, P. D., and Craft, D. C. (1984a). *IEEE J. Lightwave Technol.* **LT-2**, 160.

Dutta, N. K., Wilt, D. P., and Nelson, R. J. (1984b). *IEEE J. Lightwave Technol.* **LT-2,** 201.
Dutta, N. K., Wilt, D. P., Befomi, P., Dautremont-Smith, W. C., Wright, P. D., and Nelson, R. J. (1984c). *Appl. Phys. Lett.* **44,** 483.
Greene, P. D., and Henshall, G. D. (1979). *IEE J. Solid-State Electron Devices* **3,** 174.
Haug, A. (1983). *Appl. Phys. Lett.* **42,** 512.
Henry, C. H., Petroff, P. M., Logan, R. A., and Merritt, F. R. (1979). *J. Appl. Phys.* **50,** 3721.
Henry, C. H. Levine, B. F., Logan, R. A., and Bethea, C. G. (1983a). *IEEE J. Quantum Electron.* **QE-19,** 905.
Henry, C. H. Logan, R. A., Merritt, F. R., and Luongo, J. P. (1983b). *IEEE J. Quantum Electron.* **QE-19,** 947.
Henshall, G. D., and Thompson, G. H. B. (1982). *Proc. Int. IEEE Semicond. Laser Conf., 8th, 1982,* Paper 29.
Hirao, M., Tsuji, S., Mizuishi, K., Doi, A., and Nakamura, M. (1980). *J. Opt. Commun.* **1,** 10.
Horikoshi, Y., and Furukawa, Y. (1979). *Jpn. J. Appl. Phys.* **18,** 809.
Hsieh, J. J., and Shen, C. C. (1978). *Fiber Integr. Opt.* **1,** 357.
Imai, H., Ishikawa, H., Tanahashi, T., and Takusagawa, M. (1981). *Electron. Lett.* **17,** 17.
Ishikawa, H., Imai, H., Tanahashi, T., Nishitani, Y., and Takusagawa, M. (1981). *Electron. Lett.* **17,** 465.
Itaya, Y., Katayama, S., and Suematsu, Y. (1979). *Electron. Lett.* **15,** 123.
Iwashita, K., and Nakagawa, K. (1983). *J. Quantum Electron.* **QE-19,** 1173.
Joyce, W. B. (1980). *J. Appl. Phys.* **51,** 2394.
Joyce, W. B. (1982). *J. Appl. Phys.* **53,** 7235.
Kaiser, P. C., Cheung, N. K., Tomita, A., Ogawa, K., Cheng, S. S., Lipson, J., Stakelon, T. S., and Nelson, R. J. (1983). *Conf. Integr. Op. Opt. Fiber Commun., 1983, Tokyo,* Pap. 27C2-4.
Kaminow, I. P., Stulz, L. W., Ko, J. S., Dentai, A. G., Nahory, R. E., DeWinter, J. C., and Hartman, R. L. (1983a). *IEEE J. Quantum Electron.* **QE-19,** 1312.
Kaminow, I. P., Stulz, L. W., Ko, J. S., Miller, B. I., Feldman, R. D., DeWinter, J. C., and Pollack, M. A. (1983b). *Electron. Lett.* **19,** 877.
Kawaguchi, H., Takahei, K., Suzuki, Y., and Iwane, G. (1981a). *Appl. Phys. Lett.* **38,** 957.
Kawaguchi, H., Takahei, K., Toyoshima, Y., Nagai, H., and Iwane, G. (1981b). *IEEE J. Quantum Electron.* **QE-18,** 469.
Kishino, K., Aoko, S., and Suematsu, Y. (1982). *IEEE J. Quantum Electron.,* **QE-18,** 343.
Linke, R. A. (1984). *J. Lightwave Technol.* **LT-2,** 40.
Murrell, D. L., Phil, M., Walling, R. H., Hobbs, R. E., and Devlin, W. J. (1982). *IEEE Proc.* **6,** 262.
Nahory, R. E., Pollack, M. A., Johnston, W. D., and Burns, R. L. (1978). *Appl. Phys. Lett.* **33,** 659.
Nelson, R. J. (1979). *Appl. Phys. Lett.* **35,** 654.
Nelson, R. J., and Dutta, N. K. (1980). *Appl. Phys. Lett.* **37,** 769.
Nelson, R. J., and Dutta, N. K. (1983). *J. Appl. Phys.* **54,** 2923.
Nelson, R. J., and Sobers, R. G. (1978a). *Appl. Phys. Lett.* **32,** 761.
Nelson, R. J., and Sobers, R. G. (1978b). *J. Appl. Phys.* **49,** 6103.
Nelson, R. J., Wright, P. D., Barnes, P. A., Brown, R. L., Cella, T., and Sobers, R. G. (1980). *Appl. Phys. Lett.* **36,** 358.
Nelson, R. J., Wilson, R. B., Wright, P. D., Barnes, P. A., and Dutta, N. K. (1981). *IEEE J. Quantum Electron.* **QE-17,** 202.
Nelson, R. J., Dutta, N. K., Wilson, R. B., Wilt, D. P., Besomi, P., Wright, P. D., Shen, T. M., and Flynn, E. J. (1984). *Top. Meet. Op. Fiber Commun., New Orleans, 1984,* p. WJ5.
Nishi, H., Yano, M., Nishitani, Y., Akita, Y., and Takusagawa, M. (1979). *Appl. Phys. Lett.* **35,** 232.

Olsen, G. H., Neuse, C. J., and Ettenberg, M. (1979). *Appl. Phys. Lett.* **34**, 262.
Oomura, E., Murotani, T., Higuchi, H., Namizaki, H., and Susaki, W. (1981). *IEEE J. Quantum Electron.* **QE-17,** 646.
Oomura, E., Higuchi, H., Hirano, R., Sakaibara, Y., Namizaki, H., and Susaki, W. (1983). *Electron. Lett.* **19,** 407.
Razeghi, M., Hersee, S., Hirtz, P., Blondeau, R., de Cremoux, B., and Duchemin, J. P. (1983). *Electron. Lett.* **19,** 336.
Rode, D. L. (1974). *J. Appl Phys.* **45,** 3887.
Sermage, B., Eichler, H. J., Heritage, J. P., Nelson, R. J., and Dutta, N. K. (1983). *Appl. Phys. Lett.* **42,** 259.
Shah, J., Nahory, R. E., Leheny, R. F., and Temkin, H. T. (1981). *Appl. Phys. Lett.* **39,** 618.
Stubkjaer, K., Suematsu, Y., Asada, M., Arai, S., and Adams, A. R. (1980). *Electron. Lett.* **16,** 894.
Su, C. B., Schlafer, J., Manning, J., and Olshansky, R. (1982). *Electron. Lett.* **18,** 1108.
Sugimura, A. (1981). *IEEE J. Quantum Electron.* **QE-17,** 627.
Tamari, N., Oron, M., Shtrikman, H., and Ballman, A. A. (1982). *Top. Meet. Op. Fiber Commun., Phoenix, Arizona, 1982,* p. 56.
Tauc, J. (1959). *J. Phys. Chem. Solids* **8,** 219.
Thompson, G. H. B. (1980). "Physics of Semiconductor Laser Devices." Wiley, New York.
Thompson, G. H. B. (1983). *Electron. Lett.* **19,** 154.
Thompson, G. H. B., and Henshall, G. D. (1980). *Electron. Lett.* **16,** 42.
Tsang, W. T., Reinhardt, F. K., and Ditzenberger, J. A. (1982). *Appl. Phys. Lett.* **41,** 1094.
Tsang, W. T., Logan, R. A., Olsson, N. A., Temkin, H., Van der Ziel, J. P., Kaminow, I. P., Kasper, B. L., Linke, R. A., Mazuvczyk, V. J., Miller, B. I., and Wagner, R. E. (1983). *Top. Meet Opt. Fiber Commun., New Orleans,* Paper PD No. 9.
Turley, S. E. H. (1982). *Electron. Lett.* **18,** 590.
Turley, S. E. H., Henshall, G. D., Greene, P. D., Knight, V. P., Moule, D. M., and Wheeler, S. A. (1981). *Electron. Lett.* **17,** 869.
Ueno, M., Fakuma, I., Furufe, T., Matsumoto, Y., Kawano, H., Ide, Y., and Matsumoto, S. (1981). *IEEE J. Quantum Electron.* **QE-17,** 1930.
Ueno, M., Lang, R., Matsumoto, S., Kawano, H., Furuse, T., and Sakuma, I. (1982). *IEE Proc.* **129,** 218.
Uji, T., Iwamoto, K., and Lang, R. (1981). *Appl. Phys. Lett.* **38,** 193.
Uji, T., Iwamoto, K., and Lang, R. (1983). *IEEE Trans. Electron Devices* **ED-30,** 316.
Wada, O., Yamakoshi, S., Abe, M., Akita, K., and Toyama, Y. (1979). *Opt. Commun. Conf., 1979.* Paper 4.6-1.
Wright, P. D., Rezak, E. A., and Holonyak, N. (1977). *J. Cryst. Growth* **41,** 254.
Wright, P. D., Joyce, W. B., and Craft, D. C. (1982). *J. Appl. Phys.* **53,** 1364.
Yano, M., Nishi, H., and Tukusagawa, M. (1981). *J. Appl. Phys.* **52,** 3172.
Yano, M., Morimoto, M., Nishitani, T., and Takusagawa, M. (1982). *Appl. Phys. Lett.* **41,** 390.
Yuasa, T., Onabe, K., Ide, Y., Isoda, Y., Hayashi, J., Furuse, T., Sakuma, I., and Matsumoto, Y. (1983). *J. Appl. Phys.* **54,** 50.

CHAPTER 2

Mode-Stabilized Semiconductor Lasers for 0.7–0.8- and 1.1–1.6-μm Regions

N. Chinone and M. Nakamura

CENTRAL RESEARCH LABORATORY
HITACHI, LTD.
TOKYO, JAPAN

I.	INTRODUCTION	61
II.	TRANSVERSE-MODE CONTROL	62
	1. *Transverse-Mode Instability*	62
	2. *Index-Guided Lasers*	67
	3. *Gain-Guided Lasers*	71
III.	LONGITUDINAL MODE CONTROL	72
	4. *Longitudinal Mode of Transverse-Mode-Stabilized Lasers* .	72
	5. *Longitudinal-Mode-Stabilized Lasers*	81
IV.	CONCLUSION	88
	REFERENCES	89

I. Introduction

It has been widely recognized that lasing-mode stabilization, especially transverse-mode stabilization, is one of the most important technologies for semiconductor lasers for realizing a stable light source in optical communication systems. Transverse-mode instability is the origin of so-called kinks in light-output–current characteristics, which are accompanied by anomalous lasing behaviors such as excess optical noise, beam-direction shifts, and deterioration of modulation characteristics. The mode instability has been attributed to the deformation of the laser gain profile and of the carrier-induced refractive index profile. In order to stabilize the mode, there are two approaches: (1) introduction of rigid waveguide structure, and (2) elimination of gain-profile deformation by narrowing the current-injection path.

Longitudinal mode structure of the transverse-mode-stabilized lasers depends on the guiding mechanism of the transverse mode. Usually, lasers with a rigid waveguide have a single longitudinal mode, and only those with a narrow current path have multimodes. However, even in those single longitudinal mode lasers, the longitudinal mode easily becomes unstable by temperature change or by reflected laser light, and it also becomes multimode under high-speed modulation. In order to realize a stable light source for long-distance single-mode fiber communication systems, the stabilized single longitudinal mode operation is required. Several methods of stabilizing the longitudinal mode have been proposed: (1) mode selection by composite cavity, (2) grating feedback inside the device structure, etc.

The purpose of this chapter is to review achievements on mode stabilization. In Part II, the origin of transverse-mode instability and the method of mode stabilization are briefly discussed and characteristics of mode-stabilized lasers described. In Part III, longitudinal mode behaviors and the method of their stabilization are described.

II. Transverse-Mode Control

1. TRANSVERSE-MODE INSTABILITY

a. Experimental Results

An introduction of stripe geometry into semiconductor lasers was first reported by Dyment *et al.* (1972), where a current-injection region was defined using an oxide mask. It proved useful to reduce threshold current and also to realize single-transverse-mode operation. Similar results were also obtained in proton bombardment (D'Asaro, 1973), zinc diffusion (Yonezu *et al.*, 1973), and mesa structure (Chinone *et al.*, 1976). Typically, the stripe width was 10–20 μm. In these lasers, light is confined to the current-injection region due to gain distribution. When these lasers are excited above threshold, power saturation or kinking occurs in light–current characteristics. An experimental result of a low-mesa-structure laser is shown in Fig. 1. A kink is clearly seen at point A in Fig. 1a. The near- and far-field patterns of this laser are shown in Fig. 1b and c. It was found that the laser oscillated in the fundamental transverse mode just above threshold. However, as current was increased, the mode deformed substantially, showing a kink in the light–current characteristic. With a further increase in the applied current above point B, the lasing mode deformed and centered at a different position. The displacement of the near-field profile was typically 2–4 μm. The mode deformation was accompanied by a beam-direction shift of 3–5°. At a kink, anomalous transient behavior and huge noise were observed, as shown in Fig. 2. Such anomalous behaviors as shown in Figs. 1, and 2 were first thought to originate from defects or nonuniformity in the lasers. They were therefore expected to disappear with improvements in fabrication technology. That is partially true. If macroscopic defects exist in a laser cavity, stable single-mode operation is hardly expected. However, the reproducibility of the unstable mode behavior has strongly suggested that the observed instability originates from a more basic aspect of laser action.

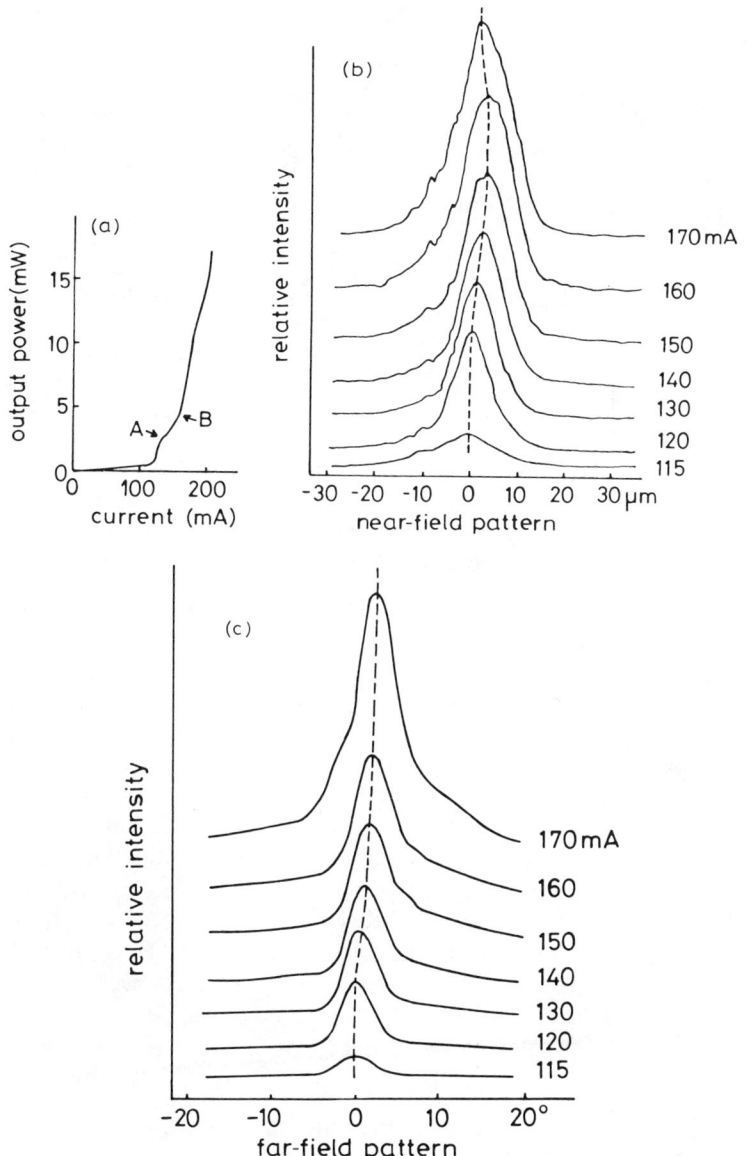

FIG. 1. Mode patterns of a low-mesa-structure gain-guided laser. The stripe width was 20 μm: (a) light–current characteristics; (b) near-field patterns; (c) far-field patterns.

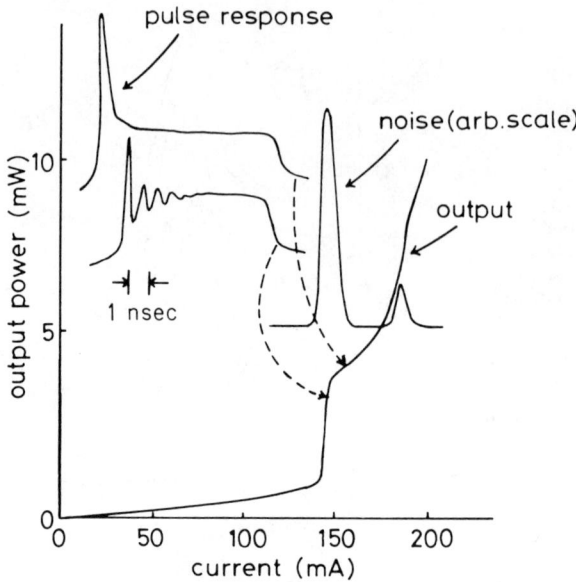

FIG. 2. Pulse response and noise at a kink. The sample is a low-mesa-structure laser with a 20-μm-stripe width.

b. Theoretical Models for the Transverse-Mode Instability

Theoretical analysis of the transverse-mode instability was first attempted by Chinone (1977) and by Seki et al. (1979), who assumed that the transverse mode was guided only by the gain profile. On the other hand, theoretical analysis by Thompson et al. (1978) and Lang (1979) included the carrier-induced refractive index profile, showing that it plays a more important role. Buus (1978) also analyzed the instability, taking into account this factor.

In lasers already mentioned, gain profile can be easily changed by the strong stimulated emission. When the intensity of the lowest order transverse mode increases, the carriers are exhausted at the peak position of the mode profile, and this causes a dip in the carrier-density profile (spatial hole burning). The local carrier consumption results in the following:

(1) A local decrease of gain g; the gain coefficient g is approximately related to the carrier density n by

$$g = an - b, \qquad (1)$$

where a and b are constant. At lasing wavelength, these are $\sim 4 \times 10^{-16}$ cm^2 and ~ 300 cm^{-1}, respectively, for lightly doped GaAs (Stern, 1976).

(2) A local increase of refractive index n_r; due to the free-carrier plasma

effect and the abnormal dispersion of the band edge, the refractive index depends on carrier density.

It has been reported that the latter effect is one order of magnitude larger than the former. The dependence is expressed by (Henry et al., 1981),

$$dn_r/dn \sim -2 \times 10^{-20} \quad cm^3. \tag{2}$$

The deformation of the gain profile results in the nonlinearity of the light–current characteristics. However, the clear kinking is shown to originate from the local variation of both the refractive index and the gain profile as follows: The local increase of the refractive index causes the mode narrowing. Since the mode is confined by a refractive index profile such as formed by the mode-intensity profile itself, this mode can easily move by the slight asymmetry of, for instance, the device structure. The mode shift into the lower gain region results in the clear kinking.

For quantitative discussion, one has to simultaneously solve the wave equation and the carrier diffusion equation:

$$(\partial^2 E/\partial x^2) + (k^2\varepsilon - \beta^2)E = 0, \tag{3}$$

$$D_a \frac{\partial^2 n}{\partial x^2} + \frac{J}{qd} - \frac{n}{\tau_s} - \frac{cg}{n_r}|E|^2 = 0, \tag{4}$$

where E is the electric field amplitude; $k = 2\pi/\lambda$, where λ is the wavelength; ε is the complex dielectric constant, β the propagation constant, D_a the ambipolar diffusion constant, J the injection current density, d the thickness of the active layer, τ_s the carrier lifetime, and c the light velocity. The complex dielectric constant ε is related to the refractive index n_r and the gain coefficient g by

$$\varepsilon = [n_r + ig/2k]^2. \tag{5}$$

The dielectric constant ε depends on the carrier density, as indicated by Eqs. (1), (2), and (5). In order to obtain carrier-density and light-intensity profiles, Eqs. (3) and (4) have to be solved self-consistently. Examples of the calculated carrier-density and light-intensity profiles are presented in Figs. 3 and 4 (Lang, 1979), respectively, where the stripe width is assumed to be 12 μm. In the self-focusing model mentioned before, the index-change ratio R, defined by

$$R = \frac{\partial \operatorname{Re}(\varepsilon)/\partial n}{\partial \operatorname{Im}(\varepsilon)/\partial n}, \tag{6}$$

is a key parameter. If R is negative and the absolute value is large, the self-focusing effect becomes more dominant. In the figures, R is assumed to be -1. More recent works have shown that R is approximately -6 (Henry et

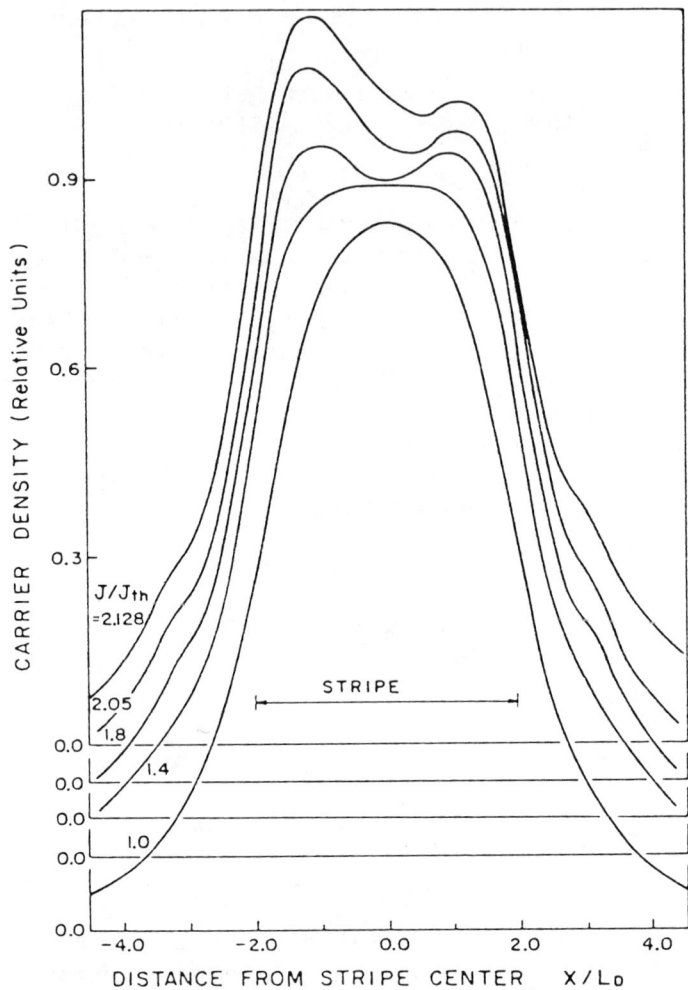

FIG. 3. Calculated carrier-density profiles for a 12-μm-wide gain-guided laser. Parameters: $S = 12$ μm, $L_D = 3$ μm, $R = -1$, $\varepsilon = 10^{-5}$. [From Lang (1979). © IEEE 1979.]

al., 1981). This factor is important for various laser characteristics. It is also assumed that the dielectric constant initially has an asymmetry that can be brought about by the device structure or material composition nonuniformity. It is seen in Figs. 3 and 4 that the light-intensity profile is gradually trapped into a dip in the carrier-density profile. As the current is increased further, the emission spot (and the dip) moves to the higher refractive index side. The accumulation of carriers in the opposite side of the stripe enhances the lateral shift. The displacement of the mode from the gain peak reduces

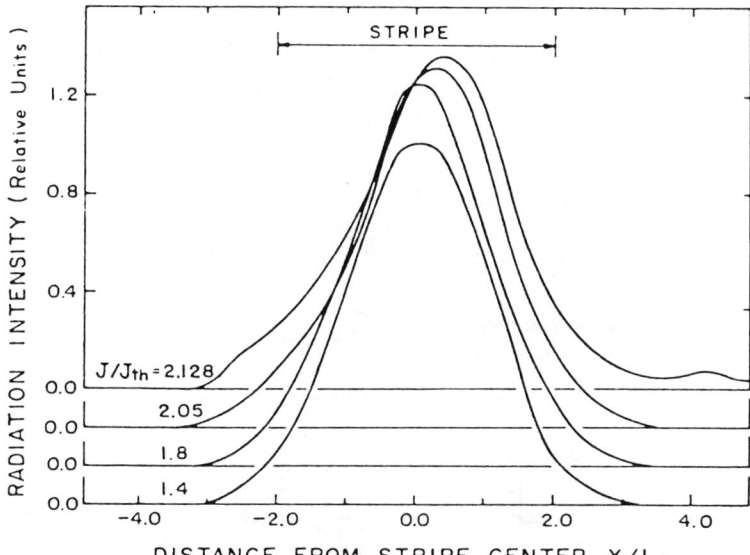

FIG. 4. Calculated radiation-intensity profiles for a 12-μm-wide gain-guided laser. Parameters: $S = 12$ μm, $L_D = 3$ μm, $R = -1$, $\varepsilon = 10^{-5}$. [From Lang (1979). © IEEE 1979.]

the mode gain, resulting in the kinking. Such mode behaviors agree well with experimental results.

There are two basic approaches to eliminate such anomalous behaviors of the transverse mode: (1) introduction of rigid-waveguide structure, and (2) narrowing the current-injection path. Lasers with a transverse mode guided by a rigid waveguide are called index-guided lasers, and those with a narrow current path are called gain-guided lasers. Such a difference in guiding mechanisms affects various laser characteristics.

2. INDEX-GUIDED LASERS

The transverse mode can be stabilized by introducing a rigid waveguide. Various structures have been successfully demonstrated. Some of them are schematically illustrated in Fig. 5. A structure typical of index-guided lasers is that of a buried heterostructure (BH) first developed by Tsukada (1974). A mesa-shaped double heterostructure is buried by a second epitaxial growth. In this laser, the GaAlAs active region is completely embedded by GaAlAs with more Al content. Therefore the spatial hole burning does not cause the mode shift described previously. In order to analyze the mode characteristics, the effective refractive index approximation can be used. First, the mode perpendicular to the junction plane is calculated in the stripe region by using Eq. (3). From the propagation constant β in Eq. (3), the effective refractive

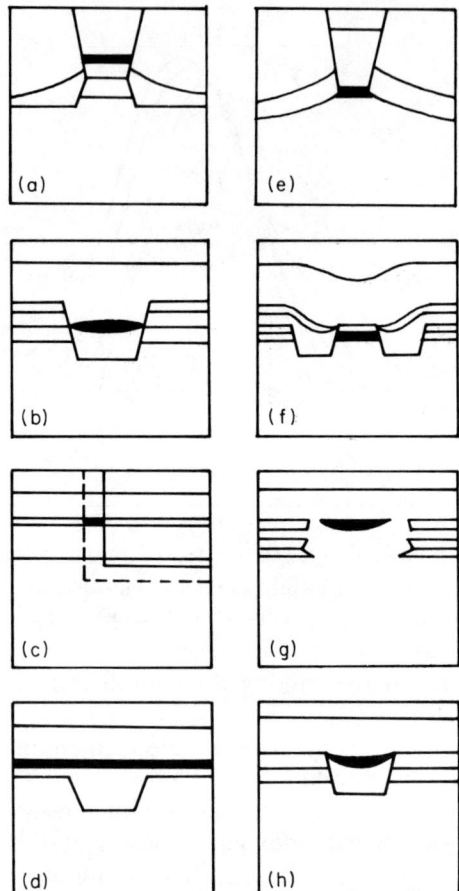

FIG. 5. Structures of index-guided lasers: (a)–(d) GaAlAs/GaAs lasers; (e)–(h) InGaAsP/InP lasers. (a) BH–BOG (Chinone et al., 1979); (b) BCW (Shima et al., 1982); (c) TJS (Kumabe et al., 1979); (d) CSP (Aiki et al., 1978); (e) BH (Hirao et al., 1980); (f) DC–PBH (Mito et al., 1982); (g) buried-crescent laser (Murotani et al., 1982); (h) VSB (Ishikawa et al., 1982).

index in the stripe region can be obtained. By using this value and the refractive index of the burying layer, the mode parallel to the junction plane is calculated (Ito and Saito, 1980). Usually, the stripe width is not very narrow, because facet degradation by high optical density is more serious as stripe width narrows. Therefore the higher order mode can propagate the waveguide. When spatial hole burning is serious, especially at high output powers, the higher-order-mode gain becomes larger than that of the lowest-order-mode (Hakki, 1975), resulting in the multi-transverse-mode oscillation. In order to eliminate this kind of instability, the stripe width should be shorter than the carrier diffusion length (~3 μm), and the mode width should be made as wide as possible by optimizing the refractive index step height. In the buried-heterostructure–buried-optical-guide (BH–BOG) laser shown in Fig. 5a, the stripe width and the Al content of the burying layer were optimized in this respect and the optical guide layer was introduced in order to have higher output power. A characteristic feature of the BH laser is its low-threshold current by superior current confinement as well as its mode stability. Light–current characteristics and far field patterns of a BH–BOG laser are shown in Fig. 6 (Chinone *et al.*, 1979). Perfectly linear light–current characteristics and low-threshold current were obtained. This laser was used in analog optical transmission experiments because of its superior linearity (Nagano *et al.*, 1979). Another example of GaAlAs BH lasers is the strip-buried-heterostructure (SBH) laser (Tsang *et al.*, 1978). Linear light–current characteristics and a stable transverse mode were obtained at an output power of above 200 mW.

Buried-heterostructure lasers can be obtained by using a grooved substrate. By liquid-phase epitaxy, an active layer is embedded in the groove

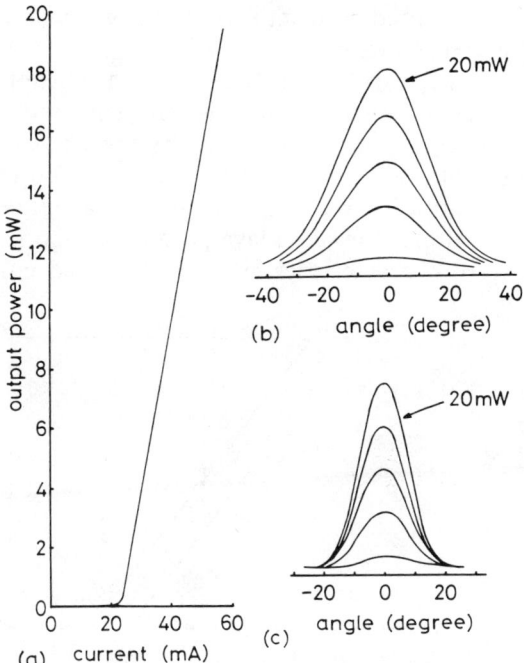

FIG. 6. Light–current characteristic (a), and far-field patterns perpendicular (b) and parallel (c) to the junction plane of a BH–BOG laser at 295 K, in the dc mode. [From Chinone *et al.* (1979).]

under an appropriate condition. A typical structure is the buried-convex-waveguide structure shown in Fig. 5b. Other examples are the channeled-substrate BH (Kirkby and Thompson, 1976) and the etched BH (Burnham and Scifres, 1975) lasers. Low-threshold currents and stable transverse modes were also achieved in these structures.

In InGaAsP/InP lasers, the facet degradation is not so serious as in GaAlAs/GaAs lasers, and the temperature dependence of threshold current density is large. Therefore the BH structure is more useful for InGaAsP/InP lasers, because of its very low threshold current and resultant high-temperature operation. Experimental results of maximum temperatures for cw operation of InGaAsP/InP BH lasers are shown in Fig. 7, where maximum temperatures are plotted both for lasers mounted in p-side-down configurations on a Si heat sink and for those in p-side-up configurations on a SiC ceramic heat sink (Tsuji et al., 1982). It is seen that maximum temperature increases as threshold current decreases. Even in p-side-up configurations, cw operation was achieved at temperatures above 100°C in spite of poor temperature characteristics of InGaAsP/InP lasers. Other examples of InGaAsP/InP BH structures are the double-channeled planar BH (DC–PBH) shown in Fig. 5f, the buried crescent shown in Fig. 5g, and the V-grooved substrate-BH (VSBH) shown in Fig. 5h.

A refractive index step can be obtained by a change in doping level due to the free-carrier effect. A typical example of this is the transverse-junction-stripe (TJS) laser shown in Fig. 5c. In TJS lasers, the active layer is Te doped with a carrier concentration of $\sim 2 \times 10^{18}$ cm^{-3}. The Zn is selectively diffused by a two-step diffusion method. As a result, carrier density has a dip around the $p-n$ junction in the active layer, resulting in optical confinement. Linear light–current characteristics and low-threshold currents were ob-

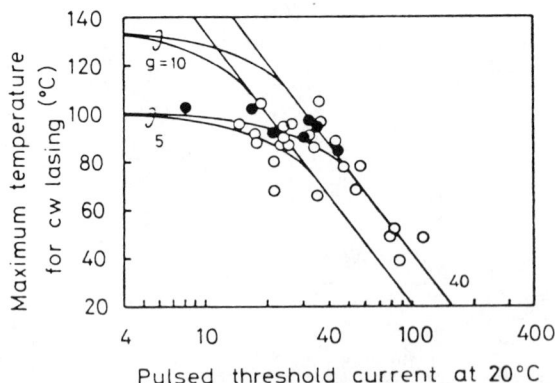

FIG. 7. Maximum temperature for cw operation of InGaAsP/InP BH lasers with $T_0 = 50$ K, $R_s = 6.0$ Ω, $V_j = 0.95$ V, $R_{th} = 180$°C/W. ●, p-side up on SiC; ○, p-side down on Si. [From Tsuji et al. (1982).]

tained. Another example is the deep-Zn-diffused planar-stripe (DPS) laser, where Zn is diffused down to the active layer (Ueno and Yonezu, 1979).

The third method for realizing stable transverse-mode operation is to grow double-heterostructure layers on a grooved substrate. A typical example is the channeled-substrate planar (CSP) structure shown in Fig. 5d. It has a flat active layer in contrast to the embedded structure. By complex index steps originating from the optical loss of the substrate, the mode is confined to the substrate groove. Linear light–current characteristics were obtained up to 50–60 mW. Other examples of laser structures formed on grooved substrates are the planoconvex waveguide (PCW) (Ide et al., 1980); the constricted-double-heterostructure (CDH) (Botez and Zory, 1978); the V-channeled-substrate inner stripe (VSIS) (Hayakawa et al., 1982); and the terraced-substrate (TS) (Sugi et al., 1979) structures.

3. GAIN-GUIDED LASERS

In contrast to the index-guided laser, no rigid waveguide is formed around the active layer in gain-guided lasers. In order to eliminate the mode instability, the injection-current path is made sufficiently narrow. Therefore the dip of the carrier-density profile is easily smoothed out by carrier diffusion. In this case, the carrier-density profile has a peak at the center of the stripe. Due to the carrier-induced refractive index change, the refractive index is small at the center and large around the edge of the stripe. Therefore the mode is gain guided but anti-index guided. This effect makes the mode width wider than that of the pure gain-guided mode and suppresses the dip of the carrier-density profile. Some narrow-stripe gain-guided lasers are schematically illustrated in Fig. 8. A typical example is the V-grooved laser shown in Fig. 8a. The current is limited to the 3–4-μm-wide region of the bottom of the V-groove. Light–current characteristics and far-field patterns are shown in Figs. 9 and 10, respectively (Marschall et al., 1979). Linear light–current characteristics were obtained by narrowing the stripe width. Far-field patterns parallel to the junction plane were much wider than those of the

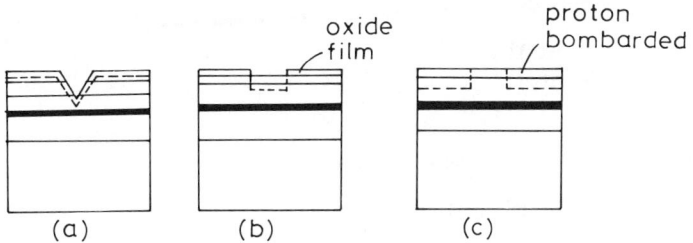

FIG. 8. Structures of narrow-stripe gain-guided GaAlAs lasers. (a) V-grooved laser (Marschall et al., 1979); (b) proton-bombarded laser (Dixon et al., 1976); (c) oxide stripe laser (Wolf et al., 1981).

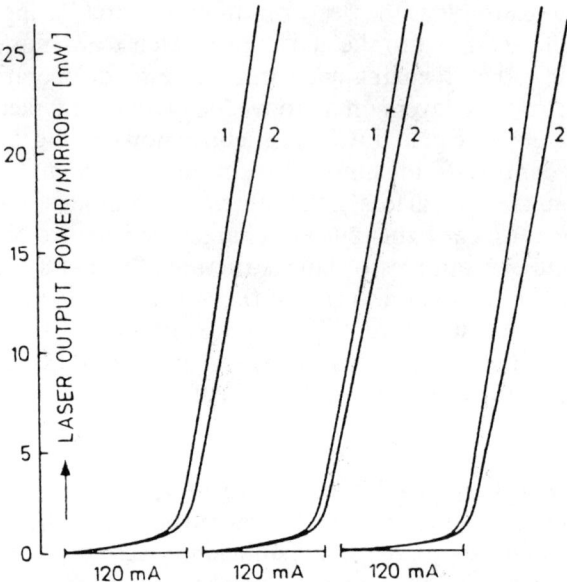

FIG. 9. Light-current characteristics of GaAlAs V-grooved lasers: (1) pulsed; (2) cw. [From Marschall et al. (1979)].

index-guided laser, due to anti-index guiding. When the stripe width is narrower than ~ 3 μm, the anti-index-guiding effect becomes prominent, and two peaks appear in a far-field pattern (Kobayashi et al., 1977). These peaks reduce the coupling efficiency between a laser and an optical fiber. Therefore the stripe width should be optimized to reduce both the mode instability and the far-field peaks.

Due to the guiding mechanism, the laser beam from gain-guided lasers has astigmatism, because the beam waist parallel to the junction plane is virtually within the device, and that perpendicular to the junction plane is at the mirror facet (Cook and Nash, 1975). The distance between the position of the beam waist parallel and perpendicular to the junction plane is usually 20–30 μm for 5-μm-wide striped lasers. It tends to be smaller the larger is the far field width (Biesterbos et al., 1982). Such astigmatism should be eliminated by using proper optics in order to collimate the laser beam.

III. Longitudinal Mode Control

4. Longitudinal Mode of Transverse-Mode-Stabilized Lasers

Semiconductor lasers with cleaved facets have a wide variety of spectral behaviors, depending on laser structure. However, there are several proper-

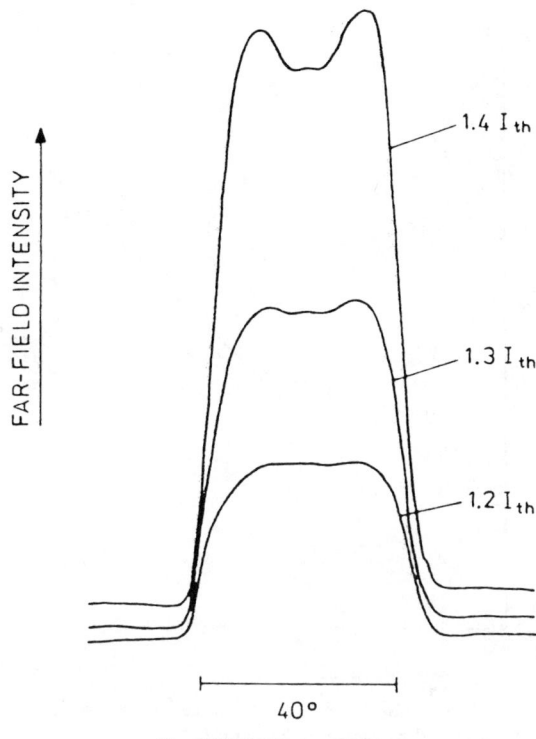

FIG. 10. Far-field patterns of GaAlAs V-grooved lasers. [From Marschall et al. (1979).]

ties inherent to semiconductor lasers that affect their spectral behaviors; the spectral gain profile and refractive index of active layers easily change with temperature or bias current. Therefore oscillation-mode hopping or spectral peak shifts are frequently observed. The spontaneous emission factor in a lasing mode is usually very large. This fact enhances multilongitudinal modes. On the other hand, intraband relaxation of carriers is very fast, and spectral hole burning is not obvious. This fact tends to suppress multimodes. Such spectral behaviors are very important for practical applications from the aspect of the laser noise.

a. Lasing Spectra of Index-Guided Lasers

Usually, index-guided lasers have a single-longitudinal mode above ~ 1.1 times threshold. Spectral narrowing is evidence of the homogeneous broadening character of the semiconductors. It is consistent with the estimation that the intraband carrier relaxation time is very small, less than ~ 1 psec. The linewidth, i.e., the coherence of single-longitudinal mode laser light, has been investigated by using a carefully arranged Fabry–Perot system (Taka-

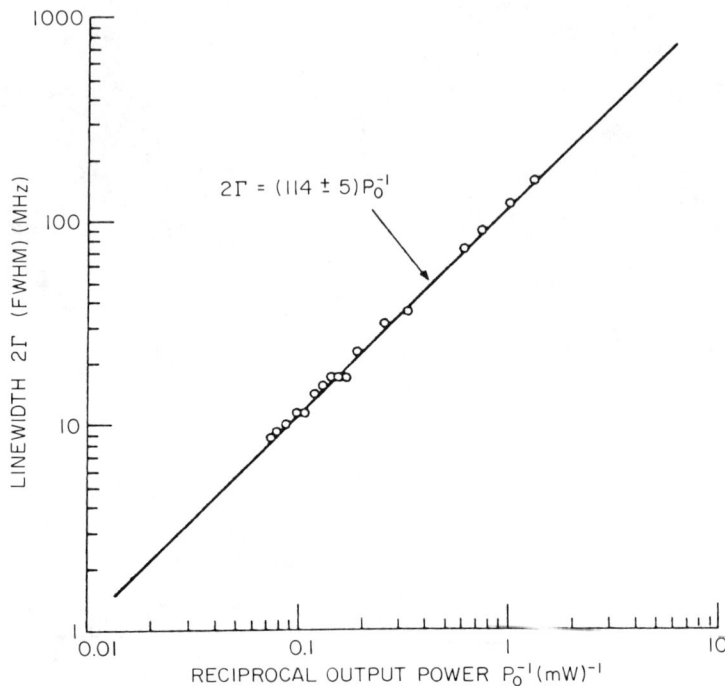

FIG. 11. Linewidth of a CSP laser at room temperature as a function of reciprocal output power. [From Fleming and Mooradian (1981).]

hara *et al.*, 1980; Fleming and Mooradian, 1981); a delay self-heterodyne method (Okoshi *et al.*, 1980); and an optical heterodyne method (Saito and Yamamoto, 1981). The half-spectral width δv is theoretically given by

$$\delta v = [2\pi h v (\Delta v_c)^2]/P, \qquad (7)$$

where hv is the laser photon energy, Δv_c the half-linewidth of the resonator, and P the output power. Figure 11 shows the experimental results of the linewidth of a CSP laser (Fleming and Mooradian, 1981). For the measurement, reflected light from the external optics into a laser device must be carefully eliminated, otherwise the linewidth tends to be narrowed by the external cavity. The observed linewidth is broader than that estimated from Eq. (7). Further, a power-independent linewidth was observed (Welford and Mooradian, 1982). A theory utilizing the index-change ratio R, defined by Eq. (6) has been proposed for the anomalous line broadening (Henry, 1982) and injected-carrier fluctuation for the power-independent linewidth (Welford and Mooradian, 1982). However, experimental results of the linewidth have been different among various laboratories (Yamamoto *et al.*, 1983).

The physical mechanism and the exact linewidth are still under investigation. A long-term frequency drift has been also investigated. The drift of less than 10 MHz over 40 min was obtained by a frequency-stabilized circuit (Tsuchida et al., 1980). Such capability of high-coherence frequency-stabilized operation of semiconductor lasers suggests the possibility of an optical heterodyne communication. Experiments on heterodyne communication systems have been demonstrated (Saito et al., 1983).

Spectral narrowing as device current increases in various mode-stabilized lasers is shown in Fig. 12 (Ikegami, 1978). It is qualitatively consistent with the simple spectral-narrowing model derived from the Statz–deMars rate equation. The details of longitudinal mode behavior were investigated by using CSP lasers (Nakamura et al., 1978). Figure 13 shows the spectra of a CSP laser under cw operation, where the intensity scales are the same throughout the five spectra shown. As shown in Fig. 13, most of the lasing power is concentrated in a single longitudinal mode whose intensity is off the scale at output powers of 1 mW or more. The drift in the lasing wavelength is due to temperature increase. An interesting finding on longitudinal mode

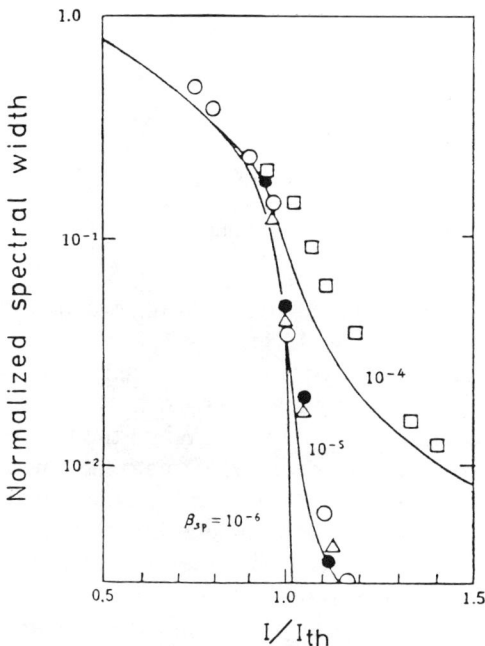

FIG. 12. Spectral width of semiconductor lasers as a function of normalized current; TJS (○), CSP (●), and DPS (△) are index guided, and NPS (□) is a narrow-planar-stripe gain-guided laser; β_{sp} is the spontaneous-emission factor. [From Ikegami (1978).]

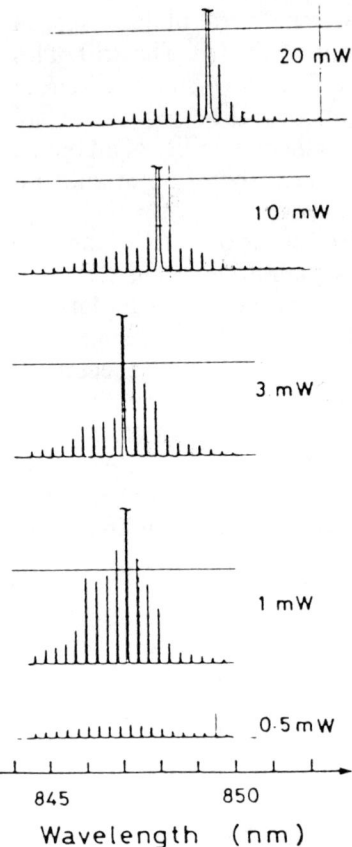

FIG. 13. Spectra of a CSP laser under cw operation at various output powers. [From Nakamura et al. (1978).]

behavior is a hysteresis observed in the lasing wavelength–temperature characteristics, as shown in Fig. 14. Here the dc current was held constant at 80 mA, while the heat-sink temperature was varied from 22 to 29°C. As the temperature was raised, the lasing wavelength continued to jump to longer wavelengths as a result of band-gap shift. When the temperature was lowered, the lasing wavelength took a different path. A similar hysteresis was also observed in lasing wavelength–current characteristics, as shown in Fig. 15. In Fig. 15, spectra at different current levels are also included. In this case, the junction temperature changed with the dc current at a rate of ~0.06 K mA^{-1}. Therefore the hysteresis in Fig. 15 is also due to temperature change. There have been several papers published on the mechanism of longitudinal mode stability in semiconductor lasers. Yamada and Suematsu (1979) analyzed it by using a density matrix formalism, and the theoretical result on the hysteresis phenomenon agreed well with the experimental result by proper choice of the intraband carrier relaxation time. Kazarinov *et*

2. MODE-STABILIZED SEMICONDUCTOR LASERS

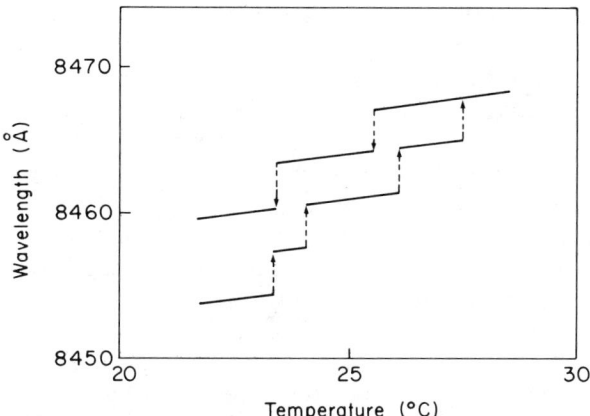

FIG. 14. Hysteresis phenomenon observed in the lasing wavelength versus temperature characteristic under a constant dc current of 80 mA. [From Nakamura *et al.* (1978).]

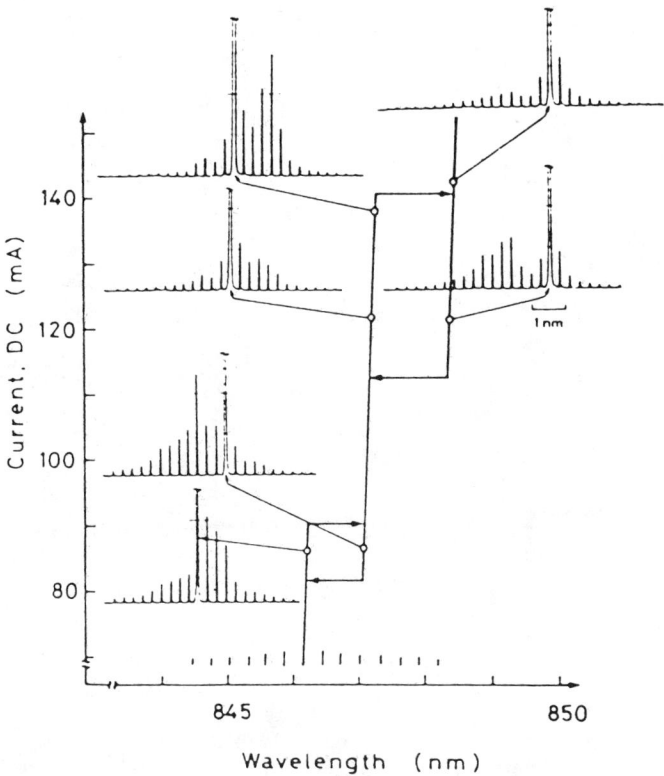

FIG. 15. Hysteresis observed in the dc-current versus wavelength characteristic under constant heat–sink temperature. [From Nakamura *et al.* (1978).]

al. (1982) also used a density matrix formalism but considered the modulation of the inverted population by the beating of the field of lasing and nonlasing modes. Lang (1982) also included the effect of the nonlasing mode as a perturbation. Much effort may be still needed to understand the complete system of the carrier and multimode laser field. On the other hand, Copeland (1980) suggested that deep traps acting as saturable absorbers, such as the DX center in AlGaAs, can cause mode stability. A laser mode penetrating the AlGaAs cladding layer suffers absorption loss of the DX centers. This loss is saturated by the standing wave of the dominant oscillating mode, and the loss grating is formed. Such loss is small for the dominant mode and large for the nonlasing modes. Thus, the mode is stabilized by the DX center.

When the mode jumps toward the next mode by the temperature or bias-current change, the anomalous noise was observed, as shown in Fig. 16 (Chinone *et al.*, 1982). Here, the output power is kept constant at 3 mW, and the ambient temperature was changed from 0 to 50°C. Then, the maximum noise between 2 and 12 MHz and the longitudinal mode were recorded. The noise level is expressed by the *relative intensity noise* (RIN), defined as

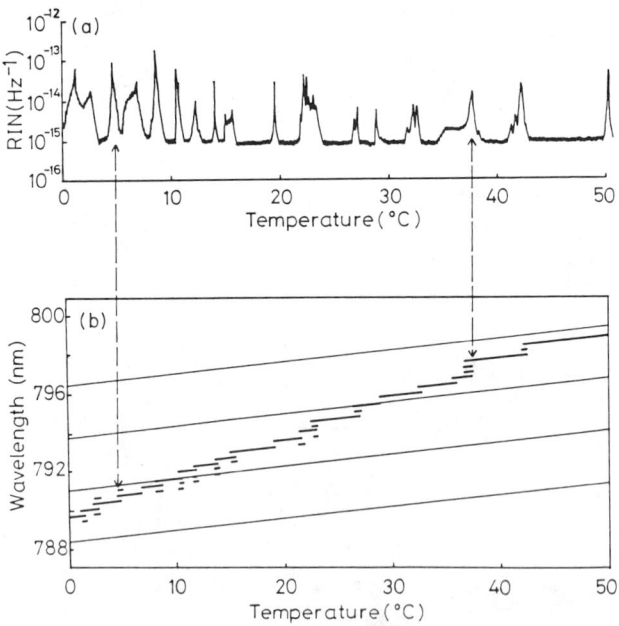

FIG. 16. (a) Mode-hopping noise and (b) longitudinal mode characteristic of an index-guided laser as a function of temperature. Output power was kept constant at 3 mW, $f = 2-12$ MHz, $\Delta f = 300$ kHz. [From Chinone *et al.* (1982). © IEEE 1982.]

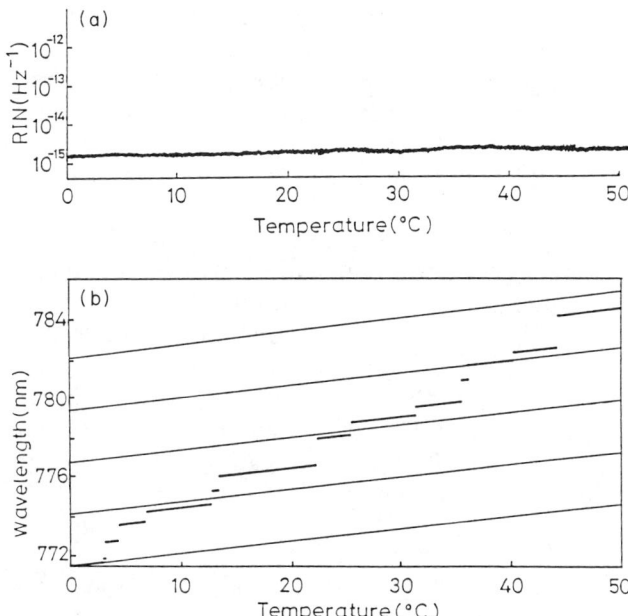

FIG. 17. (a) Mode-hopping noise and (b) longitudinal mode characteristics of an index-guided laser whose *n*-type cladding layer was highly doped with Te. Laser parameters are as given in Fig. 16. [From Chinone *et al.* (1982). © IEEE 1982.]

RIN $\equiv \langle \delta p \rangle^2 / p^2 \, \Delta f$, where δp is the output power fluctuation, p is the dc output power, and Δf is the bandwidth. Temperatures at which the noise peak appeared precisely correspond to those at which the mode jumped to the next mode. When the noise peak appeared, random switching among several modes was observed. By precise measurement of the output power, its magnitude was different among different modes. Such differences of the output power and the random switching are supposed to cause the laser noise. This mode-hopping noise was found to be suppressed by highly doping Te in the *n*-type cladding layer adjacent to the active layer, as shown in Fig. 17. In this case, Te acts as a saturable absorber. The mode selectivity produced by the saturable absorber prevents the mode competition at the mode jumping and eliminates the mode-hopping noise (Kuroda *et al.*, 1982). This fact proved the validity of the Copeland's model in a certain case.

Index-guided lasers are usually very sensitive to optical feedback from external optics, such as optical fibers, because of the superior coherency of their emitted light. When the feedback is weak, the longitudinal mode is almost single. However, laser noise at the laser resonance frequency increases (Lang and Kobayashi, 1980), and even intensity pulsation occurs

(Fye, 1982) or low-frequency noise appears, due to hopping of modes of an external cavity consisting of the laser facet and the external optics (Chinone *et al.*, 1983; Kawaguchi *et al.*, 1983). When the feedback increases, external cavity modes become multimodes and generate the laser noise by random self-mode locking. When the feedback increases further, longitudinal mode hopping or incomplete multimodes appear. The laser noise is then further increased.

Such sensitivity of index-guided lasers to external feedback is a serious drawback for optical fiber communications. However, when the laser current is rf modulated around a dc bias current that is below or equal to the threshold current, it is seen that many longitudinal modes simultaneously oscillate. This is because the gain grows above the cavity loss during the transient time, and longitudinal modes within the gain–energy profile have the gain enough to oscillate. This phenomenon reduces the feedback noise in cases such as multimode fiber communications. Otherwise, an optical isolater should be used to eliminate the feedback light.

b. Lasing Spectra of Gain-Guided Lasers

Gain-guided lasers usually have multilongitudinal modes. A typical lasing spectrum of a gain-guided laser is shown in Fig. 18 (Petermann and Arnold, 1982). Such multimode behavior has been attributed to the large spontane-

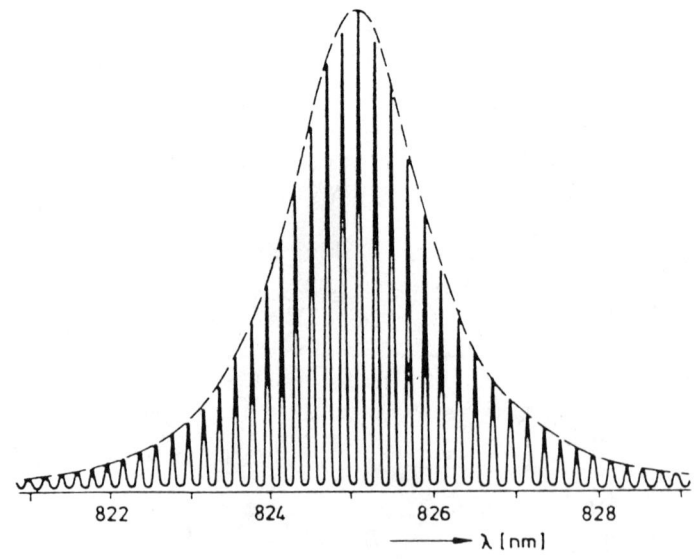

FIG. 18. Lasing spectra of a V-grooved laser. [From Petermann and Arnold (1982). © IEEE 1982.]

ous emission factor in the lasing mode. The spontaneous emission factor is the ratio of spontaneous emission power in a lasing mode to the total spontaneous emission power. It is expressed as (Petermann, 1979),

$$\beta = K[\lambda^4/(4\pi^2 n_r^3 V \Delta\lambda)], \tag{8}$$

where λ is the lasing wavelength, n_r the refractive index, V the active volume, $\Delta\lambda$ the spectral width of the spontaneous emission, and K the astigmatism parameter calculated taking into account the transverse-mode guiding mechanism. For pure index-guided mode, $K = 1$, and β was calculated to be approximately 1×10^{-5}. Using these values, the calculated spectral width agreed well with the experimental one, as shown in Fig. 12. For pure gain-guided mode, $K = 2$. However, in the usual gain-guided lasers, the mode is anti-index guided, and K can be larger than 10 (Streifer et al., 1981). Quantum mechanical analysis confirmed the validity of this estimation (Marcuse, 1982). Such a large spontaneous emission factor brings about the multi-longitudinal-mode oscillation frequently observed in gain-guided lasers. This property is advantageous for multimode fiber communications, because the poor coherency of the laser beam makes the feedback-light-induced noise and the fiber speckle-pattern noise (modal noise) less serious.

5. Longitudinal-Mode-Stabilized Lasers

In index-guided lasers, a single longitudinal mode is observed under dc operation. However, as mentioned in Section 4, many longitudinal modes appear when the current is modulated with pulse signal. In this case, since spontaneous emission is randomly distributed among laser modes, the large intensity fluctuation is observed when one mode is selected, for instance, by a monochromator. it is also the case for multimode gain-guided lasers. Such mode-competition noise seriously degrades signal-to-noise ratio in long-distance fiber communications, because of the wavelength dispersion of the fiber material (Arnold and Petermann, 1980). Therefore a longitudinal-mode-stabilized laser is needed for long-distance communications. In order to achieve a stable single mode, there are four approaches: (1) short-cavity lasers, (2) multiple-mirror lasers, (3) external light injection, and (4) lasers with grating feedback.

a. Short-Cavity Lasers

The longitudinal mode spacing $\Delta\lambda$ is expressed as

$$\Delta\lambda = \frac{\lambda^2}{(n_{\text{eff}} - \lambda \, dn_{\text{eff}}/d\lambda)2L}, \tag{9}$$

where λ is the lasing wavelength, n_{eff} the effective refractive index, and L the cavity length. When the cavity length is short, the spacing is large, and the mode jumping or multimodes are not likely to occur. As one fabrication method of the short-cavity laser, the microcleaving process has been developed (Blauvelt et al., 1982; Levine et al., 1982). A region other than the stripe is selectively etched underneath the double heterostructure, leaving a cantilever structure. By ultrasound vibrator or mechanical stress, this cantilever is cleaved off. The cleaved facet thus obtained has almost the same quality as that obtained by the conventional cleaving process. Since the selectively etched region is defined by the photoresist mask, the cavity length can be made short; 40–60-μm-long cavity lasers have been realized by this method. Even in such lasers, multimodes appeared under pulsed modulation. Therefore a shorter cavity is necessary for complete mode stabilization. In order to realize a very short cavity laser, a surface-emitting laser (Soda et al., 1982) is one of the candidates. In this device, the upper surface of the epitaxial layer and the rear metal electrode formed a Fabry–Perot cavity. The cavity length was approximately 10 μm. Only one longitudinal mode was observed in this laser. However, the threshold current was very high at room temperature, because the laser light has the gain only while it propagates perpendicularly through the thin active layer. Laser operation has been thus far obtained at liquid-nitrogen temperatures.

b. *Multiple-Mirror Lasers*

If an additional mirror is placed in or out of the Fabry–Perot cavity, longitudinal mode selection takes place by coupled-cavity effect. Longitudinal modes can be stabilized by this method to some extent. In this simple approach, an external mirror is placed close to the laser facet. By using a combination of this method and the short-cavity laser, a single longitudinal mode was observed even under pulsed modulation (Lin and Burrus, 1983). The additional mirror can be monolithically fabricated by introducing an etched groove perpendicular to the stripe (Coldren et al., 1982). One side of the groove acts as a mirror of the laser device, and the other side is an external mirror. Longitudinal mode stabilization by the external mirror thus fabricated was also observed. An interferometric laser (Fatah and Wang, 1982) can be classified in the present method. An additional cleaved facet connected with a branched stripe stabilized the longitudinal modes. A longitudinal mode was observed to be stable against a temperature change of about 20°C.

c. *External Light Injection*

When two laser diodes are coupled with each other and the oscillation wavelength of one laser is tuned to that of the other, the longitudinal mode is stabilized at this wavelength. Even under high-speed modulation, single-mode operation was obtained (Kobayashi et al., 1980; Malyon and McDonna, 1982). In addition to the single-mode oscillation, the linewidth of a mode can be also narrowed by the injection of narrow-spectrum laser light; 1-GHz linewidth of isolated laser spectrum was narrowed to 1.5 MHz by He–Ne laser-injection locking (Wyatt et al., 1982). However, precise temperature control is required for wavelength tuning in this method.

d. Lasers with Grating Feedback

By methods mentioned previously, realization of a pure single mode at various temperatures is very difficult. Such single-mode operation has been demonstrated by introducing a wavelength-selective grating inside the laser device. Distributed feedback (DFB) lasers have a grating all through the laser cavity. On the other hand, distributed Bragg reflector (DBR) lasers have a grating outside of a pumped region. In both lasers, single-mode operation was realized. The period of the grating Λ is expressed by

$$\Lambda = l\lambda_B/2n_r, \qquad (10)$$

where λ_B is the Bragg wavelength, l an integer, and n_r the effective refractive index. In the DFB laser, oscillation takes place at a wavelength λ_0, which is expressed by

$$\lambda_0 = \lambda_B \pm [(q + \tfrac{1}{2})\lambda_B^2/2n_r L] \qquad (11)$$

where L is the length of the grating region and q is an integer. From Eq. (11), the oscillation wavelength is not the same as the Bragg wavelength, and many longitudinal modes exist. However, two longitudinal modes closest to the Bragg wavelength ($q = 0$) have the lowest threshold. A single-mode oscillation is then obtained at one of the lowest threshold modes.

The DFB and DBR lasers were first realized in GaAlAs/GaAs double heterostructures (Casey et al., 1975; Aiki et al., 1975; Reinhart et al., 1975). For long-distance fiber optical communications, InGaAsP/InP DFB or DBR lasers have been extensively studied. Since fiber loss is small and their material dispersion is considerable for wavelengths of approximately 1.5 μm, pure single-mode lasers are especially required. From a fabrication point of view, the long-wavelength region is more attractive because of the longer grating period. For GaAlAs/GaAs lasers, a second- or third-order grating ($l = 2, 3$) was used, but for 1.5-μm InGaAsP/InP lasers, a first-order grating ($l = 1$) is generally used.

An InGaAsP/InP DFB laser was first realized at a wavelength of 1.1 μm (Doi et al., 1979). However, the maximum temperature for laser operation was 180 K. Room-temperature cw operation was achieved using a BH at a wavelength of approximately 1.5 μm (Utaka et al., 1981; Matsuoka et al., 1982). Several structures developed so far are schematically illustrated in Fig. 19. The grating is usually fabricated by holographic lithography using a

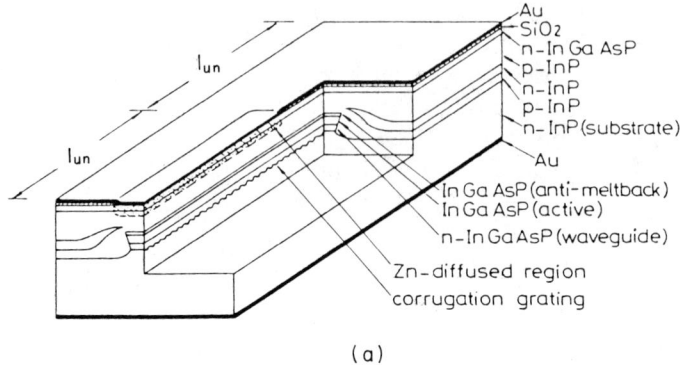

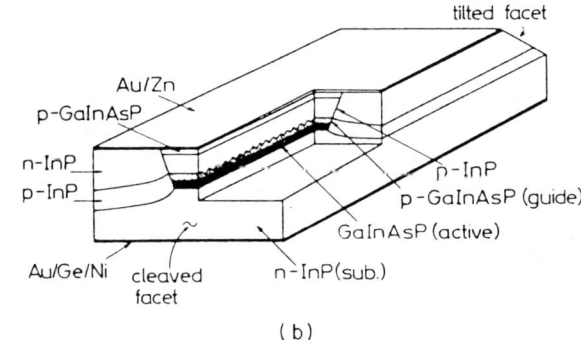

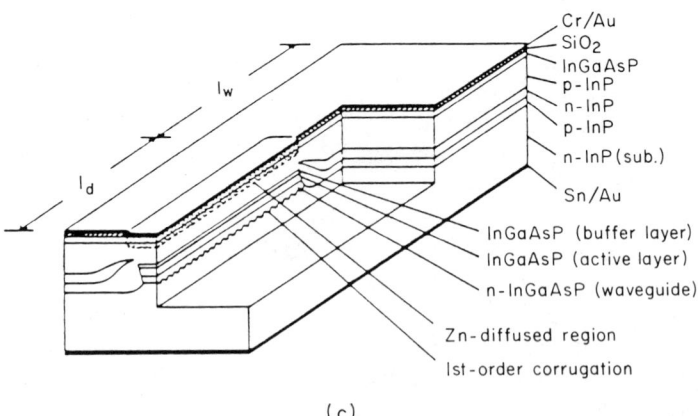

FIG. 19. Structures of InGaAsP/InP DFB–BH lasers: (a) Utaka *et al.* (1981); (b) Itaya *et al.* (1982); (c) Utaka *et al.* (1982).

short-wavelength gas laser. Electron-beam lithography by a computer-controlled scanning electron microscope has been also attempted (Westbrook *et al.*, 1982). In InGaAsP/InP lasers, InP substrate can be used as a cladding layer. Therefore the grating is usually fabricated on a surface of the InP substrate, and an InGaAsP optical guide layer is grown on the substrate following the growth of the active layer, as shown in Fig. 19a. Otherwise, the grating is formed on the InGaAsP optical guide layer grown on the active layer and is buried by the second epitaxial growth of an InP cladding layer, as shown in Fig. 19b. For $\lambda = 1.5$ μm, $\Lambda = 0.214$ μm ($l = 1$) and 0.428 μm ($l = 2$), which is about twice the grating period in GaAlAs/GaAs lasers. The grating height is typically 300–600 Å. In order to hold the shape of the grating during the epitaxial growth, low-temperature growth of approximately 600°C is usually required.

In DFB lasers, the two laser modes closest to Bragg frequency have the same threshold gain. However, reflection from one of end facets of the laser device makes their gain differ. In the actual device, this effect enhances the single-mode oscillation. In order to intentionally introduce mode selection, several methods were proposed, such as asymmetrical device structure (Streifer *et al.*, 1975) or chirped grating (Suzuki and Tada, 1980). In DFB lasers, two end facets of the device form a Fabry–Perot cavity, which affects lasing characteristics. In order to eliminate this effect, the attempt was made to bury one end facet, as shown in Fig. 19c. It was reported that the lasing mode was well stabilized by this method (Utaka *et al.*, 1982). Pure single mode was obtained even under pulsed modulation in well-fabricated DFB lasers, as shown in Fig. 20, and no mode hopping was observed at temperatures 20–70°C, as shown in Fig. 21 (Utaka *et al.*, 1981). In DFB–BH lasers, a threshold current of approximately 50 mA was obtained at room tempera-

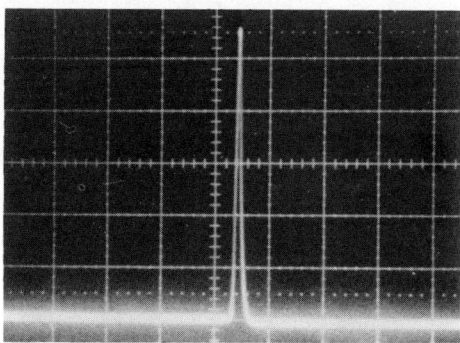

FIG. 20. Spectrum of a DFB–BH laser measured under 500-Mbit/sec^{-1} pseudorandom-pulse modulation; horizontal scale, 28 Å div^{-1}. $I_{dc} = 53$ mA, $I_p = 42$ mA. [From Utaka *et al.* (1981).]

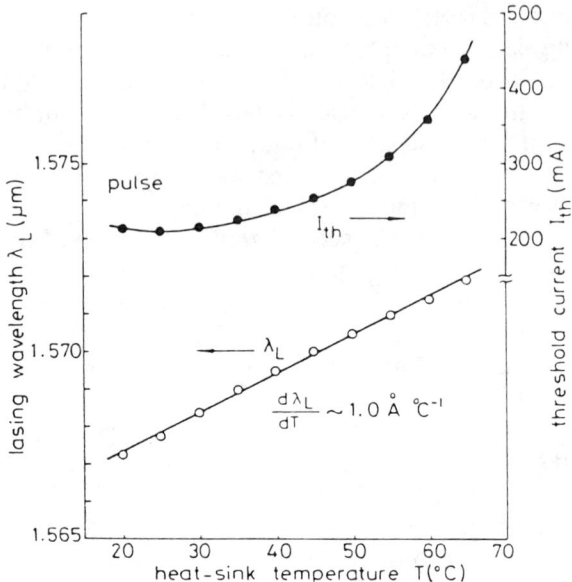

FIG. 21. Threshold current I_{th} and lasing wavelength λ_L of DFB-BH laser in pulsed operation as a function of heat-sink temperature; $l_{ex} = 360$ μm, $l_{un} = 220$ μm. [From Utaka et al. (1981).]

ture (Matsuoka et al., 1982). It is comparable to conventional Fabry–Perot lasers.

In contrast to DFB lasers, DBR lasers have grating outside the pumped region. Therefore, fabrication is easier and is applicable to optical integrated devices. In DBR lasers, the laser oscillation can take place at the Bragg wavelength λ_B. However, by the temperature change, this lowest threshold mode moves apart from the Bragg wavelength, and the adjacent mode has the lowest threshold. Then, the mode hops toward the adjacent mode. In order to avoid this phenomenon throughout a wide range of temperature, precise control of the length of the pumped region is necessary. Several structures that have been developed are schematically illustrated in Fig. 22. In DBR lasers, single-mode oscillation under pulsed modulation was also obtained (Koyama et al., 1981). Oscillation-mode spectra under high-frequency modulation are shown in Fig. 23 (Koyama et al., 1981). Even if modulation efficiency is 100%, single-mode oscillation is observed. However, as modulation frequency increases, the linewidth of the mode increases and becomes the largest at a modulation frequency close to the laser resonance frequency. It is due to laser oscillation frequency variation induced by carrier-induced refractive index variation. This phenomenon would limit

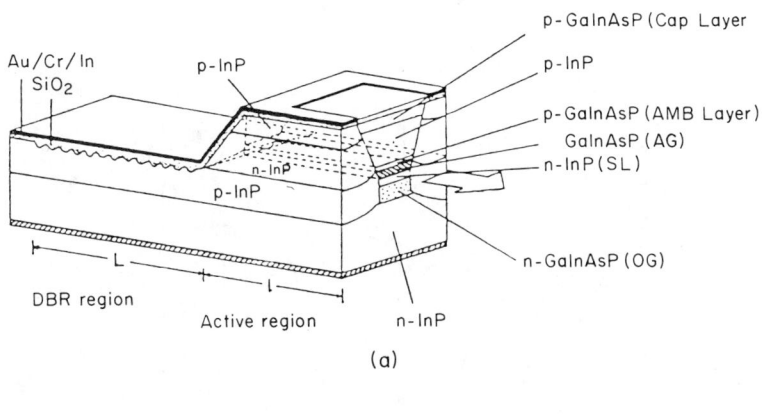

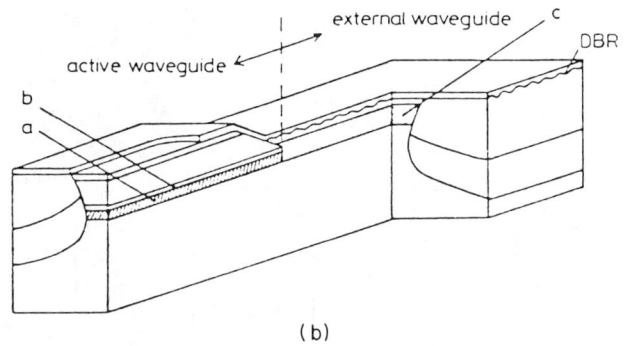

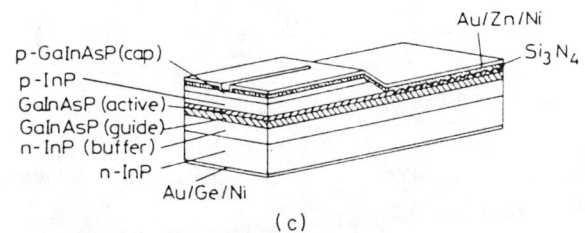

FIG. 22. Structures of InGaAsP/InP DBR lasers: (a) integrated-twin-guide (Tanbun-ek *et al.*, 1981); (b) butt-jointed (Abe *et al.*, 1982); *a*, active layer; *b*, antimelt-back layer; *c*, guide layer; (c) built-in optical waveguide (Mikami *et al.*, 1982).

the bandwidth of optical fiber communications, even if the fiber loss is very small. Room-temperature cw operation was also achieved in DBR lasers (Abe *et al.*, 1982). However, threshold current is generally approximately 100 mA, not as small as DFB lasers. If the optical coupling between the pumped region and the grating region were improved, this drawback would

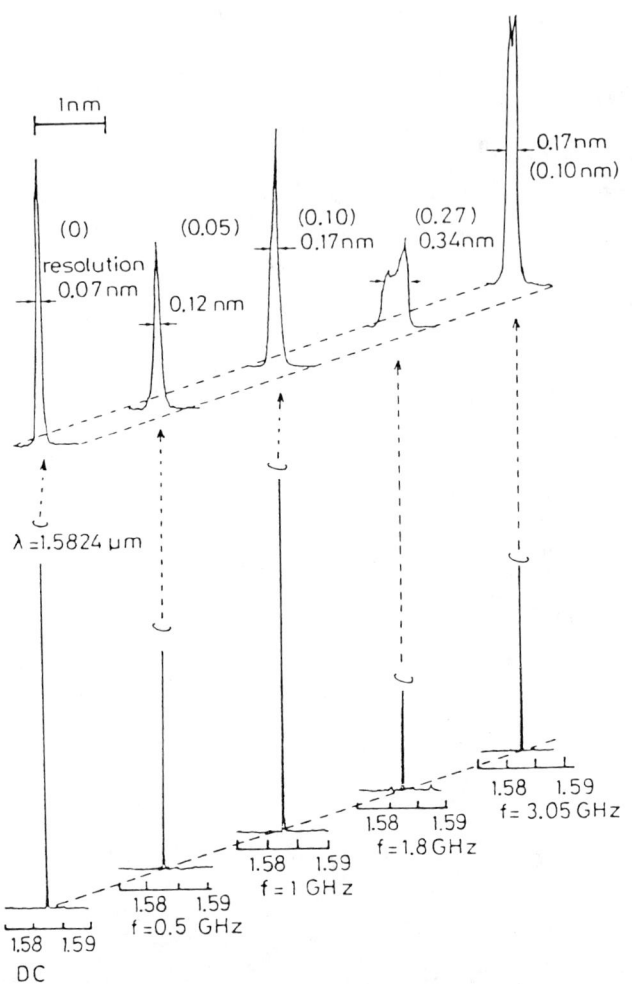

FIG. 23. Lasing spectra of a DBR laser under dc operation and high-speed modulation with modulation depth of 100%. [From Koyama et al. (1981).]

be eliminated. Several waveguide structures have been proposed, as shown in Fig. 22.

IV. Conclusion

Technology and theory for semiconductor laser mode control have been reviewed. Transverse-mode control has been successfully demonstrated for output powers up to several 10 mW. In transverse-mode-stabilized lasers, longitudinal mode control is necessary for long-distance fiber communications. Various methods have been reported for longitudinal mode control. Among them, lasers with grating feedback seem to be the most promising. However, more effort is needed to reproducibly fabricate pure single-mode lasers. Longitudinal mode and related noise behaviors have been intensively studied. However, they have not been fully clarified. In semiconductor lasers, the interaction between carriers and laser fields make this behavior complicated. Much effort is also needed to analyze them.

REFERENCES

Abe, K., Koshino, K., Tanbun-Ek, T., Arai, S., Koyama, F., Matsumoto, K., Watanabe, T., and Suematsu, Y. (1982). *Electron. Lett.* **18**, 410.
Aiki, K., Nakamura, M., Umeda, J., Yariv, A., Katziv, A., and Yen, H. W. (1975). *Appl. Phys. Lett.* **27**, 145.
Aiki, K., Nakamura, M., Kuroda, T., Umeda, J., Ito, R., Chinone, N., and Maeda, M. (1978). *IEEE J. Quantum Electron.* **QE-14**, 89.
Arnold, G., and Petermann, K. (1980). *Opt. Quantum Electron.* **12**, 207.
Biesterbos, J. W. M., Brouwer, R. P., Valster, A., de Poorter, J. A., and Acket G. A. (1982). *Proc. Int. IEEE Semicond. Laser Conf., 8th, 1982,* Paper 45, p. 122.
Blauvelt, H., Bar-Chaim, N., Fekete, D., Margalit, S., and Yariv, A. (1982). *Appl. Phys. Lett.* **15**, 289.
Botez, D., and Zory, P. (1978). *Appl. Phys. Lett.* **32**, 261.
Burnham, R. D., and Scifres, D. R. (1975). *Appl. Phys. Lett.* **27**, 510.
Buus, J. (1978). *Electron. Lett.* **14**, 127.
Casey, H. C., Jr., Somekh, S., and Ilegems, M. (1975). *Appl. Phys. Lett.* **27**, 142.
Chinone, N. (1977). *J. Appl. Phys.* **48**, 3237.
Chinone, N., Ito, R., and Nakada, O. (1976). *J. Appl. Phys.* **47**, 785.
Chinone, N., Saito, K., Ito, R., and Aiki, K. (1979). *Appl. Phys. Lett.* **35**, 513.
Chinone, N., Takahashi, K., Kajimura, T., and Ojima, M. (1982). *Proc. Int. IEEE Semicond. Laser Conf. 8th, 1982,* Paper 25, p. 74.
Chinone, N., Oishi, A., and Nakamura, M. (1983). *Proc. Spring Meet. Jpn. Appl. Phys. Soc.,* Paper 6a-H-9, p. 131 (in Japanese).
Coldren, L. A., Furuga, K., Miller, B. I., and Rentschler, J. A. (1982). *IEEE J. Quantum Electron.* **QE-18**, 1679.
Cook, D. D., and Nash, R. R. (1975). *J. Appl. Phys.* **46**, 1660.
Copeland, J. A. (1980). *IEEE J. Quantum Electron.* **QE-16**, 721.
D'Asaro, L. A. (1973). *J. Lumin.* **7**, 310.
Dixon, R. W., Nash, F. R., Hartman, R. L., and Hepplewhite, R. J. (1976). *Appl. Phys. Lett.* **29**, 372.
Doi, A., Fukuzawa, T., Nakamura, M., Ito, R., and Aiki, K. (1979). *Appl. Phys. Lett.* **35**, 441.
Dyment, J. C., D'Asaro, L. A., North, J. C., Miller, B. I., and Ripper, J. E. (1972). *Proc. IEEE* **60**, 726.
Fatah, I., and Wang, S. (1982). *Appl. Phys. Lett.* **41**, 112.
Fleming, M. W., and Mooradian, A. (1981). *Appl. Phys. Lett.* **38**, 511.
Fye, D. M. (1982). *IEEE J. Quantum Electron.* **QE-18**, 1675.
Hakki, B. W. (1975). *J. Appl. Phys.* **46**, 292.
Hayakawa, T., Miyauchi, N., Yamamoto, S., Hayashi, H., Yano, S., and Hijikata, T. (1982). *J. Appl. Phys.* **53**, 7224.
Henry, C. H. (1982). *IEEE J. Quantum Electron.* **QE-18**, 259.

Henry, C. H., Logan, R. A., and Bertness, K. A. (1981). *J. Appl. Phys.* **52**, 4457.
Hirao, M., Doi, A., Tsuji, S., Nakamura, M., and Aiki, K. (1980). *J. Appl. Phys.* **51**, 4539.
Ide, Y., Furuse, T., Sakuma, I., and Nishida, K. (1980). *Appl. Phys. Lett.* **36**, 121.
Ikegami, T. (1978). *Oyo Butsuri* **47**, 812 (in Japanese).
Ishikawa, H., Imai, H., Tanahashi, T., Hori, K., and Takahei, K. (1982). *IEEE J. Quantum Electron.* **QE-18**, 1704.
Itaya, K., Matsuoka, T., Nakano, Y., Suzuki, Y., Kuroiwa, K., and Ikegami, T. (1982). *Electron. Lett.* **18**, 1006.
Ito, R., and Saito, K. (1980). *IEEE J. Quantum Electron.* **QE-16**, 205.
Kawaguchi, H., Yoshikuni, Y., and Okegami, T. (1983). *Proc. Spring Meet. Jpn. Appl. Phys. Soc.*, Paper **40-H-5**, p. 118 (in Japanese).
Kazarinov, R. F., Henry, C. H., and Logan, R. A. (1982). *Proc. Int. IEEE Semicond. Laser Conf., 8th, 1982,* Paper 7, p. 32.
Kirkby, P. A., and Thompson, G. H. B. (1976). *J. Appl. Phys.* **47**, 4578.
Kobayashi, T., Kawaguchi, H., and Fukukawa, Y. (1977). *Jpn. J. Appl. Phys.* **16**, 601.
Kobayashi, S., Yamada, J., Machida, S., and Kimura, T. (1980). *Electron. Lett.* **16**, 746.
Koyama, F., Arai, A., Suematsu, Y., and Kishino, K. (1981). *Electron. Lett.* **17**, 938.
Kuroda, T., Ohtoshi, T., Chinone, N., Kajimura, T., and Ouchi, H. (1982). *Proc. Autumn Meet. Jpn. Appl. Phys. Soc.,* Paper **28-P-B-4**, p. 120 (in Japanese).
Kumabe, H., Tanaka, T., Namigaki, H., Takamiya, M., Ishii, M., and Susaki, W. (1979). *Jpn. J. Appl. Phys.* **18-1**, Suppl., 371.
Lang, R. (1979). *IEEE J. Quantum Electron.* **QE-15**, 718.
Lang, R. (1982). *Proc. Int. IEEE Semicond. Laser Conf., 8th, 1982,* Paper 6, p. 30.
Lang, R., and Kobayashi, K. (1980). *IEEE J. Quantum Electron.* **QE-16**, 347.
Levine, B. F., Van der Ziel, J. P., Logan, R. A., and Bethea, C. G. (1982). *Electron. Lett.* **18**, 690.
Lin, C., and Burrus, A. (1983). *Proc. Top. Meet. Opt. Fiber Commun., 1983, New Orleans,* Paper PD5-1.
Malyon, D. J., and McDonna, A. P. (1982). *Electron. Lett.* **18**, 62.
Marcuse, D. (1982). *Electron. Lett.* **18**, 920.
Marschall, P., Schlosser, E., and Wölle, C. (1979). *Electron. Lett.* **15**, 38.
Matsuoka, T., Nagai, H., Itaya, Y., Noguchi, Y., Suzuki, Y., and Ikegami, T. (1982). *Electron. Lett.* **18**, 27.
Mikami, O., Saitoh, T., and Nakagome, H. (1982). *Electron. Lett.* **18**, 458.
Mito, I., Kitamura, M., Kobayashi, Ke., and Kobayashi, Ko. (1982). *Electron. Lett.* **18**, 953.
Murotani, T., Oomura, E., Higuchi, H., Namizaki, H., and Susaki, W. (1982). *Electron. Lett.* **16**, 566.
Nagano, K., Takahashi, Y., Takasaki, Y., Maeda, M., and Tanaka, M. (1979). *Proc. Eur. Conf. Opt. Commun., 5th, 1979,* Vol. 13, p. 1.
Nakamura, M., Aiki, K., Chinone, N., Ito, R., and Umeda, J. (1978). *J. Appl. Phys.* **49**, 4644.
Okoshi, T., Kikuchi, K., and Nakayama, A. (1980). *Electron. Lett.* **16**, 630.
Petermann, K. (1979). *IEEE J. Quantum Electron.* **QE-15**, 566.
Petermann, K., and Arnold, G. (1982). *IEEE J. Quantum Electron.* **QE-18**, 543.
Reinhart, F. K., Logan, R. A., and Shenk, C. V. (1975). *Appl. Phys. Lett.* **27**, 45.
Saito, S., and Yamamoto, Y. (1981). *Electron. Lett.* **17**, 325.
Saito, S., Yamamoto, Y., and Kimura, T. (1983). *IEEE J. Quantum Electron.* **QE-19**, 180.
Seki, K., Kamiya, T., and Yanai, H. (1979). *Trans. IECE Jpn.* **E-62**, 73.
Shima, K., Hanamitsu, K., and Takusagawa, M. (1982). *IEEE J. Quantum Electron.* **QE-18**, 1688.
Soda, H., Motegi, Y., and Iga, K. (1982). *Proc. Int. IEEE Semicond. Laser Conf., 8th, 1982,* Paper 37, p. 102.

Stern, F. (1976). *J. Appl. Phys.* **47**, 5382.
Streifer, W., Burnham, R. D., and Seifres, D. R. (1975). *IEEE J. Quantum Electron.* **QE-11** 154.
Streifer, W., Seifres, D. R., and Burnham, R. D. (1981). *Electron. Lett.* **17**, 933.
Sugi, T., Wada, M., Shimizu, H., Ito, K., and Teramoto, I. (1979). *Appl. Phys. Lett.* **34**, 270.
Suzuki, A., and Tada, K. (1980). *Proc. Soc. Photo-Opt. Instrum. Eng.*
Takakura, T., Iga, K., and Tako, T. (1980). *Jpn. J. Appl. Phys.* **19**, 1725.
Tanbun-ek, T., Arai, S., Koyama, F., Kishino, K., Yoshizawa, S., Watanabe, T., and Suematsu, Y. (1981). *Electron. Lett.* **17**, 967.
Thompson, G. H. B., Lovelace, D. F., and Turley, S. E. H. (1978). *IEEE J. Solid-State Electron Devices* **2**, 12.
Tsang, W. T., Logan, R. A., and Ilegems, M. (1978). *Appl. Phys. Lett.* **32**, 311.
Tsuchida, H., Sanpei, S., Ohtsu, M., and Tako, T. (1980). *Jpn. J. Appl. Phys.* **19**, L721.
Tsuji, S., Mizuishi, K., Nakayama, Y., Shimaoka, M., and Hirao, M. (1982). *Jpn. J. Appl. Phys.* **22-1**, Suppl., 239.
Tsukada, T. (1974). *J. Appl. Phys.* **45**, 4899.
Ueno, M., and Yonezu, H. (1979). *IEEE J. Quantum Electron.* **QE-15**, 1189.
Utaka, K., Akiba, S., Sakai, K., and Matsushima, Y. (1981). *Electron. Lett.* **17**, 961.
Utaka, K., Akiba, S., Sakai, K., and Matsushima, Y. (1982). *Proc. Int. IEEE Semicond. Laser Conf., 8th, 1982,* Paper 34, p. 96.
Welford, D., and Mooradian, A. (1982). *Appl. Phys. Lett.* **40**, 560.
Westbrook, L. D., Nelson, A. W., and Dix, C. (1982). *Electron. Lett.* **18**, 863.
Wolf, H. D., Metler, K., and Zschaur, K. H. (1981). *Jpn. J. Appl. Phys.* **20**, L693.
Wyatt, R., Smith, D. W., and Cameron, K. H. (1982). *Electron. Lett.* **18**, 179.
Yamada, M., and Suematsu, Y. (1979). *Jpn. J. Appl. Phys.* **18-1**, Suppl., 347.
Yamamoto, Y., Saito, S., and Mukai, T. (1983). *IEEE J. Quantum Electron.* **QE-19**, 47.
Yonezu, H., Sakuma, I., Kobayashi, K., Kamejima, T., Ueno, M., and Nannichi, Y. (1973). *Jpn. J. Appl. Phys.* **12**, 1585.

CHAPTER 3

Semiconductor Lasers with Wavelengths Exceeding 2 μm

Yoshiji Horikoshi

MUSASHINO ELECTRICAL COMMUNICATION LABORATORY
NIPPON TELEGRAPH AND TELEPHONE PUBLIC CORPORATION
TOKYO, JAPAN

I.	INTRODUCTION	93
II.	SEMICONDUCTOR MATERIALS	94
	1. III–V *Compound Semiconductors*	94
	2. IV–VI *Compound Semiconductors*	98
	3. II–VI *Compound Semiconductors*	101
III.	CRYSTAL GROWTH AND LASER FABRICATION	103
	4. III–V *Compound Semiconductors*	103
	5. IV–VI *Compound Semiconductors*	110
IV.	LASER-DIODE CHARACTERISTICS	120
	6. *Introduction*	120
	7. III–V *Compound Semiconductor Lasers*	121
	8. *Lasing Characteristics for* IV–VI *Semiconductor Lasers*	126
	9. *Output Properties and High-Temperature Operation*	140
	REFERENCES	147

I. Introduction

Semiconductor lasers with emission wavelengths exceeding 2 μm have been studied for almost 20 years. The first demonstration of GaAs p–n-junction lasers was in 1962 (Nathan *et al.*, 1962). Since that time, laser action in narrow-energy-gap semiconductor materials has been reported not only for direct energy-gap III–V semiconductors such as InAs and InSb, but also for direct energy-gap lead chalcogenide materials that have band extrema at the Brillouin zone edge in the [111] direction rather than at the Γ point, as with direct energy-gap III–V semiconductors. In 1963, Melngailis observed 3.1-μm laser emission from an InAs p–n-junction diode, whereas Phelan *et al.* (1963) and Guillaume and Lavallard (1963) reported InSb diode lasers having a 5.3-μm emission wavelength. Laser action at much longer wavelengths was reported by Butler *et al.* in 1964 with PbTe and PbSe p–n-junction laser diodes, which stimulated investigations of lasers with ternary mixed IV–VI compound semiconductors. In contrast to 0.8–1.6-μm-wavelength III–V semiconductor lasers, only a few reports, are available for laser diodes of narrow energy-gap III–V semiconductors. However, many advances have so far been made concerning IV–VI semiconductor lasers toward high-temperature operation, improved tuning characteristics, and temperature-cycle stability.

Although no useful applications have so far been reported for the lasers of narrow-gap III–V compound semiconductors, those of lead chalcogenide materials have found applications in various areas of technology as well as in basic research. Some of these include applications to high-resolution gas spectroscopy, air pollution monitoring, and infrared heterodyne detection. Most recently, much attention has been paid to ultra-low-loss optical fiber materials in the near- and middle-infrared regions, such as thallium bromide and chalcogenide glass materials. Pinnow *et al.* (1978) estimated the minimum transmission loss for TlBr and KRS-5 within the spectral range of 4–5.5 μm to be as low as 10^{-2}–10^{-5} dB km^{-1}. This means that the narrow-gap semiconductor lasers mentioned earlier may be much more useful for providing extremely low-loss optical communication links. Nevertheless, unlike the development of lasers with AlGaAs/GaAs and InP/InGaAsP double heterostructures, which has been stimulated by progress in optical fibers, semiconductor lasers with wavelengths exceeding 2 μm are now only at a very early stage in their development.

In this chapter, we summarize semiconductor materials and laser fabrication technology appropriate for laser diodes having emission wavelengths exceeding 2 μm, including those with lattice-matched double heterostructure configurations. Narrow-gap semiconductor lasers suffer from excessive loss mechanisms, such as those due to tunneling current, Auger recombination, and free-carrier absorption. These raise threshold currents and limit the highest temperature for laser oscillation. Therefore, in the later half of this chapter, we discuss the factors determining threshold current density, and compare the lasers fabricated from different semiconductor materials, taking a look at their threshold characteristics.

II. Semiconductor Materials

1. III–V Compound Semiconductors

Semiconductors with direct energy gap corresponding to an emission wavelength exceeding 2 μm include III–V, IV–VI, and II–VI compound semiconductors. First, III–V compound semiconductors are discussed. Taking Al, Ga and In from among group III elements, and P, As, and Sb from group V elements, we have 15 quaternaries that comprise 9 III$_x$III$'_{1-x}$V$_y$V$'_{1-y}$-type mixed crystals, three III$_x$III$'_y$III$''_{1-x-y}$–V-type, and three III–

$V_x V_y V''_{1-x-y}$-type mixed crystals. They, of course, include many binaries and ternaries at their boundaries.

Based on Vegard's law and the formula proposed by Moon *et al.* (1974) and by Williams *et al.* (1978), lattice constants and energy gaps for the quaternaries just mentioned have been interpolated from available binary and ternary values that include bowing parameters (Glisson *et al.*, 1978). As for $InAs_{1-x-y}P_xSb_y$, the composition dependence of the energy gap was obtained from photoresponse measurement of metallo-organic chemical vapor deposition (MOCVD)-grown $InAs_{1-x-y}P_xSb_y$ (Fukui and Horikoshi, 1981). From these results, the expected emission wavelength ranges corresponding to the direct energy gap at room temperature were obtained, as shown in Fig. 1. The thick portion of each bar denotes the wavelength range covered by the composition lattice matched to a stable binary substrate such as (a) GaAs, (b) InP, (c) GaSb, and (d) InAs. The shaded areas denote miscibility gaps, although the exact extent of these is unknown for most of the quaternaries. In Fig. 1, AlAsPSb is not shown, because this quaternary has a minimum indirect energy gap regardless of the crystal composition.

Many quaternaries have solid-solution compositions with a direct energy

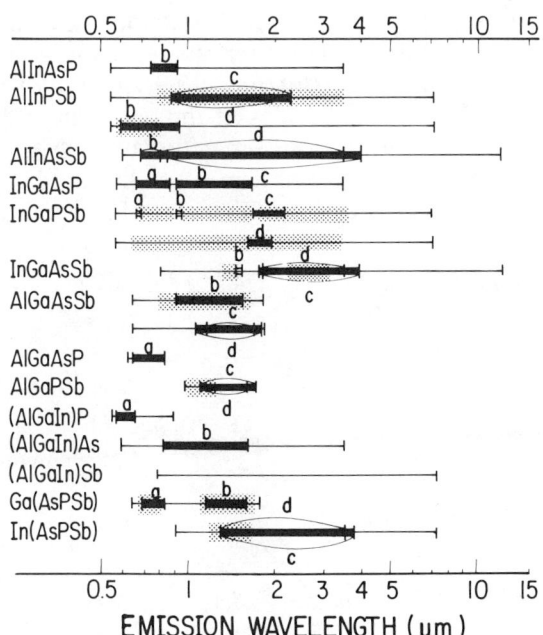

FIG. 1. Expected emission wavelengths for various III–V quaternaries at 300 K. The thick portion denotes the compositions lattice matched to stable binary crystals: (a) GaAs, (b) InP, (c) GaSb, (d) InAs; shaded areas correspond to the miscibility gaps.

gap corresponding to emission wavelengths exceeding 2 μm. Although with most of these quaternaries there are difficulties in finding stable substrates, there are three exceptions: AlInAsSb, InGaAsSb, and InAsPSb. These have compositions lattice matched to GaSb and InAs substrates. Liquid-phase epitaxial growth of uniform AlInAsSb is very difficult, however, because of severe Al segregation. Therefore, in the following, the remaining two quaternaries, InGaAsSb and InAsPSb, are described. Figure 1 also suggests that III–V semiconductors lattice matched to binary crystals cannot be applied to wavelengths exceeding 4 μm, where lead salt materials become important.

Figure 2 indicates the x–y compositional plane of an InAs–InSb–GaAs–GaSb pseudoquaternary system, where the compositional dependence of direct energy gaps and the lattice constants are shown by solid and dashed curves, respectively. $In_{1-x}Ga_xAs_ySb_{1-y}$ lattice matched to GaSb can be obtained along the $y = 0.91x$ composition line, whereas that lattice matched to InAs can be obtained along the $y = 0.92x + 0.08$ composition line. The extent of the miscibility gap is also shown in Fig. 2 according to the experimental result by Nakajima et al. (1977). Although this system suffers from a wide miscibility gap in the central portion, both GaSb and InAs-enriched single-phase compositions can be grown by liquid-phase epitaxy. The former corresponds to emission wavelengths of approximately 2 μm, whereas the latter covers 3–4-μm emission. However, a simple interpolation method based on a single effective oscillator model (Wemple and DiDomenico, 1971) suggests that the refractive index of this quaternary decreases with decreasing energy gap along the lattice-matched composition line. Therefore, to achieve efficient photon confinement within the InGaAsSb active region, it is necessary to use material with a lower refractive index for confin-

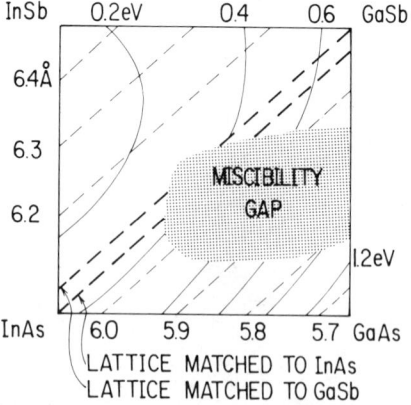

FIG. 2. Composition plane for InGaAsSb: (——) equi-energy-gap lines; (– – –) equi-lattice-constant lines.

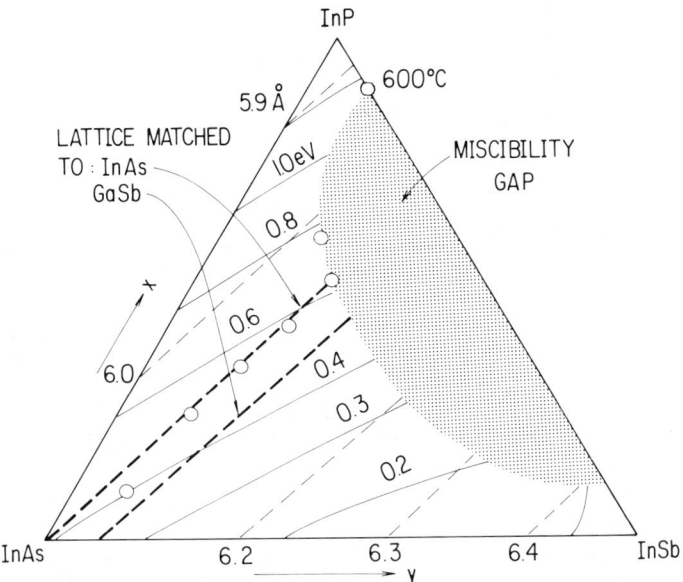

FIG. 3. Compositional plane for InAsPSb. (———) equi-energy-gap lines; (– – –) equi-lattice-constant lines; (○) solid compositions grown by MOVPE. [From Fukui and Horikoshi, (1981).]

ing layers. Fortunately, AlGaAsSb that is lattice matched to GaSb or InAs shows a much lower refractive index than GaSb or GaSb-enriched In-GaAsSb.

Figure 3 shows the x–y compositional plane of the InAs–InP–InSb pseudoternary system. Dashed curves denote equi-lattice-constant lines, and solid curves show equi-energy-gap lines deduced from the photoresponse measurement at 77 K. In this system, $InAs_{1-x-y}P_xSb_y$ alloys lattice matched to the InAs substrate have solid compositions on the $y = 0.47x$ line, where $x + y \leq 1$. This quaternary also suffers from a miscibility gap in the InPSb ternary and penetration into the quaternary. Open circles in Fig. 3 denote compositions with stable solid solutions grown at 600°C by MOCVD (Fukui and Horikoshi, 1981). Figure 3 suggests that the energy gap of $InAs_{1-x-y}P_xSb_y$ with a composition lattice matched to InAs or GaSb increases with increasing x. The refractive index, as estimated by a single effective oscillator model, showed a monotonic decrease with increasing x along the lattice-matched composition line. Therefore, efficient photon and carrier confinement can be realized by combination of different x crystals. Figure 4 shows optical confinement factors for the fundamental TE mode in double heterostructures with $InAs_{0.63}P_{0.25}Sb_{0.12}$ confining layers. The curves in Fig. 4 correspond to various active-region compositions.

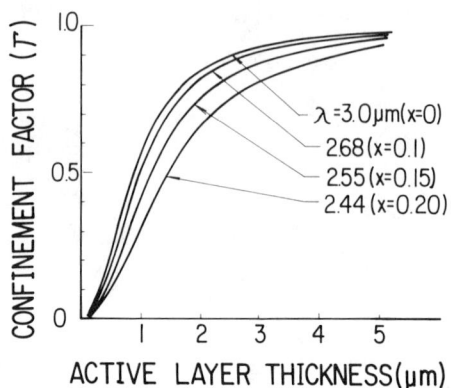

FIG. 4. The optical confinement factors for the fundamental TE mode as a function of the active-region thickness for $InAs_{1-x-y}P_xSb_y$ DH structures with $InAs_{0.63}P_{0.25}Sb_{0.12}$ confining layers.

2. IV–VI Compound Semiconductors

Laser action has so far been observed in p–n-junction diodes with a number of lead- and lead–tin-based binary and ternary IV–VI semiconductors. These include PbTe, PbSe, PbS, PbSnTe, PbSnSe, and PbSSe, which have been covered by earlier review articles (Harman, 1971; Calawa, 1972; Melngailis, 1968; Preier, 1979). Due to the well-known band-inversion effect (Dimmock et al., 1966) in $Pb_{1-x}Sn_xTe$ and $Pb_{1-x}Sn_xSe$, the energy gap approaches zero at certain values of x. This leads, in principle, to laser action up to very long wavelengths in these materials. However, the shortest wavelength is limited by the energy gap of PbS or PbTe, to approximately either 4 or 5.5 μm at 77 K.

Therefore, several attempts have been made to shorten the emission wavelength by employing other narrow-gap semiconductor materials. Some of these include ternary IV–VI semiconductors incorporated with Ge, such as PbGeTe (Wrobel et al., 1973) for 4.4–6.5-μm wavelengths, PbGeS (Melngailis and Harman, 1973) for 3.4–4.3-μm wavelengths, and mixtures of IV–VI and II–VI compounds, such as PbCdS (Nill et al., 1973), for the 2.5–4.3-μm region. Wavelength regions covered by the laser diodes made of these materials are summarized in Fig. 5. The emission wavelength range for each ternary was deduced from the available energy-gap data for binaries and ternaries (Harman and Melngailis, 1974; Kucherenko et al., 1976; Adler et al., 1973; Anderson, 1977). More recently, IV–VI semiconductors incorporated with rare-earth compounds were found to be useful for shorter-wavelength laser materials (Partin, 1983), and laser action in the 2.7–6.6-μm wavelength range has been demonstrated with DH laser diodes having

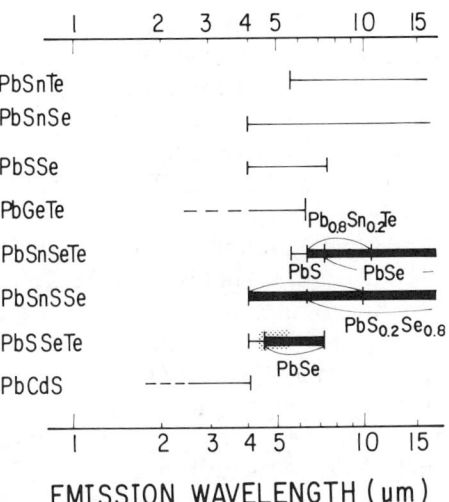

FIG. 5. Expected emission wavelengths for various IV–VI mixed crystals at 77 K. The thick portion denotes the lattice-matched compositions and the shaded area, the miscibility gap.

PbEuSeTe active regions lattice matched to PbTe substrate (Partin and Thrush, 1984). In Fig. 5, examples of the expected emission wavelength for quaternary IV–VI semiconductors, such as PbSnSeTe, PbSnSSe, and PbSSeTe, are also shown. The thick portion corresponds to the wavelengths covered by the quaternaries that are lattice matched to PbSe and $Pb_{0.8}Sn_{0.2}Te$ substrates for PbSnSeTe, to PbS and $PbS_{0.2}Se_{0.8}$ substrates for PbSnSSe, and to a PbSe substrate for the PbSSeTe system.

The refractive index of binary crystals and its variation with composition has been studied by many workers (Dalven, 1969; Dionne and Wooley, 1972; Kucherenko et al., 1977; Bauer, 1978). It has been summarized by Preier (1979) for various ternary systems. In each system, the refractive index increases with decreasing energy gap, which seems to be a phenomenon favorable for constructing double heterostructures with good optical confinement, even though there is a lattice-mismatch problem.

Although investigations to date have largely dealt with homojunction and single-heterojunction-diode lasers (Grisar et al., 1979; Linden et al., 1977), there have been expectations that the introduction of a double heterostructure (DH) would improve laser performance, as in AlGaAs/GaAs and InP/InGaAsP DH lasers. Indeed, the PbSnTe/PbTe DH configuration has reduced the threshold current density J_{th} at relatively high temperatures and made cw operation possible above 77 K (Groves et al., 1974; Walpole et al., 1976a; Tomasetta and Fonstad, 1974a,b; Yoshikawa et al., 1977). However, at lower temperatures, PbSnTe/PbTe DH lasers exhibit a very high J_{th} when compared with their diffused homostructure and grown-junction homostructure counterparts (Lo *et al.*, 1976; Oron and Zussman, 1980). Moreover, PbSnTe/PbTe DH lasers suffer from degradation due to thermal cycling. All of these problems have been attributed to the extremely high dislocation density in the vicinity of the heterojunctions that is caused by the difference in the lattice constant between PbSnTe and PbTe (Tamari and Shtrikman, 1979; Yoshikawa *et al.*, 1979).

A lattice-matched DH configuration can be realized by combination of ternary mixed crystals. For example, in $Pb_{1-x}Sn_xTe/PbSe_yTe_{1-y}$ the system forms lattice-matched heterojunctions when $x = 2.52y$, whereas for the $Pb_{1-x}Sn_xTe/PbS_yTe_{1-y}$ system, lattice-matched combinations are possible when $x = 3.93y$. Kasemset and Fonstad (1979b) reported a remarkably reduced J_{th} and an increased temperature range for laser oscillation in lattice-matched $Pb_{0.8}Sn_{0.2}Te/PbSe_{0.08}Te_{0.92}$ DH lasers, indicating the usefulness of a lattice-matched DH structure in IV–VI semiconductor lasers. However, in these lattice-matched heterojunctions, different lattice-constant substrates are required, depending upon the emission wavelength. This problem can be solved by employing quaternary alloys for active- or confining-layer materials.

Although there are many possible IV–VI quaternaries, most of them suffer from widely spread miscibility gaps in their solid compositions. Among these, such quaternaries as the PbSnSeTe, PbSnSSe, and PbSSeTe systems yield NaCl-type cubic crystals within fairly wide solid compositions (Abrikosov *et al.*, 1969) and have lattice-matched regions, as shown in Fig. 5. Figures 6 and 7, respectively, show the dependence on the solid composition

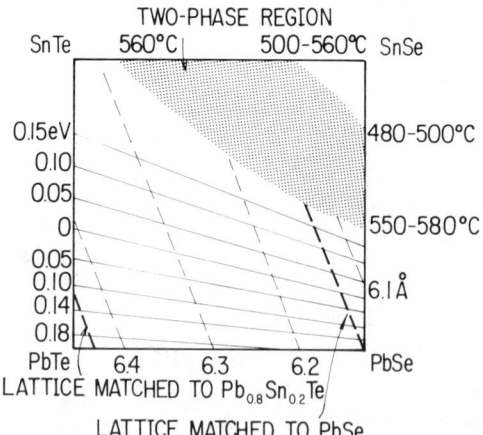

FIG. 6. Compositional plane for PbSnSeTe: (——) equi-energy-gap lines at 77 K; (– – –) equi-lattice-constant lines; shaded area shows the miscibility gap.

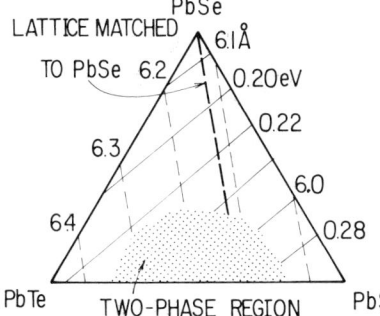

FIG. 7. Compositional plane for PbSSeTe: (——) equi-energy-gap lines at 77 K; (- - -) equi-lattice-constant lines; shaded area shows the miscibility gap.

of the fundamental energy gap measured at 77 K and the lattice constant for PbSnSeTe and PbSSeTe, as constructed using linear interpolation from previously reported ternary (Harman and Melngailis, 1974) and quaternary data (Abrikosov et al., 1969; Kennedy and Linden, 1970; Starik et al., 1978; Zayachuk and Starik, 1980). The solid and dashed lines represent the equi-energy-gap and equi-lattice-constant lines, respectively. In Fig. 6, the shaded area near the SnSe-enriched corner indicates the miscibility gap, and the material beyond this area has an orthorhombic B29 lattice. In the PbSSeTe system, a two-phase region occupies the central portion of the PbSTe ternary and limits the stable cubic area that is lattice matched to PbSe. $Pb_{1-x}Sn_xSe_yTe_{1-y}$ alloys with NaCl-type cubic lattices have the same lattice constant as the $Pb_{1-z}Sn_zTe$ substrate when $x = -2.52y + z$. Therefore, for a $Pb_{0.8}Sn_{0.2}Te$ substrate, the lattice-matched quaternary has solid compositions expressed by $x = -2.52y + 0.2$. Under these conditions, the energy gap of $Pb_{1-x}Sn_xSe_yTe_{1-y}$ was calculated (Kasemset et al., 1981) as a function of the solid composition, as is shown in Fig. 8. Although the refractive index for PbSnSeTe at the emission wavelength is not known, linear interpolation from available data concerning the optical dielectric constant for related binary and ternary compounds has been used to estimate the optical confinement factors for PbSnSeTe/PbSeTe DH structures (Horikoshi et al., 1981) as a function of the PbSnSeTe active-region thickness. Results for the $Pb_{0.85}Sn_{0.15}Se_{0.02}Te_{0.98}/PbSe_{0.08}Te_{0.92}$ DH structure are presented in Fig. 9.

3. II–VI Compound Semiconductors

Among the other semiconductor materials with a direct energy gap corresponding to emission wavelengths exceeding approximately 2 μm, mercury chalcogenides and ternary II–VI alloys incorporated with mercury chalcogenide are perhaps the most important, because, except for the CdS/HgS system, they form continuous series of substitutional solid solutions (Abrikosov et al., 1969). Among these, HgCdTe has been most extensively inves-

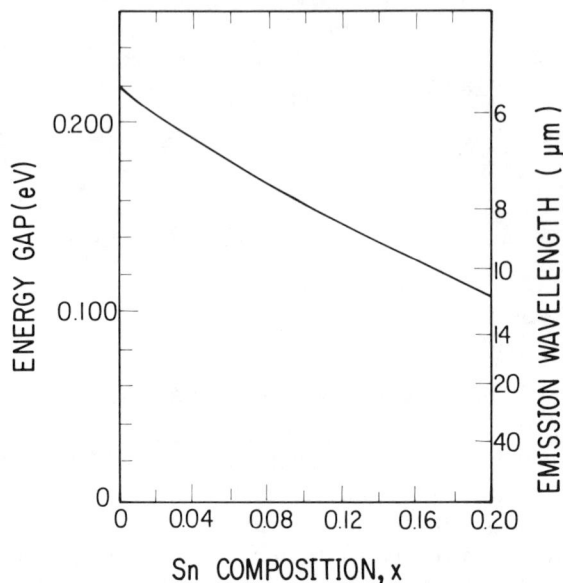

FIG. 8. The energy gaps for $Pb_{1-x}Sn_xSe_yTe_{1-y}$ under lattice-matching conditions, with $y = 0.396(0.20 - x)$, $T = 77$ K. [From Kasemset et al. (1981).]

tigated as an intrinsic infrared detector material. Laser action at 3.8 and 4.1 μm and spontaneous emission in the 3–15-μm wavelength range have been observed (Melngailis and Strauss, 1966) in $Hg_xCd_{1-x}Te$ crystals excited optically by the radiation from a GaAs diode laser. Photoluminescence and cathode luminescence from HgCdTe having various solid compositions have also been reported (Ivanov-Omskii et al., 1978; Hunter et al., 1980).

It should be noted that the equivalent threshold current estimated for the laser action mentioned before is considerably lower than that observed in optically pumped InSb and InAs lasers. Although spontaneous emission

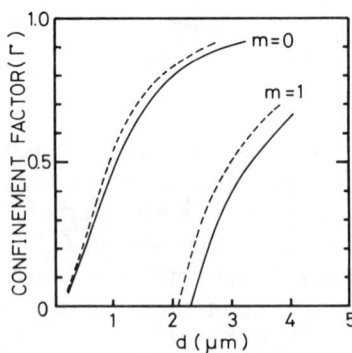

FIG. 9. The optical confinement factor as a function of the active-region thickness d for the TE mode of the $Pb_{0.85}Sn_{0.15}Se_{0.02}Te_{0.98}$/$PbSe_{0.08}Te_{0.92}$DH laser at 4.2 K and 77 K.

from $Hg_xCd_{1-x}Te$ $p-n$ junctions (Verie and Granger, 1965) has also been observed, no successful work has so far been reported with diode lasers. The HgCdTe/CdTe system is not a lattice-matching system. However, good-quality heterojunctions can be fabricated if the interdiffusion between the neighboring layers is suppressed by careful preparation. This is because the difference in the lattice constants between HgTe and CdTe is as small as 0.3%. Therefore, high-quality HgCdTe epitaxial films and heterostructures have been grown on CdTe substrates using liquid-phase epitaxy (Harman, 1979; Lanir *et al.*, 1979; Wang *et al.*, 1980; Chu *et al.*, 1980; Schmit and Bowers, 1979) and molecular-beam epitaxy (Faurie and Million, 1981, 1982). In a succeeding section, the expected lasing temperature range for HgCdTe/CdTe DH structures is briefly discussed in comparison with the other semiconductor materials.

III. Crystal Growth and Laser Fabrication

4. III–V Compound Semiconductors

a. Introduction

Preparation of heterostructure lasers with III–V compound semiconductors requires the development of heteroepitaxial growth processes. These include liquid-phase epitaxy (LPE), chemical vapor deposition (CVD), molecular-beam epitaxy (MBE), and the recently developed CVD method that utilizes metallo-organic compounds (MOCVD). Among these, LPE has been most extensively used for heterostructure fabrication. Of particular importance for LPE is the finding of phase equilibria between the solid and liquid phases in the III–V compound semiconductor systems. Phase equilibria in binary, ternary, and in some quaternary systems have been described in several review works (Panish and Ilegems, 1972; Panish and Hayashi, 1974; Casey and Panish, 1978) and are not reproduced here. Instead, this part is mainly devoted to phase-diagram considerations and LPE for two quaternary III–V compound semiconductors that are of interest: InGaAsSb and InAsPSb.

Both MBE and MOCVD methods are highly nonequilibrium techniques where it is expected that phase-equilibria constraints will be somewhat relaxed. Therefore, it is conceivable that these techniques will be useful for growth of III–V quaternaries such as InGaAsSb, InAsPSb, and AlInAsSb, since these suffer from miscibility gaps or severe segregation. Although no applications of MBE to these quaternaries have yet been reported, the MOCVD method has been used for growth of InAsPSb and related ternary compounds (Fukui and Horikoshi, 1981). The results showed that this MOCVD method is useful for the growth of high-quality InAsPSb. However, no applications to diode lasers have been reported up to the present.

b. *LPE Growth of* InGaAsSb

The existence of a miscibility gap in the central portion of the GaAs–GaSb–InAs–InSb pseudoquaternary system has been demonstrated experimentally as shown in Fig. 2. This was also predicted theoretically on the basis of spinodal curve calculations (Onabe, 1982; Stringfellow, 1982). Except for in the unstable composition range mentioned earlier, the LPE growth of InGaAsSb on either a GaSb or InAs substrate has been reported (Sankaran and Antypas, 1976; Nakajima *et al.*, 1977; Dolginov *et al.*, 1978a; Kobayashi *et al.*, 1979), and DH lasers with emission wavelengths around 2 μm have been fabricated that employ a GaSb-enriched quaternary active region (Dolginov *et al.*, 1978b; Kobayashi *et al.*, 1980).

In the solid compositions enriched with GaSb, lattice-matched growth conditions have been determined by a source-dissolution–successive-growth technique that is composed of the saturation step of an undersaturated In–Ga–As–Sb mixture on a GaSb source crystal and the epitaxial growth of InGaAsSb on a GaSb substrate (Kobayashi *et al.*, 1979, 1980). The growth melt composition was determined from the initial charge and the weight loss of the GaSb source wafer, whereas the composition of the grown $In_xGa_{1-x}As_ySb_{1-y}$ layer was determined by electron probe microanalysis (EPMA) measurement. The lattice-matching condition was confirmed by x-ray double-crystal diffraction.

Figure 10 shows the atomic fractions of Ga (X^l_{Ga}), As (X^l_{As}) and Sb (X^l_{Sb}) in the liquid phase, where they are given as a function of the atomic fraction of

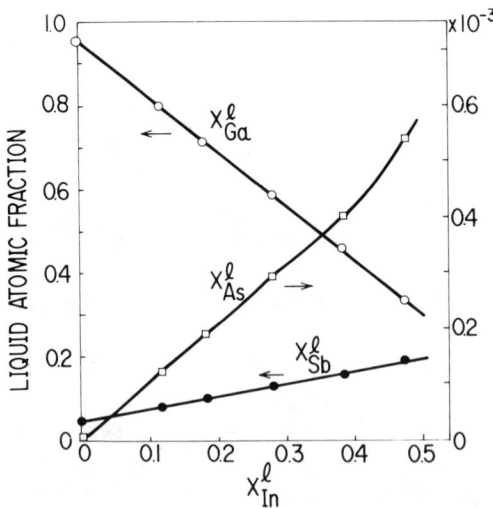

FIG. 10. The liquid atomic fractions of X^l_{Ga}, X^l_{As}, and X^l_{Sb} for (001) GaSb at 527°C as a function of X^l_{In} under lattice-matching conditions.

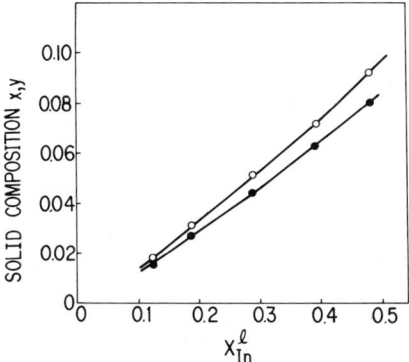

FIG. 11. The $In_xGa_{1-x}As_ySb_{1-y}$ solid compositions at 527°C as a function of X_{In}^l under conditions of lattice-matching to a (001) oriented GaSb substrate. ○, x; ●, y.

In in the liquid phase (X_{In}^l), which gives a lattice-matched condition with a GaSb substrate at 527°C. The corresponding quaternary solid composition x, y as lattice matched to the GaSb substrate, is shown in Fig. 11 as a function of X_{In}^l.

Applying these results, DH wafers with an $In_xGa_{1-x}As_ySb_{1-y}$ active region have been grown using a conventional sliding boat. As was discussed in the previous section, GaSb does not provide efficient confining layers because of its large refractive index. Therefore, AlGaAsSb lattice matched to GaSb was used instead. Figure 12 shows an optical micrograph of the stained cleaved edge of a grown DH wafer composed of $Al_{0.2}Ga_{0.8}As_{0.02}Sb_{0.98}$ confining layers, an undoped $In_{0.05}Ga_{0.95}As_{0.04}Sb_{0.96}$ active layer, and a p-GaSb cap layer successively grown on a [100]-oriented GaSb substrate. In this structure, the energy-gap difference and refractive index difference between the active layer and the confining layers are estimated to be about 0.26 eV and 2.6%, respectively. A 20-μm-wide stripe-geometry electrode was formed on the outermost p-GaSb layer through utilizing a native oxide film as an insulator. The laser diodes thus obtained showed a relatively low threshold current density of about 5 kA cm^{-2} with a 1.8-μm emission wavelength at room temperature.

As has been mentioned, $In_xGa_{1-x}As_ySb_{1-y}$ with larger x,y values covers emission wavelength around 3–4 μm. It also has compositions lattice matched to GaSb or InAs, as shown is in Fig. 2. However, it is very difficult to grow this quaternary on a GaSb substrate while employing a near-equilibrium LPE method, since the growth solutions for larger x,y alloys tend to dissolve severely the GaSb substrate. Therefore, investigations concerning

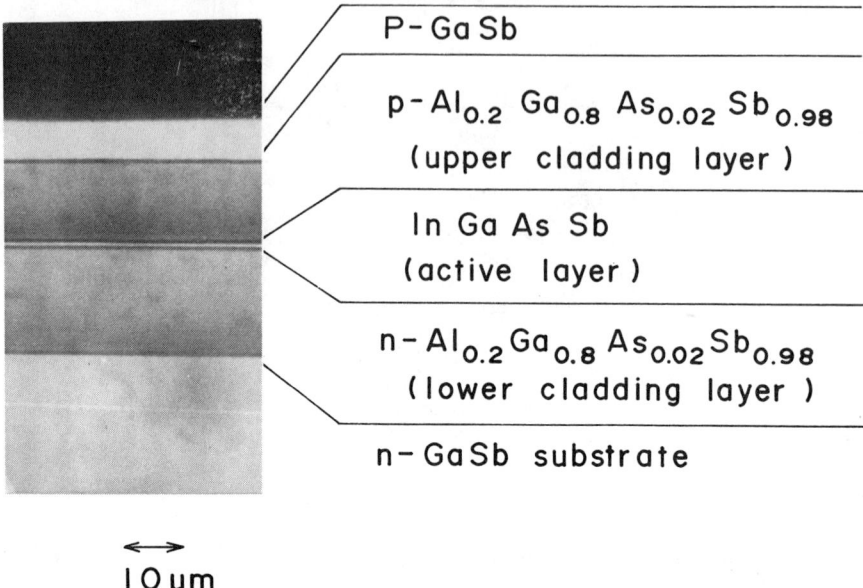

Fig. 12. An optical micrograph of a cleaved AlGaAsSb/InGaAsSb DH layer grown on [100]-oriented GaSb substrate.

LPE growth of this quaternary were performed mainly on an InAs substrate by again applying a source-dissolution–successive-growth technique. The subsequently defined relationship between solid and liquid compositions has already been reported. Experimental solidus isotherms obtained for pseudoquaternary solutions by various authors agree with each other very well (Nakajima et al., 1977; Kobayashi and Horikoshi, 1982). Accordingly, LPE growth of quaternary alloys enriched with InAs through use of an InAs substrate has now become well established. However, the interface instability mentioned earlier, for example, as seen in the interface problem between this quaternary and the GaSb substrate, has also been observed between GaSb- and InAs-enriched solid solutions. This makes it difficult to fabricate DH wafers with efficient carrier and photon confinement. Consequently, no DH lasers with an active region employing this quaternary have been reported up to the present.

Phase-diagram calculations have also been reported on this quaternary (Dolginov et al., 1978a; Kobayashi and Horikoshi, 1982). The pseudoquaternary phase diagram was calculated on the basis of a model suggested by Ilegems and Panish (1974) by applying available thermodynamic parameters. However, resulting solidus isotherms do not agree well with the experimental results (Kobayashi and Horikoshi, 1982). This discrepancy may

c. LPE Growth of InAsPSb

It has already been mentioned that the InAsPSb quaternary also possesses compositions that are lattice matched to InAs or GaSb substrates and cover 3–4-μm emission wavelengths. With this system, however, there are difficulties in growing uniform, high-quality crystals due to the existence of a miscibility gap and a very large distribution coefficient of P. Epitaxial growth of InAsPSb on an InAs substrate has been performed using both a conventional LPE method that involves equilibrium-cooling, supercooling, and step-cooling techniques (Gertner et al., 1977; Dolginov et al., 1978c; Kobayashi and Horikoshi, 1981) and a steady-state LPE method that employs a vertical temperature gradient in the growth solution (Gertner et al., 1977). The equilibrium and supercooling methods result in compositionally graded layers, whereas both the step-cooling method and steady-state LPE method yield uniform composition layers.

In order to determine suitable LPE conditions, especially growth temperatures, an InAs–InP–InSb pseudoternary phase diagram has been calculated by applying a model developed by Ilegems and Panish (1974), in which the liquid is taken to be a simple solution of four elements: In, As, P, and Sb, and the solid is assumed to be a regular solution of three binary compounds. Figure 13 shows the calculated InAs–InP–InSb pseudoternary solid isotherms, together with the experimental results obtained by the source-dissolution–successive-growth technique (Kobayashi and Horikoshi, 1981). Figure 13 shows that the agreement between the experimental and calculated results is fairly good and that the solution temperature for an $InAs_{1-x-y}P_xSb_y$ quaternary lattice matched to InAs decreases as the x or y composition increases. Moreover, it has been suggested that the pseudoternary solid isotherms predict the maximum P and Sb concentrations for this quaternary solid under lattice-matching conditions (Kobayashi and Horikoshi, 1981). Therefore, LPE growth of InAsPSb alloys with a higher P and Sb concentration requires lower solution temperatures. This limits the solid compositions grown practically by LPE to relatively smaller x and y values, since the solubility of P and As in the solution decreases steeply with decreasing temperature.

The source-dissolution–successive-growth technique has also been used for determining lattice-matching conditions. Figure 14 shows lattice mismatch $\Delta a/a$ as a function of the atomic fraction X_P^l of P in the solution, where a is the lattice constant of InAs. By applying the lattice-matching condition thus obtained, DH wafers have been grown with $InAs_{0.63}P_{0.25}Sb_{0.12}$ confining layers. A cleaved section of InAs/

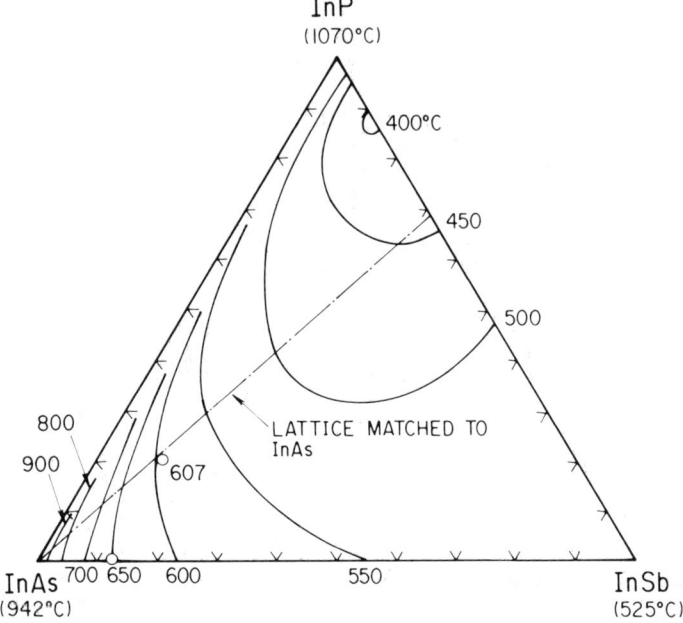

FIG. 13. The calculated InAs–InP–InSb pseudoternary solid isotherms; (O) experimental results. [From Kobayashi and Horikoshi (1981).]

$InAs_{0.63}P_{0.25}Sb_{0.12}$ DH wafer is shown in Fig. 15. Stripe-geometry laser diodes have been fabricated from DH wafers with differing active-region compositions that correspond to the emission wavelength between 2.6 and 3.0 μm.

It was mentioned previously that LPE growth with equilibrium-cooling and super-cooling modes results in epitaxial layers with graded solid compositions. This is because of the large distribution coefficient of P and As, which leads to lattice-constant variation along the growth direction. Therefore, these layers can be used as a buffer layer between the different lattice constant partners. Figure 16 shows examples of lattice-constant variation in the InAs–InP–InSb compositional plane. As was expected, lattice-constant variation depends strongly upon growth-solution thickness. With application of the buffer layer mentioned earlier, high-quality $InAs_{1-y}Sb_y$ layers have been grown on an InAs substrate (Gertner et al., 1977; N. Kobayashi, private communication, 1982) for use as long-wavelength lasers and detectors. The latter author observed a 3.56-μm-wavelength laser emission from p–n junctions of $InAs_{0.92}Sb_{0.08}$ grown on an $InAs_{1-x-y}P_xSb_y$ buffer layer.

3. SEMICONDUCTOR LASERS WITH WAVELENGTHS $>2\,\mu m$

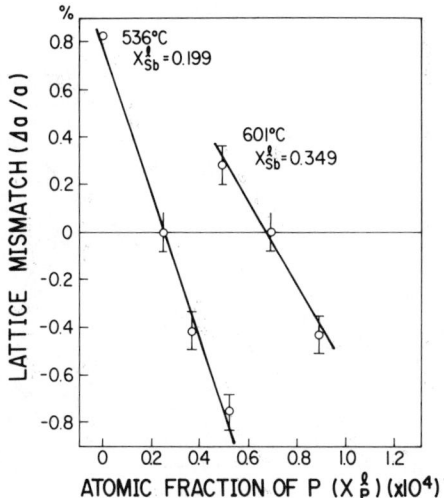

FIG. 14. The lattice mismatch between InAsPSb and InAs substrates as a function of the liquid atomic fraction of P. [From Kobayashi and Horikoshi (1981).]

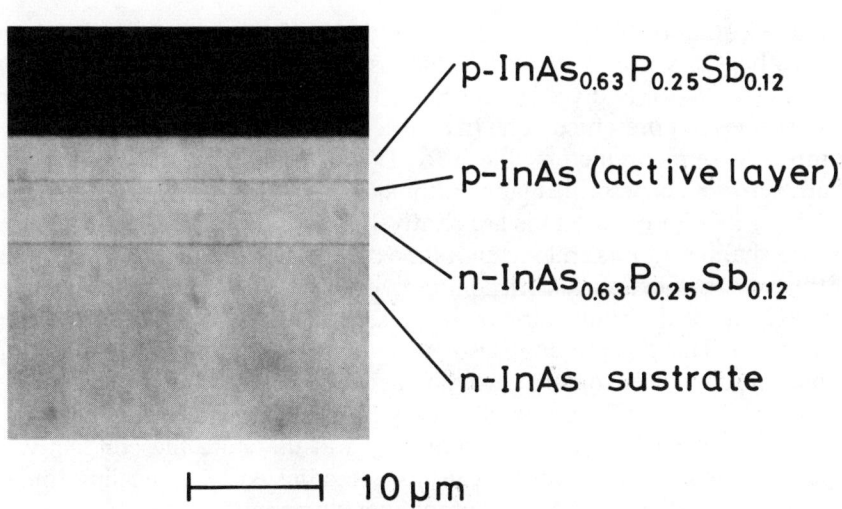

FIG. 15. A cleaved section of InAs/InAs$_{0.63}$P$_{0.25}$Sb$_{0.12}$ DH wafer.

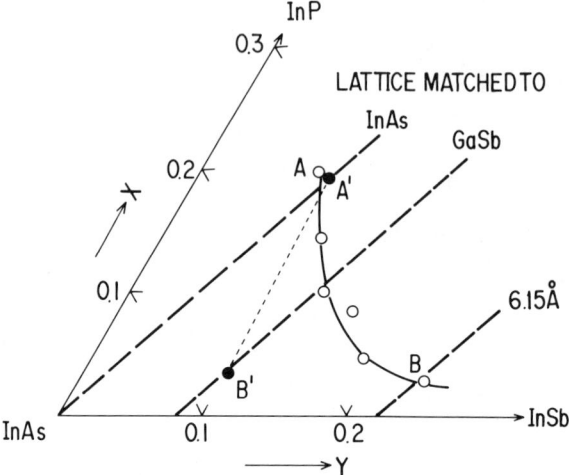

FIG. 16. The lattice-constant variation during growth in the InAs–InP–InSb compositional plane. Data was obtained by EPMA analysis on step-etched surfaces of the grown layer with (O) 1-mm-thick growth solution and (●) 10-mm-thick solution; (A, A') initial compositions; (B, B') surface compositions. [From Gertner et al. (1977).]

5. IV–VI Compound Semiconductors

a. Introduction

Phase equilibria for IV–VI compound semiconductors have been reviewed by Abrikosov et al. (1969), Novoselova et al. (1972), Harman and Melngailis (1974), and Casey and Panish (1978), and therefore a detailed discussion is not presented here. In contrast to the phase equilibria in III–V compound semiconductors, those for IV–VI compounds show peculiar characteristics, as described later. Although departures from stoichiometry in binary IV–VI compounds are relatively small, the lattice vacancies behave as shallow donors and acceptors and determine the electrical properties of the crystals. Therefore, both conductivity type and carrier concentration can be controlled by annealing in the presence of an excess of a group IV or VI element. This phenomenon also makes it possible to establish the solidus nonstoichiometry boundary through carrier-concentration measurement. Although the phase diagrams of ternary IV–VI semiconductors have not been investigated in detail, stoichiometry and therefore the conductivity type and carrier concentration can also be adjusted by annealing under appropriate partial pressures of component elements.

Vapor-phase characteristics in an equilibrium with binary crystals also needs to be pointed out. Binary IV–VI compound semiconductors (MX) have a high gaseous-molecule dissociation energy. Therefore, they are present mainly in the form of MX molecules in the vapor phase. For example, in the PbTe system, the partial pressure of PbTe(g) is greater than the pressure of Pb over the entire composition range and greater than the Te_2 pressure in the Pb-rich region. This nature is very important when these pressure relationships are applied to vapor transport epitaxy and MBE.

b. Growth of Bulk Ternary Crystals

Various techniques have been investigated for the growth of binary and ternary IV–VI compound semiconductors. The most important techniques for the growth of ternary materials that are in use today are the Bridgman, Czochralski, and vapor growth methods. Although the crystals grown by the Bridgman and Czochralski methods are large in size, they include compositional inhomogeneities in the growth direction, relatively high dislocation densities, and high carrier concentrations. The origin of the carriers is attributed to lattice-point defects, which can be reduced by long-time annealing under controlled conditions of temperature and vapor pressure for various alloy constituents. The relationships between the annealing conditions and resulting carrier concentrations have become well established for some ternary alloy systems, such as PbSnTe, PbSnSe, and PbSSe (Harman, 1973), through isothermal annealing experiments (Brebrick and Allgaier, 1960). Unlike the inhomogeneities observed along the growth direction, both Bridgman and Czochralski crystals show excellent transverse homogeneity in their alloy composition. Therefore, they can serve as a substrate for the subsequent growth of epitaxial layers.

Among the growth techniques shown in Table I, vapor growth techniques have been particularly useful for preparation of materials in the fabrication of laser diodes. Because these techniques yield uniform and low-dislocation-density crystals with carrier concentrations controllable down to the 10^{17} cm^{-3} range even with no application of a long-term annealing (Maier *et al.*, 1976; Daniel *et al.*, 1977), they can be directly used for fabrication of diffused or single-heterostructure lasers.

c. Liquid-Phase Epitaxy

A variety of epitaxial growth techniques have been investigated for laser-diode fabrication. The most important techniques reported so far are shown in Table I. These techniques have been applied not only to the PbTe/PbSnTe system but also to other ternary materials, such as PbSe/PbSnSe and PbS/PbSSe. Among these, the majority of investigations to date have dealt with LPE, MBE, and hot-wall vapor deposition (HWVD). Therefore, this section will be devoted to these three growth techniques. It should be noted, however, that highly uniform PbSnTe layers have been grown by open-tube

TABLE I

Method of epitaxy	Laser application	Reference
Liquid-phase epitaxy	PbTe/PbSnTe	Tomasetta and Fonstad (1974a,c), Groves (1977)
	PbSnTe/PbSeTe	Kasemset and Fonstad (1979a,b), Shinohara et al. (1982)
	PbSnTe/PbSnSeTe	Horikoshi et al. (1982a)
Molecular-beam epitaxy	PbTe/PbTe	Walpole et al. (1977), Partin and Lo (1981)
	PbSnTe/PbSnTe	Walpole et al. (1977)
	PbTe/PbSnTe	Walpole et al. (1973a, 1976a)
Hot-wall vapor deposition	PbS/PbSSe	Preier et al. (1976, 1977)
	PbSe/PbSnSe	Grisar et al. (1979)
Vapor-transport epitaxy		

vapor-transport epitaxy during detector array fabrication (Bellavance and Johnson, 1976).

Liquid-phase epitaxy will be considered first. The LPE method for semiconductor materials is an established technique in the preparation of high-quality crystalline layers for device applications. In this connection, it has been shown that the LPE method is one of the best techniques to produce PbSnTe layers with low carrier-concentration (Longo et al., 1971; Andrews et al., 1972) and efficient radiative recombination (Tomasetta and Fonstad, 1974a,b) characteristics. Early work on PbSnTe LPE growth (Longo et al., 1972) suggested that achieving a large vertical temperature gradient in the growth solution is essential for obtaining uniform epitaxial growth. However, Tomasetta and Fonstad (1974a) have successfully grown PbTe/PbSnTe DH wafers with shiny surfaces that have no intentional vertical temperature gradients.

The Pb–Sn–Te phase diagram required for $Pb_{1-x}Sn_xTe$ LPE growth has been investigated both experimentally and theoretically by several workers (Laugier et al., 1974; Harris et al., 1975; Ilegems and Pearson, 1975; Muszynski et al., 1979). The solidus lines calculated by Ilegems and Pearson (1975) are shown in Fig. 17, together with the experimental results from Harris et al. (1975). Figure 17 predicts the existence of a solid composition range insensitive to the atomic fraction of Sn in the solution (X_{Sn}^l). This indicates that solid-composition control is very easy when $0.18 < x < 0.20$, because Sn/Pb ratios in the solution do not have to be controlled accurately. Figure 17 also suggests that the solid composition x is almost proportional to X_{Sn}^l when $x < 0.18$. One major difficulty encountered in DH-diode prepara-

EPITAXIAL GROWTH METHODS FOR LASER AND DETECTOR FABRICATION WITH IV–VI COMPOUND SEMICONDUCTORS

Detector and others applications	Reference
PbTe/PbTe	Wagner and Thompson (1970)
PbSnTe/PbSnTe	Wang and Lorenzo (1977)
PbSnSe/PbSnSe	Kasai and Bassett (1974)
BaF_2/PbSnTe	Smith and Pickhardt (1976)
BaF_2/PbSnSe	Hohnke et al. (1976)
PbS/PbSSe	Schoolar et al. (1977)
PbSe/PbSnSe	Schoolar and Jensen (1977)
BaF_2/PbSnTe	Lopez-Otero (1978)
KCl/PbSnTe	Kasai et al. (1975)
PbTe/PbSnTe	Rolls et al. (1970)
PbSnTe/PbSnTe	Bellavance and Johnson (1976)

tion by LPE is the reduced carrier-injection efficiency due to diffusion of the p–n junction out of the active region into the substrate during epitaxy steps. This problem has, however, been overcome by using highly Tl-doped substrates or by doping the epitaxial layer with Tl and Bi (Tomasetta and Fonstad, 1974b).

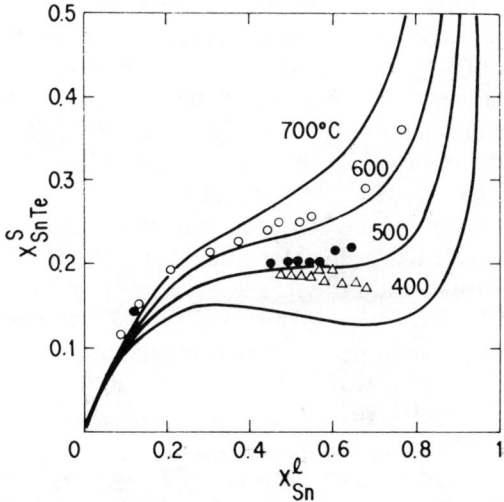

FIG. 17. $Pb_{1-x}Sn_x$Te solidus versus liquidus composition along several isotherms calculated by Ilegems and Pearson (1975): ○, 640°C; ●, 550°C; △, 475°C. Experimental results are from Harris et al. (1975). [From Ilegems and Pearson (1975).]

Although the majority of investigations have been devoted to PbTe/PbSnTe heterostructures, this system includes an inherent lattice mismatching, which causes such laser performance areas as threshold current and reliability to deteriorate. As was described earlier, there are numerous lattice-matching combinations with IV–VI ternary and quaternary materials. The importance of the PbSnTe/PbSeTe system has been noticed by Walpole *et al.* (1973a). Kasemset and Fonstad(1979b) reported the first successful LPE growth of lattice-matched DH wafers composed of a $Pb_{0.8}Sn_{0.2}Te$ active layer and $PbSe_{0.08}Te_{0.92}$ confining layers and demonstrated a reduced threshold current and higher-temperature operation in lattice-matched DH lasers.

Application of the PbSnSeTe quaternary to DH configurations is much more advantageous, since this quaternary has compositions that are lattice matched to such binary crystals as PbSe and SnTe, as well as to the related ternary alloys PbSnTe and PbSeTe that have arbitrary solid compositions. Lattice-matched DH layers with a PbSnSeTe active layer and $PbSe_{0.08}Te_{0.92}$ confining layers have been grown on $Pb_{0.8}Sn_{0.2}Te$ substrates by a conventional LPE method. The growth temperature range was from 560 to 530°C and cooling rates from 1 to 3°C min^{-1} (Horikoshi *et al.*, 1982a). The LPE growth condition for $PbSe_{0.08}Te_{0.92}$ confining layers has been determined by a source-dissolution–successive-growth technique with a PbTe source wafer. The lattice mismatch $\Delta a/a$, obtained by an x-ray double-crystal diffraction measurement, where a represents the lattice constant of $Pb_{0.8}Sn_{0.2}Te$, is shown in Fig. 18 as a function of the Se/Te atomic fraction ratio (X^l_{Se}/X^l_{Te}) in the growth solution. It should be noted that, within the range $0.07 < X^l_{Se}/X^l_{Te} < 0.09$, closely lattice-matched layers were grown regardless of the values of X^l_{Se}/X^l_{Te}, as in the LPE growth of ternary III–V compound semiconductors (Wu and Pearson, 1972; Stringfellow, 1972). Therefore, it is very easy to grow high-quality PbSeTe layers that are closely lattice matched to a $Pb_{0.8}Sn_{0.2}Te$ substrate.

The growth condition for PbSnSeTe quaternaries lattice matched to a $Pb_{0.8}Sn_{0.2}Te$ substrate has been determined using the growth conditions for ternary PbSnTe. First, the growth solution composition of $Pb_{1-x}Sn_xTe$, where $0 < x < 0.2$, was obtained from available phase-diagram data (Harris *et al.*, 1975), and an accurate saturation composition was determined by employing a PbTe source wafer. The value of x for the grown layer was confirmed by x-ray diffraction measurement. A small amount of Se was added to the solution thus obtained, and a $Pb_{1-x}Sn_xSe_yTe_{1-y}$ layer was grown from this solution. The amount of Se was varied until the x-ray diffraction measurement showed complete lattice matching between the grown quaternary and the substrate. The composition of the growth solution thus obtained was somewhat oversaturated. Therefore, an additional saturation adjustment was required, because the growth of the active region re-

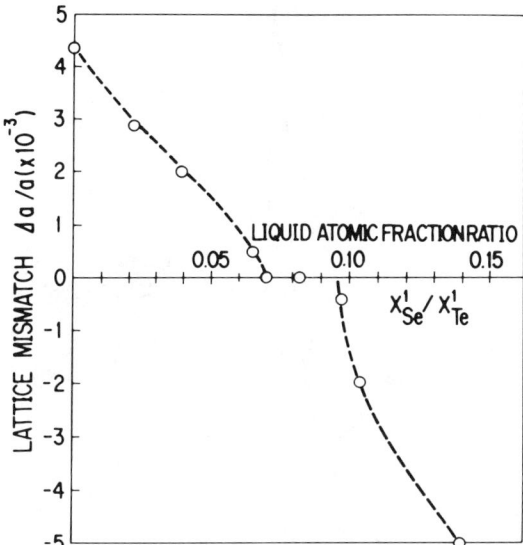

FIG. 18. The lattice mismatch as a function of X^l_{Se}/X^l_{Te}, which shows lattice-matching conditions between PbSeTe and $Pb_{0.8}Sn_{0.2}Te$.

quires precise thickness control. This was done by changing the amount of Te and Se in the growth solution while keeping the ratio X^l_{Se}/X^l_{Te} constant. It was found that the solid composition of Sn, $x\,(=X^s_{Sn})$ is not sensitive to the addition of Se in the growth solution, thus making it easy to find the growth condition for PbSnSeTe quaternary layers.

Figure 19 shows a view of an etched cleaved section of a typical PbSnSeTe/PbSeTe DH wafer composed of a Tl-doped p-$Pb_{0.8}Sn_{0.2}Te$ substrate ($\sim 1 \times 10^{19}$ cm^{-3}), Tl-doped p-$PbSe_{0.08}Te_{0.92}$ ($p \sim 2.5 \times 10^{18}$ cm^{-3}), a nondoped PbSnSeTe active layer, and In-doped n-$PbSe_{0.08}Te_{0.92}$

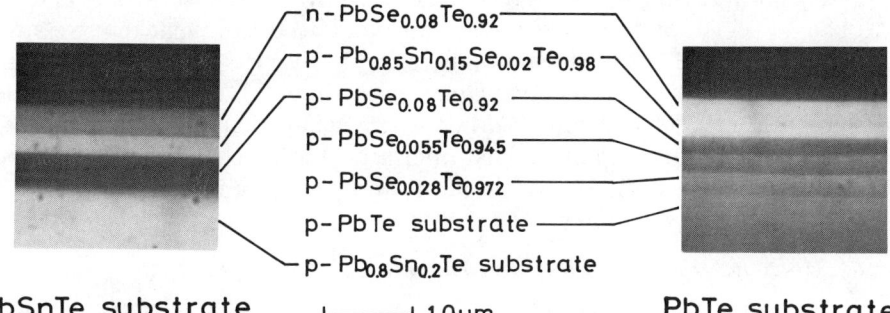

FIG. 19. Etched-cleaved surface of PbSnSeTe/PbSeTe DH layers grown on a $Pb_{0.8}Sn_{0.2}Te$ or PbTe substrate.

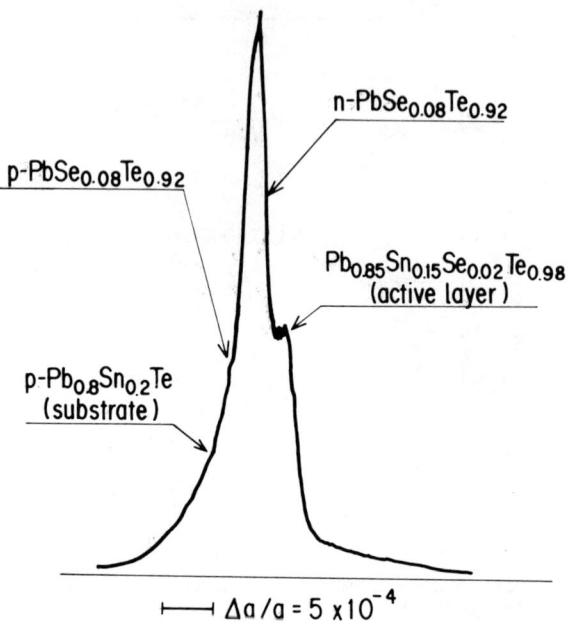

FIG. 20. X-ray rocking curve for $Pb_{0.85}Sn_{0.15}Se_{0.02}Te_{0.98}/PbSe_{0.08}Te_{0.92}$ DH wafer with $Pb_{0.8}Sn_{0.2}Te$ substrate.

($n \sim 1 \times 10^{18}$ cm^{-3}). The resulting DH wafers exhibited shiny, flat surfaces throughout the wafer area of 12×10 mm^2. An x-ray double-crystal rocking curve measured for a DH wafer is shown in Fig. 20. Here the diffraction signal from each layer was assigned through x-ray measurement made on successively etched wafer surfaces. The PbSnSeTe/PbSeTe DH layers mentioned above have also been grown on PbTe substrates. In this case, however, two additional layers with intermediate solid compositions, $PbSe_{0.028}Te_{0.972}$ and $PbSe_{0.055}Te_{0.945}$, were grown between the PbTe substrate and p-$PbSe_{0.08}Te_{0.92}$ confining layer. This was done to circumvent the stress resulting from the lattice constant difference. In Fig. 19, a typical view of an etched cleaved section of this wafer is also shown. During growth of the step-graded layers mentioned earlier, a somewhat greater cooling rate, 3°C min^{-1}, was required to obtain flat interfaces. Lasing characteristics for the resulting DH laser diodes are similar to those observed for DH lasers with lattice-matched substrates.

d. Molecular Beam Epitaxy and Hot-Wall Vapor Deposition

The LPE growth mentioned previously requires relatively high substrate temperatures (500–600°C) to achieve sufficient solubility in the growth solution and better wetting between the solution and substrate surface. Epitaxial growth by the MBE method, on the other hand, is possible at very low temperatures (~ 300°C). Therefore, this latter technique seems to be particularly well suited to the fabrication of abrupt heterojunctions and sharp impurity profiles with IV–VI compound semiconductors, because both the impurity diffusion and interdiffusion coefficients are very large in these materials.

Indeed, Partin and Lo (1981) demonstrated steep carrier-concentration profiles in a PbTe homojunction grown by MBE at 300°C, as is shown in Fig. 21. By applying the MBE method, PbTe homojunction (HJ) (Walpole *et al.*, 1977), PbTe/PbSnTe single-heterostructure (SH) (Walpole *et al.*, 1973a), and DH lasers (Walpole *et al.*, 1976a) have been fabricated with relatively low substrate temperatures (300–425°C). Each layer has usually been grown from a single polycrystalline source with the desired alloy compositions. Layers grown from undoped sources have very low carrier concentrations ranging from 10^{16} to 10^{17} cm^{-3} (Walpole *et al.*, 1976a; Ralston *et al.*, 1975; Smith and Pickhardt, 1978). Although both n- and p-type doping are achievable by stoichiometry adjustment using coevaporated Pb and Te, respectively, only ~ 0.1% of each can be incorporated (Smith and Pickhardt, 1978). Therefore, it is very important to establish impurity doping conditions for the fabrication of laser diodes. Smith and Pickhardt (1978) found that more than half the incident Bi and Tl evaporated from separate sources became incorporated as n and p dopants, respectively, to levels exceeding 10^{19} cm^{-3}. Walpole *et al.* (1976a) controlled the carrier type and concentration in PbTe and PbSnTe by adding, respectively, Bi and Tl chalcogenide to the sources for the n- and p-type impurities. The MBE method was also applied to grow DH layers composed of PbEuSeTe quaternaries lattice matched to PbTe by using PbTe, Eu, Te, PbSe, Tl$_2$Te (for p-type dopant) and Bi$_2$Te$_3$ (for n-type dopant) source ovens (Partin, 1983).

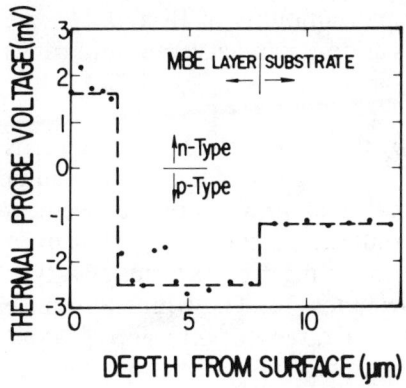

FIG. 21. Thermal probe voltage as a function of depth through MBE-grown p^+–p–n^+ PbTe HJ structure. [From Partin and Lo (1981).]

Hot-wall vapor deposition (HWVD) is a slight variation of vacuum evaporation, or the MBE method. However, it is a comparatively new method that differs from previous methods in its near-equilibrium growth conditions. Although this method has been used for studying stoichiometry control in PbTe and PbSnTe(Lopez-Otero, 1978; Bis et al., 1972; Lowney and Cammack, 1974; Kasai et al., 1975; Jakobus and Hornung, 1978), its application to laser structure fabrication has mainly been restricted to such materials as PbSSe (Preier et al., 1976, 1977; McLane and Sleger, 1975; Duh and Preier, 1975; Bleicher et al., 1977) and PbSnSe (Grisar et al., 1981) rather than PbSnTe, as can be seen in Table I. In HWVD growth, the conduction type and carrier concentrations can be controlled by changing the starting material compositions and temperature of the chalcogenide reservoir set independently in the vacuum system. However, variation of the reservoir temperature causes changes in both carrier concentration and solid composition (Duh and Preier, 1975). Therefore, careful determination of the starting material compositions and chalcogenide reservoir temperature is required. Moreover, the carrier-concentration ranges achievable with stoichiometry adjustment are restricted to those predicted by the phase diagram (Harman, 1973). In some IV–VI compounds, only carrier concentrations below 10^{18} cm^{-3} may be possible when the substrate temperature is relatively low (300–400°C). Therefore, doping with foreign impurities is useful for fabrication of well-controlled $p-n$ junctions.

Among the works published on HWVD, it is noteworthy that Prier et al. (1977) demonstrated Peltier-cooled PbS/PbSe DH lasers operating at temperatures of up to 230 K with emission wavelengths of around 5.5 μm. The PbSe active layer and PbS upper confining layer have been grown on a [100]-oriented PbS substrate by employing separate furnace arrangements that are placed inside a vacuum system for the growth of epitaxial films having different compositions and carrier types. Consective layers were grown by moving the substrate from one furnace to the other. In order to achieve higher carrier concentrations, small amounts of Bi and Tl were respectively added to the source materials as donor and acceptor impurities

e. Laser-Diode Fabrication

Unlike preparation of laser structures by the variety of epitaxial methods just described, direct application of bulk crystals to laser structures require creation of a $p-n$ junction inside the crystal. The surface carrier concentration of high-quality vapor-grown bulk crystals decreases during cooling from the growth temperature, and sometimes the cooling process forms an n-type skin at the cryatal surface and creates $p-n$ junctions. These junctions have been used for laser structures. In addition, a variety of diffusion, proton-bombardment, and ion-implantation processes have been investigated. Among these, diffusion processes may be the most important, since low-threshold and high-output-power lasers have already been fabricated through these methods.

Self-diffusion in 1% metal- or chalcogen-rich powder causes a conversion of the conduction type of the surface region, and leads to formation of an interdiffused $p-n$ junction (Melngailis and Harman, 1970; Lo and Swets, 1978). Diffusion of foreign impurities such as Sb (Antcliffe and Wrobel, 1970; Lo et al., 1976) and Cd (Lo, 1976; Grisar et al., 1979) has also been investigated. Lo (1977) demonstrated excellent tuning characteristics with Te-rich ingot nucleated crystals by Sb diffusion. Single heterostructures have been fabricated by annealing $PbS_{1-x}Se_x$ or $Pb_{1-x}Sn_xSe$ in the presence of PbS or PbSe powder, respectively. This process leads to the formation of heterojunctions as a product of Se or Sn out-diffusion from the surface region (Linden et al., 1977; Grisar et al., 1979), as shown in Fig. 22.

A variety of stripe geometries have been developed for lateral confinement of carriers and photons. Most of them are similar to those developed for AlGaAs/GaAs DH lasers. Corrugated structures for distributed-feedback (DFB)-mode oscillation have been fabricated by constructing a corrugation on the top of MBE-grown DH wafers (Walpole et al., 1976b). Distributed-feedback lasers have also been produced by LPE growth of $Pb_{1-x}Sn_xTe$ and PbTe layers directly on a corrugated PbTe substrate (Hsieh and Fonstad, 1979). In both cases, clean single-mode spectra and a very wide continuous tuning range have been demonstrated.

Lowering contact resistance is essential for achieving cw operation. Lo and Gifford (1980) reported that a contact resistance of 10^{-4} Ω cm^{-2} or less is necessary for cw operation at temperatures approaching 40 K. This has been attributed to the low thermal conductivity of rock-salt materials. Making

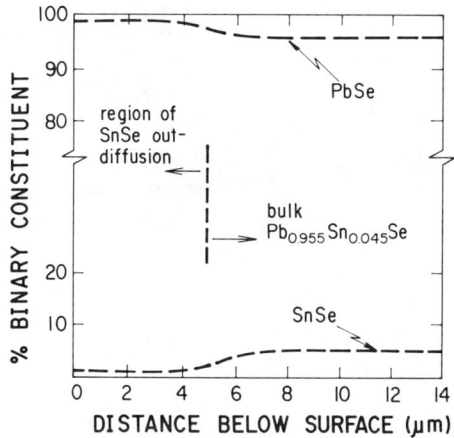

FIG. 22. Compositional interdiffusion effect of $Pb_{0.955}Sn_{0.045}Se$ wafer annealed with $Pb_{0.99}Sn_{0.01}Se$ source at 600°C for 6.5 hr [From Linden et al. (1977). © IEEE 1977.]

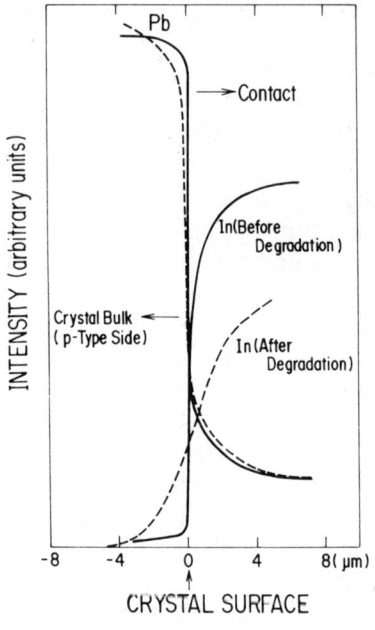

FIG. 23. Indium diffusion into a p-type bulk crystal during room-temperature storage. [From Lo and Gifford (1980). Reprinted by permission of the publisher, The Electrochemical Society, Inc.]

low-resistance contacts to n-type materials presents no difficulty because of high electron affinities in IV–VI compound semiconductors, and Au and In have proved to be the best materials. As for the p-side, Au and Pt have been successfully used for low-resistance contacts, although careful preparation of the crystal surface is required. Lo and Gifford (1980) noticed that In tends to diffuse into p-type lead-salt crystals during a few weeks of storage at room temperature, as can be seen in Fig. 23. This phenomenon increases p-type-contact resistance resulting from a reduction in the hole concentration of the surface region, because In acts as a shallow donor. Using Au–Pt or Pt–Au barriers between the In film and the crystal surface have proved to prevent In migration (Lo and Gifford, 1980). Although In yields low-resistance contacts to n-type material, as has already been mentioned, In diffusion may cause the p–n junctions or active regions of lasers to degrade when In is directly applied to the n-layer surface. This is because a few micrometers of diffusion occur even at room temperature, as shown in Fig. 23.

IV. Laser-Diode Characteristics

6. Introduction

Diode lasers with emission wavelengths exceeding 2 μm involve inherent problems when compared with near-infrared AlGaAs/GaAs and InP/

InGaAsP DH lasers. Since the energy gap is comparatively small, the importance of the band-to-band Auger process in determining the minority-carrier lifetime increases with increasing emission wavelength. This may lower the internal quantum efficiency. Therefore, low-temperature operation is probably essential for these lasers. Moreover, free-carrier absorption will increase with λ^2, where λ is the emission wavelength. Therefore, the optical gain required for laser oscillation will increase with the emission wavelength. However, unlike lasers with shorter wavelengths in which a variety of electronic and optical loss processes are competing, it is anticipated that in lasers with longer wavelengths the band-to-band Auger effect may be the determining process, especially, at high temperatures. The most important laser properties may therefore be threshold current variation as a function of heat-sink temperature, output power, and temperature range for stable laser oscillation. Therefore, these properties are discussed in the rest of this part for both the III–V and IV–VI compound semiconductor lasers described in Part III.

7. III–V Compound Semiconductor Lasers

a. *Homojunction and Heterostructure Lasers*

Laser diodes made from narrow-gap III–V compound semiconductors have been reported that employ binary InAs and InSb, as well as such mixed crystals as InAsSb, InGaAsSb, and InAsPSb. Melngailis and Rediker (1966) reported characteristics of InAs $p-n$-junction lasers having 3.1-μm emission wavelengths that were fabricated by means of Zn diffusion. They demonstrated an external quantum efficiency as high as 12% and an optical output power exceeding 100 mW at 11 K under pulsed operation. This suggested that the characteristics of InAs-diode lasers are very similar to those observed for GaAs HJ lasers. Application of a magnetic field perpendicular to the diode current greatly improved both the threshold current and the external quantum efficiency. Such a result was attributed to improved carrier confinement within the active area. This implies the usefulness of DHs in which sufficient carrier and photon confinement can be expected.

Heterojunction lasers with these wavelengths have also been investigated. Dolginov *et al.* (1977) reported spontaneous emission for a spectral range of 1.9–2.1 μm at room temperature with InGaAsSb/GaSb heterojunction laser structures. They also reported 1.9-μm laser oscillation at 90 K (Dolginov *et al.*, 1978b). Room-temperature laser operation at 1.8-μm wavelength was demonstrated for AlGaSb/GaSb (Dolginov *et al.*, 1981) and AlGaAsSb/InGaAsSb (Kobayashi and Horikoshi, 1980) DH lasers. In an InGaAsSb quaternary lattice matched to InAs or GaSb, as was described earlier, high-quality crystals can be obtained in both InAs- and GaSb-enriched regions.

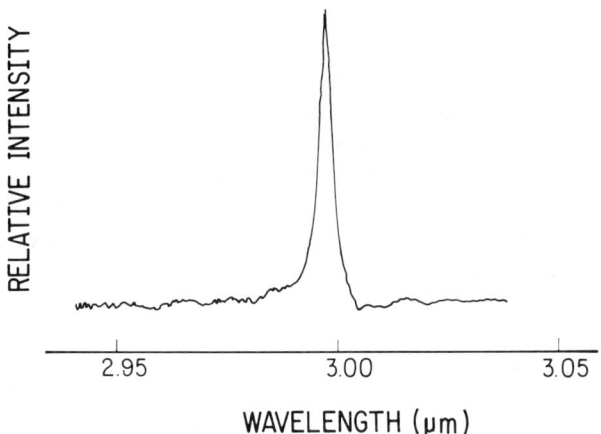

FIG. 24. Typical emission spectrum for $InAs_{0.63}P_{0.25}Sb_{0.12}/InAs$ DH laser with InAs active region at 77 K.

However, no successful work on DH-laser fabrication has so far been reported with compositions beyond the miscibility gap. Instead, heterostructure lasers have been investigated whereby InAsPSb quaternaries are grown on InAs substrates. Dolginov et al. (1978c) reported on electron-beam-pumped lasers with 3.1 – 3.7-μm emission wavelengths at 80 K and demonstrated a high efficiency of 8% as well as a high optical output power, reaching 100 W. InAsPSb/InAs lattice-matched DH-diode lasers with emission wavelengths of 2.6 – 3.0 μm at 77 K have been reported by Kobayashi and Horikoshi (1980) using the InAs or InAsPSb active regions.

Figure 24 shows a typical emission spectrum at 77 K for a $InAs_{0.63}P_{0.25}Sb_{0.12}$ InAs (active region) DH laser. However, the spectrum in Fig. 24 is the longest obtainable with this lattice-matched DH system at 77 K. Applications of solid-solution InAsSb may be useful for creating much longer-wavelength lasers if sufficiently low-dislocation-density layers are prepared. As was described earlier, low-defect-density $InAs_{0.92}Sb_{0.08}$ layers have been grown on a compositionally graded InAsPSb buffer layer. Figure 25 shows the emission spectrum measured at 4.2 K for a grown $InAs_{0.92}Sb_{0.08}$ $p-n$-junction laser thus obtained. Although this laser ceased its laser action at about 70 K, the temperature dependence of the threshold current density J_{th} was very similar to that observed for InAsPSb/InAs DH lasers.

b. Threshold Characteristics

The J_{th} values have been calculated for InGaAsSb active-region and AlGaAsSb active-region lasers through considering both the band-to-band

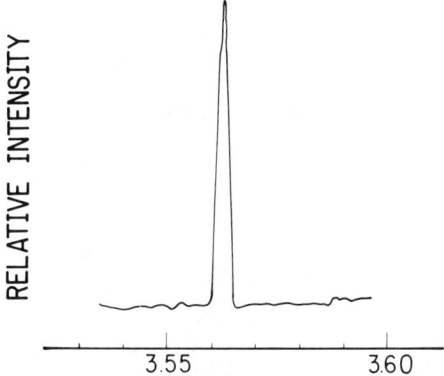

FIG. 25. Laser-emission spectrum for a $InAs_{0.92}Sb_{0.08}$ HJ laser at 4.2 K grown on the compositionally gradient InAsPSb layer. [From Kobayashi (1982).]

Auger effect and radiative recombination (Sugimura, 1982). Results for room-temperature values are shown in Fig. 26 as a function of the active-region emission wavelength, where J_{thr} represents the ideal J_{th} determined by radiative recombination, and J_{tha}^c, J_{tha}^s, and J_{tha}^l denote the threshold current components due to the band-to-band Auger effect with CCHC, CHHS, and CHHL processes, respectively. Here, in the CCHC process, two conduction-band electrons interact with one heavy hole leaving a high-energy electron in the conduction band, whereas in the other two processes, two holes interact with one conduction-band electron leaving a high-energy hole in the split-off valence band (CHHS process) or in the light-hold band (CHHL process). The calculated J_{th} values for a 1.8-μm wavelength agree well with the experimental results, which are also shown in Fig. 26. However, although not shown here, the experimental temperature dependence of J_{th} (Kobayashi and Horikoshi, 1980) is much more gradual than the calculated result (Sugimura, 1982). This discrepancy is still not well understood. Perhaps the experimental J_{th} at a lower temperature includes some additional loss components which have not been considered in the theory. Figure 26 also suggests that the definitive process for InGaAsSb lasers with emission wavelengths exceeding 2 μm is the CHCC-type Auger process, which includes two conduction-band electrons and one heavy hole.

Figure 27 shows the experimental temperature dependence of J_{th} (J_{th}-T relationship) for an $InAs_{0.63}P_{0.25}Sb_{0.12}$InAs DH laser with an emission wavelength of 3 μm at 77 K. The measurement was performed with a 100-nsec pulse current and a 100-Hz repetition. Although J_{th} at this temperature is as low as 1 kA cm^{-2}, the highest observed temperature for laser oscillation was about 150 K, due to the high-temperature sensitivity of J_{th}. Calculated J_{th} values are also shown in Fig. 27. The dashed curves denote the radiative

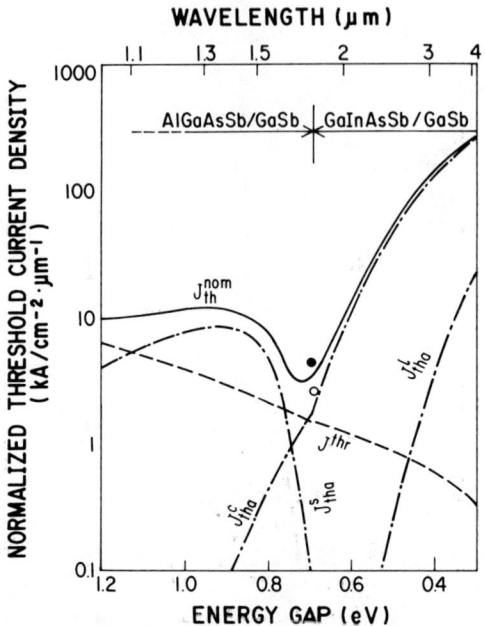

FIG. 26. Calculated nominal threshold current density for AlGaAsSb and InGaAsSb lasers: (J_{th}^{nom}) total threshold current density per 1-μm-thick active region composed of radiative J_{thr} and nonradiative Auger components J_{tha}^c, J_{tha}^s, and J_{tha}^l; Data points are from (●) Dolginov et al. (1980) and (○) Kobayashi and Horikoshi (1980). [From Sugimura (1982). © IEEE 1977.]

component of J_{th}, J_{thr}, as determined by the band-to-band direct-transition model (Stern, 1973). The solid curves represent the theoretical J_{th}, including the CHCC-type band-to-band Auger component. In Fig. 28, the maximum optical gain g_{max} used in the J_{thr} calculation is shown as a function of the injected-carrier density. For the estimation of the threshold optical gain g_{max}^{th}, which is needed for obtaining the injected-carrier concentration at threshold n_{th} (cf. Fig. 28), both free-carrier absorption and the optical confinement factor were taken into account.

In the temperature range below about 60 K, experimental results are much higher than the calculated result. This implies the existence of leakage current components that do not increase the band-edge carrier concentration. Above about 80 K, experimental and theoretical results show a similar temperature dependence. This fact suggests that the higher-temperature characteristics are mainly determined by the band-to-band Auger effect. Although J_{th} for this laser is much lower than the lowest value for the InAs HJ lasers reported to date (Melngailis and Rediker, 1966), both InAsPSb/InAs DH and InAs HJ lasers show a similar $J_{th}-T$ relationship and stop

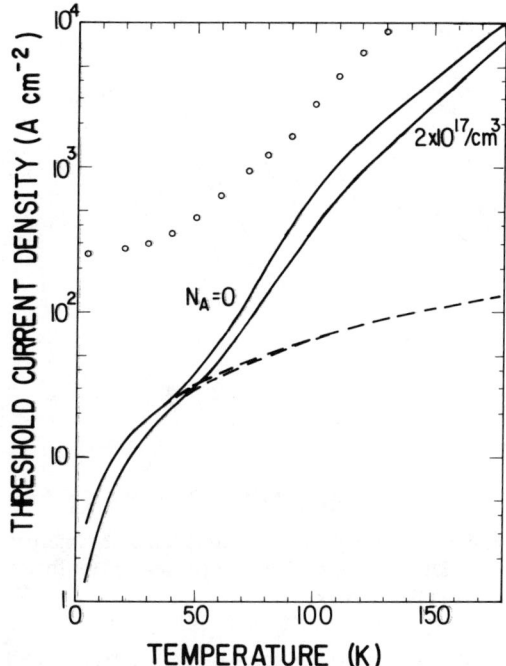

FIG. 27. The temperature dependence of J_{th} for an $InAs_{0.63}P_{0.25}Sb_{0.12}/InAs$ DH laser with a 3-μm emission wavelength. O, Experimental data; ——, $J_{th}(=J_{thr}+J_{tha})$ calculated; - - -, J_{thr} calculated.

lasing at almost the same temperature (150 K). At this temperature, the theoretical internal quantum efficiency, that is, J_{thr}/J_{th}, is about 2.5%.

Spontaneous carrier-lifetime measurements at the threshold provide useful information on the recombination processes in the active region. Although the sensitivity of detectors in the wavelength region past 2 μm is much less than with shorter-wavelength detectors, the higher optical output of lasers of narrow-energy-gap III–V semiconductors makes it possible to evaluate carrier lifetimes using such conventional measurements as the lasing time-delay method (Konnerth and Lanza, 1964) and luminescence decay-time measurements. Figure 29 shows the temperature dependence of the carrier lifetime ($\tau_s - T$ relationship) for $InAs_{0.63}P_{0.25}Sb_{0.12}/InAs$ DH and InAs HJ lasers as measured by the lasing-delay method. An HgCdTe p–n-junction detector was used for measurement. Curves are calculated results as based on a band-to-band direct-transition model and the CHCC-type Auger effect. There is considerable discrepancy in the variation of lifetime with temperature between the experimental and calculated results at low temper-

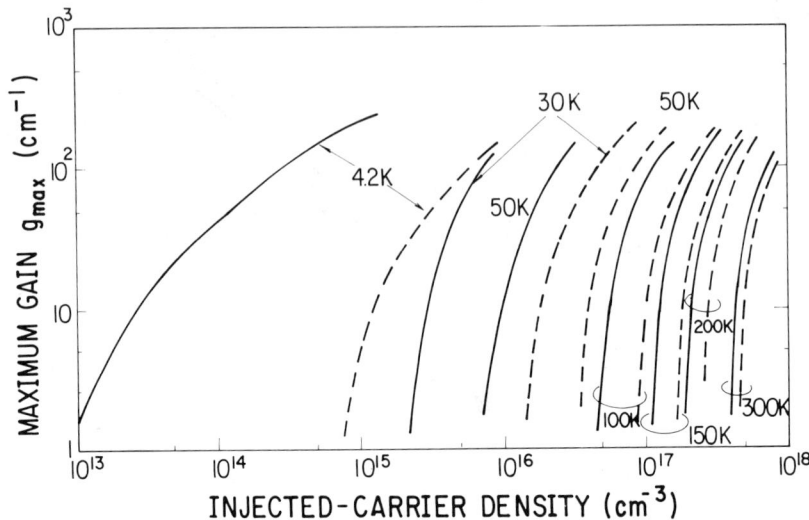

FIG. 28. The maximum optical gain g_{max} as a function of the injected-carrier density for an InAs/InAs$_{0.63}$P$_{0.25}$Sb$_{0.12}$ DH laser. Available mass parameters for InAs were used. ——, $N_A = 2 \times 10^{17}$ cm^{-3}; – – –, $N_A = 0$.

atures. Since n_{th} at low temperatures is very small, electronic transitions including tail states, rather than a simple band-to-band transition, may be responsible for the observed lifetime. Difference in lifetime between DH and HJ lasers is not well understood. However, it should be noted that a DH diode has a nondoped active region, whereas in the HJ diode, a Cd-doped active region was used. Therefore, there is a considerable difference between these lasers in the tail states in the active region.

8. Lasing Characteristics for IV–VI Semiconductor Lasers

a. Homojunction and Heterostructure Lasers

Available J_{th} data for IV–VI compound semiconductor lasers, including diffused or grown HJ, SH, and DH configurations, show that the low-temperature J_{th} values are smaller in HJ laser diodes than in SH and DH diode lasers, although some SH diodes (Walpole *et al.*, 1973a) exhibit a very low J_{th}, which is comparable with the lowest J_{th} values of HJ lasers (Oron and Zussman, 1980). As has been mentioned, J_{th} values at high temperatures have been greatly improved, and cw operation at temperatures well above 77 K has been demonstrated in heterostructure configurations in general and in DH diode lasers in particular.

Figure 30 shows typical J_{th} variations in relation to T for some low-thresh-

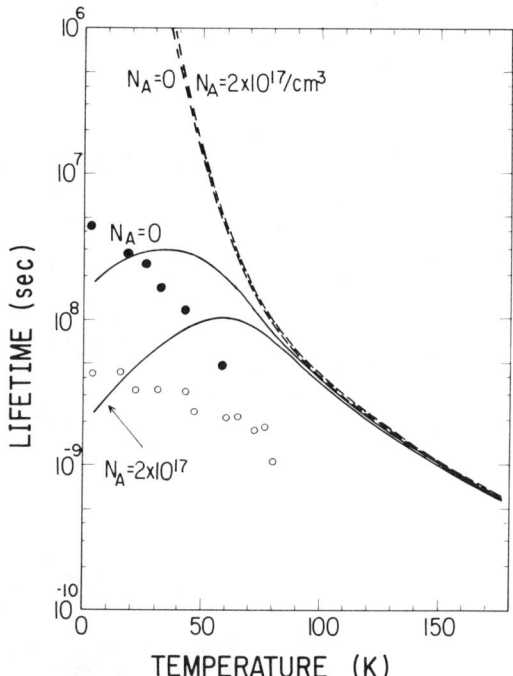

FIG. 29. Temperature dependence of the spontaneous carrier lifetime at threshold for an InAs$_{0.63}$P$_{0.25}$Sb$_{0.12}$/InAs DH laser (○) and an InAs HJ laser (●). Curves are calculated results with the band-to-band direct-transition model and the CHCC-type Auger effect. ——, τ calculated; – – –, τ_a calculated; ○, experimental data. [From N. Kobayashi (1982).]

old HJ and DH PbSnTe diode lasers, including lattice-matched DH lasers that have so far been reported on. The large J_{th} values at low temperatures in the thin active-region DH laser diodes have been attributed to interface recombination at the heterojunction, because there is a considerable lattice mismatch between the adjacent layers. Because of this effect, reduction of J_{th} with decreasing temperature tends to saturate more severely in DH and SH lasers when the active region thickness d is small. With HJ and thick-active-region SH and DH lasers, however, there is no distinct saturation (Horikoshi et al., 1982a).

In HJ laser diodes, much effort has been put into improving J_{th} and output power and into achieving high-temperature cw operation. Included here is development of a crystal-growth method that will yield low-defect-density materials (for example, Lo et al., 1976); adjustment of proper carrier concentration profiles (Walpole et al., 1977; Lo, 1976, 1977; Oron and Zussman, 1980; Partin and Lo, 1981); and development of low-resistance electrical contacts. There have been expectations of sufficient optical and carrier

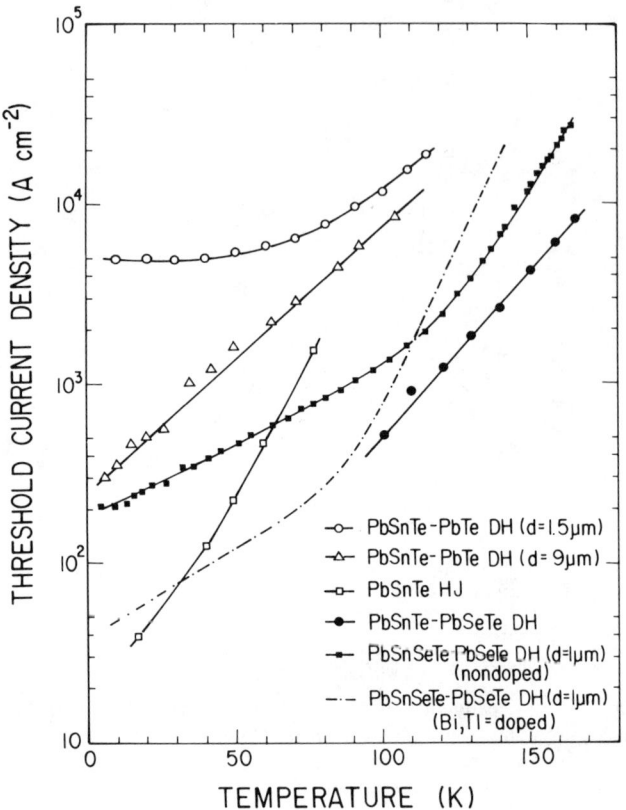

FIG. 30. Temperature dependence of low-threshold PbSnTe-based DH and HJ lasers: PbSnTe/PbTe DH with $d = 1.5$ μm from Walpole et al., (1976a); PbSnTe/PbTe DH with $d = 9$ μm from Kasemset and Fonstad (1979c); PbSnTe HJ from Oron and Zussman (1980); PbSeTe/PbSnTe DH from Kasemset and Fonstad (1979b). [From Horikoshi (1983).]

confinement with HJ lasers that have properly controlled carrier-concentration profiles. Indeed, Walpole et al. (1977) demonstrated cw operation at up to 100 K for n^+-n-p^+-structure MBE- and LPE-grown PbTe HJ lasers and for MBE-grown $Pb_{0.78}Sn_{0.22}Te$ HJ lasers. Oron and Zussman (1980) reported n^+-p-p^+ layer-structure $Pb_{0.85}Sn_{0.15}Te$ diode lasers prepared by LPE and demonstrated low J_{th} (40 A cm^{-2} at 17 K) and high external quantum efficiency (17% at 40 K).

As has been described above, a high interface recombination velocity is responsible for the high DH- and SH-laser J_{th} values at low temperatures. Interface recombination velocity for PbSnTe/PbTe DH lasers was investigated through measurement of spontaneous carrier lifetimes by Kasemset and Fonstad (1979a). They measured effective carrier lifetimes from the pulse-duration dependence of J_{th} (Gorbylev et al., 1972) for PbSnTe/PbTe DH laser diodes having active-region thicknesses ranging from 0.7 to 3.9 μm. It was shown that the interface recombination velocity is as high as 1×10^5 cm sec^{-1} at 5 K, which leads to an internal quantum efficiency reduction of about 80% when the active-region thickness is 2 μm. Since the optimum d value for the lowest J_{th} is smaller than 2 μm for 10-μm emission wavelength lasers, as will be described later, a reduction of the interface recombination velocity seems to be indispensable if low J_{th} laser diodes are to be obtained.

Lattice-matched DH lasers using the PbSnTe/PbSeTe system have been demonstrated to have a very low J_{th}, an increased temperature range of laser oscillation, and improved output power of over 400 μW (Kasemset and Fonstad, 1979b). DH lasers with lattice-matched PbSnSeTe/PbSeTe structure were also investigated, and similar improvements, especially the reduced J_{th} at 4.2 K (47 A cm^{-2} for a 500-μm cavity length) have been observed, as shown in Fig. 30. Spontaneous carrier-lifetime measurements at threshold for lattice-matched $Pb_{0.85}Sn_{0.15}Se_{0.02}Te_{0.98}/PbSe_{0.08}Te_{0.92}$ DH lasers with a 10-μm emission wavelength at 4.2 K revealed that lifetimes for lasers with unintentionally doped 1-μm-thick active regions were between 4 and 6 nsec at 4.2 K. These values are very similar to the bulk minority-carrier lifetime of 4.3 nsec (Kasemset and Fonstad, 1979a) obtained from a PbSnTe/PbTe DH laser having a similar emission wavelength as an interface-recombination free limit. Therefore, the interface recombination velocity at PbSnSeTe/PbSeTe lattice-matched heterojunctions is very slow. This enables us to discuss the doping effect on the threshold-temperature characteristics, because the interface recombination does not seem to dominate the threshold characteristics.

Some typical J_{th}–T relations for PbSnSeTe/PbSeTe lattice-matched DH lasers with Tl- and Sb-doped active regions are shown in Figs. 31 and 32, respectively. The active-region thicknesses for these lasers are between 1 and 2 μm. The amounts of Tl and Sb added per 1 g Pb solution in the LPE process are shown in milligrams in Figs. 31 and 32. The hole concentration in these Tl-doped active regions vary from 4.6×10^{17} cm^{-3} to 1.2×10^{18} cm^{-3}, whereas for the Sb-doped active regions, the electron concentration changes between 2×10^{17} and 4×10^{17} cm^{-3}. The diodes with the lowest doping levels exhibited slightly lower J_{th} at low temperatures but higher J_{th} at relatively high temperatures, compared with the nondoped active-region laser, shown by dashed curves in Figs. 31 and 32. This phenomenon in doped active-region lasers can be understood by considering the reduced carrier concentration required for population inversion at low temperatures and an enhanced band-to-band Auger recombination at higher temperatures. However, further increase in doping levels increases low-temperature J_{th} values,

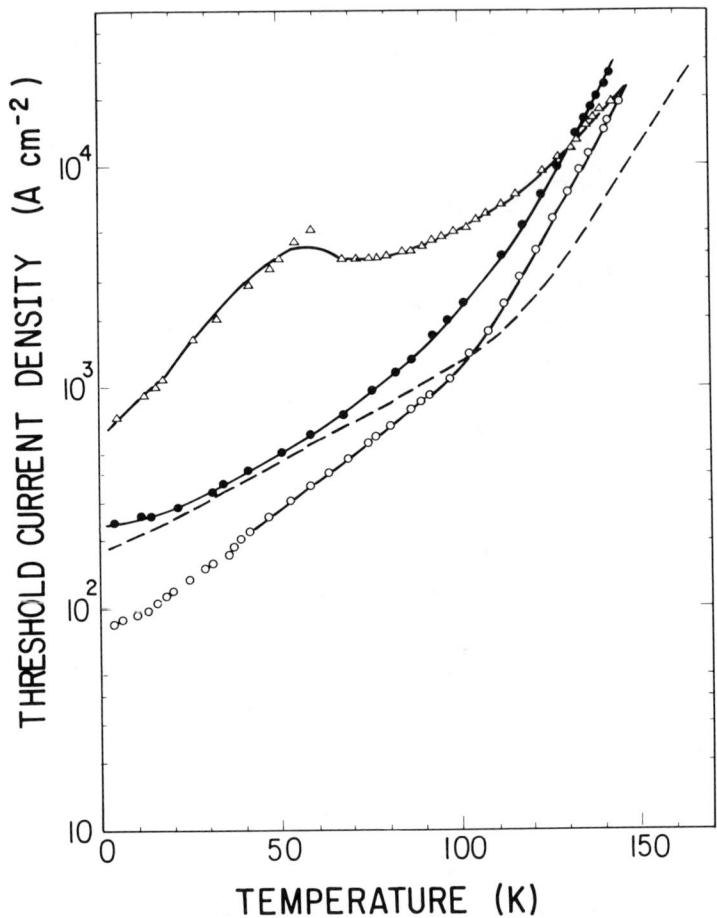

FIG. 31. Temperature dependence of laser threshold current density for Tl-doped PbSnSeTe active-region lasers. Doping levels: △, 10 mg; ●, 5 mg; ○, 1 mg; ---, undoped. [From Horikoshi (1983).]

indicating an increased nonradiative recombination rate in the active region and/or leakage-current components.

Much lower J_{th} values at low temperatures are observed in compensated active-region lasers as shown in Fig. 33, where the characteristics for the nondoped active-region laser are again shown for comparison. Compensated active regions were grown by adding small amounts of Bi and Tl in the growth solution as indicated in Fig. 33. Figure 34 shows measured lifetimes as a function of temperature, together with the theoretical results calculated at the threshold condition by using the models proposed by Anderson (1977)

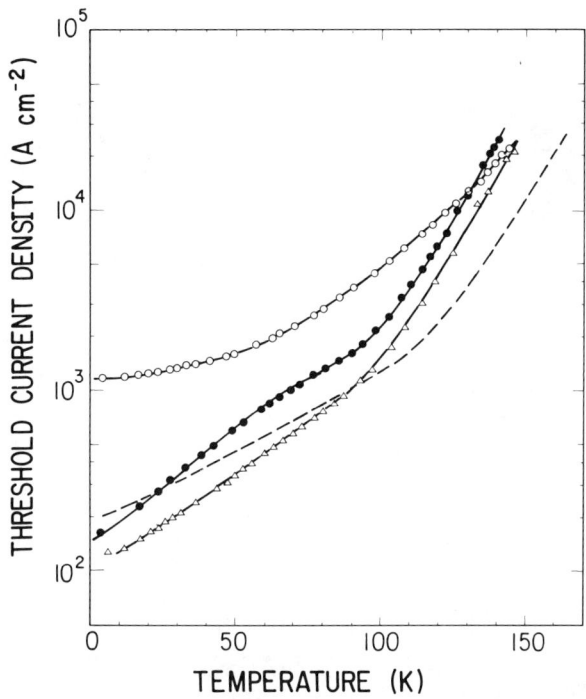

FIG. 32. Temperature dependence of laser threshold current density for Sb-doped PbSnSeTe active-region lasers. Doping levels: △, 0.7 mg; ●, 1.8 mg; ○, 3.6 mg; – – –, undoped. [From Horikoshi (1983).]

and Emtage (1976) for radiative and nonradiative Auger lifetimes, respectively.

b. *Emission Spectrum for* IV–VI *Semiconductor Lasers*

The IV–VI compound semiconductor lasers with lateral confinement structures, such as stripe geometry or mesa structures, show well-defined single longitudinal mode operation within a certain current range. Typical results for PbSnTe and PbCdS lasers are shown in Fig. 35. Spectral linewidth measurements were performed through spectrum observation of the beat frequency between the diode laser and a CO_2 laser in a heterodyne experiment (Hinkley and Freed, 1969). The researchers found that the linewidth of a mode with 240 μW of power from a $Pb_{0.88}Sn_{0.12}Te$ laser was as small as 54 kHz and was inversely proportional to the mode power. The emission wavelength of IV–VI semiconductor diodes can be tuned by temperature, magnetic field, or pressure. Among these, temperature can be easily changed by varying the injection-current level. Temperature variation causes changes in

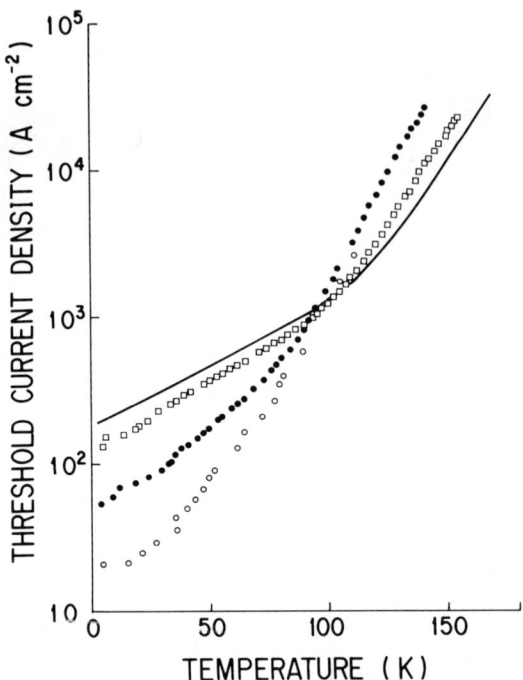

FIG. 33. Temperature dependence of laser threshold current density for both Tl-Bi-doped active-region lasers. $L = 300$ μm. ●, 1-0.5 mg; ○, 1-5 mg; □, 1-10 mg.

both energy gap and refractive index, and therefore causes shifts in the gain spectrum peak and the mode wavelength. Since the gain-peak shift rate with temperature is much greater than the mode shift rate, mode hopping occurs from one longitudinal mode to the neighboring one. In about a 10-μm wavelength range, a maximum continuous tuning range of 50 GHz has been reported (Harman and Melngailis, 1974). An improved continuous tuning range in excess of 20 cm^{-1} (600 GHz) has been reported with an LPE-grown PbSnTe DFB laser having a 12.8-μm center wavelength (Hsieh and Fonstad, 1979). The lasing-peak wavelengths are shown for this DFB laser in Fig. 36 as a function of heat sink temperature.

c. Factors Determining Threshold Current

It is expected that carrier and photon confinement to active regions is well defined in lattice-matched DH lasers, since the carrier-loss process at the heterojunctions is not important, as was mentioned before. Thus, it is now possible to discuss the intrinsic factors that determine the J_{th} of IV–VI

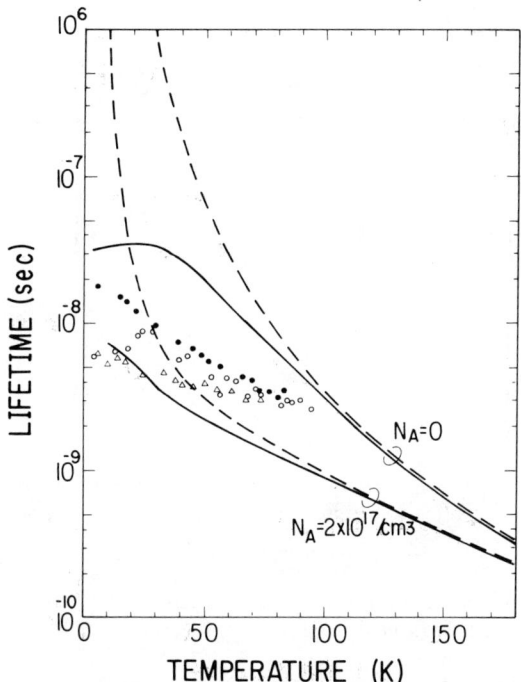

FIG. 34. Spontaneous carrier lifetimes at threshold for $Pb_{0.85}Sn_{0.15}Se_{0.02}Te_{0.98}/PbSe_{0.08}Te_{0.92}$ DH lasers with active regions: ○, nondoped; △, Tl-doped; ●, Bi-, Tl-doped. ——, Total lifetimes; – – –, Auger component of the lifetime; calculated on the basis of theory by Anderson (1977) and by Emtage (1976). [From Horikoshi (1983).]

compound semiconductor lasers. In the following discussion, predominantly lattice-matched PbSnSeTe/PbSeTe DH lasers are dealt with.

Optical gain variation as a function of injected-carrier density n is the most essential factor determining J_{th}. The maximum value of the gain spectrum g_{max} can be expressed as

$$g_{max} = \beta(\eta/d)J - \alpha_0, \qquad (1)$$

where β and α_0 represent gain constants, and η is the internal quantum efficiency. At the threshold, g_{max} is equal to the total cavity loss:

$$g_{max} = \frac{1}{\Gamma}\left(\alpha_i + \frac{1}{L}\ln\frac{1}{R}\right). \qquad (2)$$

Here, α_i represents the total internal loss, which may be dominated by free-carrier absorption, and R represents the mirror reflectivity. Thus, J_{th}

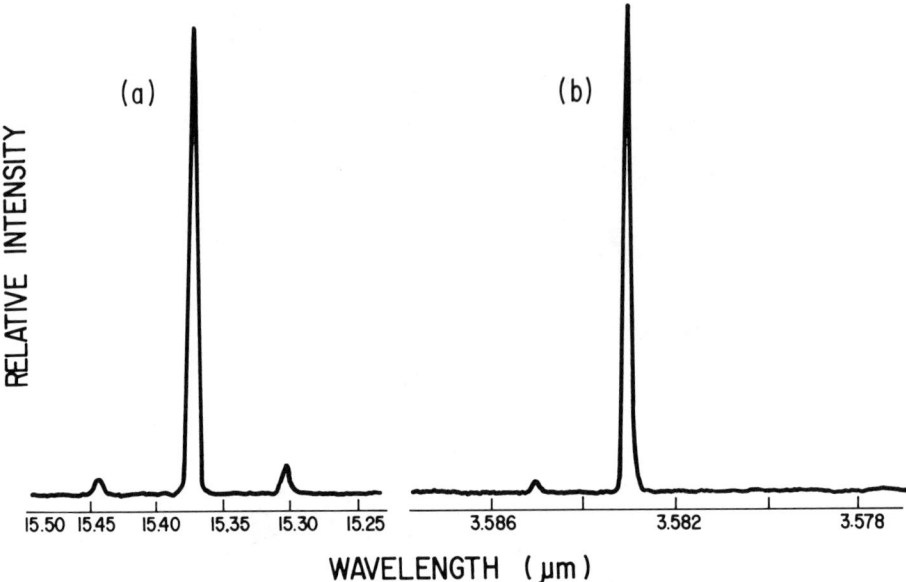

FIG. 35. Emission spectra for (a) a $Pb_{0.782}Sn_{0.218}Te/PbTe$ DH laser ($T \simeq 12$ K, $J = 5.8$ kA cm^{-2}) (Walpole et al., 1976a) and (b) a $Pb_{0.98}Cd_{0.02}S$ HJ laser ($T = 10$ K, $T_{dc} = 60$ mA) (Nill et al., 1973).

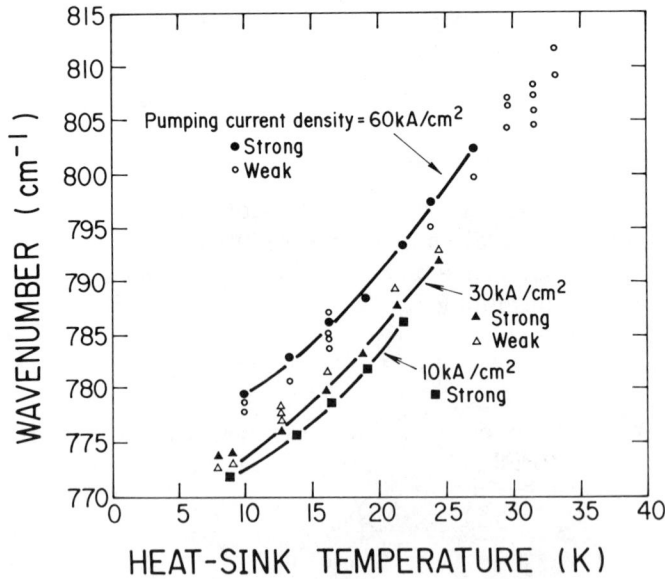

FIG. 36. Temperature dependence of lasing peak wavelength of a DFB laser grown on a corrugated PbTe substrate. [From Hsieh and Fonstad (1979). © 1979 IEEE.]

can be expressed as

$$J_{th} = \frac{d}{\eta}\left[\frac{\alpha_0}{\beta} + \frac{1}{\beta\Gamma}\left(\alpha_i + \frac{1}{L}\ln\frac{1}{R}\right)\right]. \qquad (3)$$

At low temperatures, g_{max} is proportional to n as is described later. Therefore, α_0 can be neglected in Eq. (3), and thus,

$$J_{th} = \frac{d}{\eta}\left[\frac{1}{\beta\Gamma}\left(\alpha_i + \frac{1}{L}\ln\frac{1}{R}\right)\right]. \qquad (4)$$

Equation (4) suggests that the lowest J_{th} can be obtained by choosing the d value so as to minimize d/Γ. From the Γ–d relationship shown in Fig. 9, the optimum d value can be estimated to be about 1 μm for PbSnSeTe/PbSeTe DH lasers having a 10-μm emission wavelength at 77 K.

Figure 37 shows J_{th} variation as a function of $1/L$ for $Pb_{0.85}Sn_{0.15}Se_{0.02}Te_{0.98}$/$PbSe_{0.08}Te_{0.92}$ DH lasers with Tl-doped 1-μm-thick active regions measured at 4.2 K. The plots show a linear relationship except for the two lasers with the longest cavities (shown by dots), which exhibit an extremely low J_{th} (14–15 A cm^{-2}). This low J_{th} has been interpreted in terms of the total reflection-mode excitation (Horikoshi et al., 1982b). By applying a least squares fitting to the experimental results (shown by circles), α_i and η_β can be evaluated as 2.8 cm^{-1} and 0.513 cm^2 A^{-1}, respectively, using an optical confinement factor of 0.48. It should be noted that the internal optical loss α_i is much smaller than that previously reported for PbSnTe HJ lasers (Antcliffe and Parker, 1973) and is very similar to the value for AlGaAs/GaAs DH lasers measured at 4.2 K (Hwang et al., 1978).

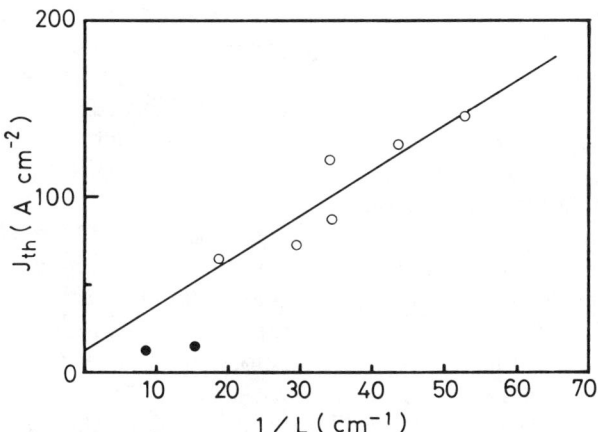

FIG. 37. J_{th}–L^{-1} plots for $Pb_{0.85}Sn_{0.15}Se_{0.02}Te_{0.98}$/$PbSe_{0.08}Te_{0.92}$ DH lasers with a Tl-doped 1-μm-thick active region at 4.2 K.

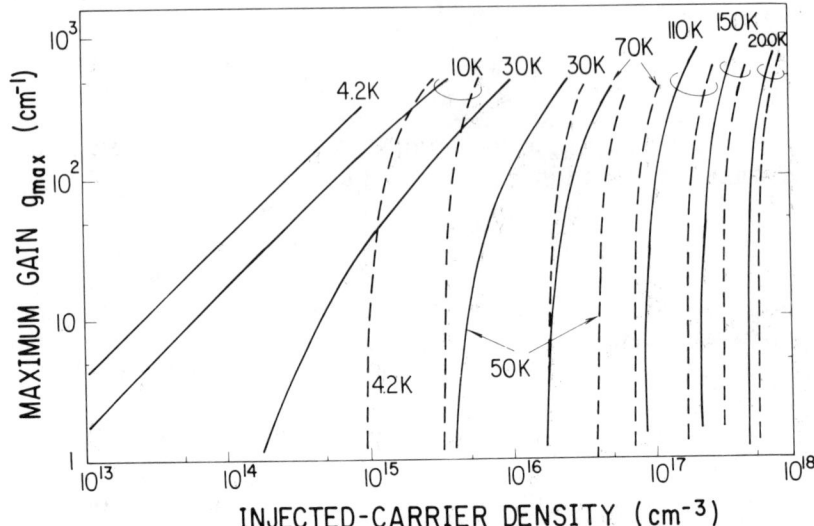

FIG. 38. Maximum optical gain for $Pb_{0.85}Sn_{0.15}Se_{0.02}Te_{0.98}/PbSe_{0.08}Te_{0.92}$ DH laser as a function of the injected-carrier concentration. ———, $N_A = 2 \times 10^{17}\,\text{cm}^{-3}$; – – –, $N_A = 0$. [From Horikoshi (1983).]

Several attempts have been made to calculate the optical gain spectrum in IV–VI compound semiconductors (Tomasetta and Fonstad, 1975; Anderson, 1977; Ziep et al., 1980) by assuming the direct allowed band-to-band transition with k conservation. Based on the gain expression presented by Anderson (1977), g_{max} and radiative lifetimes have been calculated as a function of injected-carrier concentration. This was done assuming a Fermi–Dirac distribution for injected carriers and uniform gain approximation. The calculated g_{max}–n relationships are shown in Fig. 38 at various temperatures. For a doped active region, g_{max} is proportional to n at low temperatures. This linear g_{max}–n relationship has been obtained at active-layer doping levels down to $1 \times 10^{16}\,\text{cm}^{-3}$. This is due to the fact that the n values required for population inversion are very small when the majority-carrier population is degenerate.

The radiative component of J_{th}, J_{thr} can be obtained from the threshold injected-carrier concentration n_{th} and the corresponding radiative lifetime. This can be calculated if the required optical gain for laser oscillation g_{max}^{th} is known. When determining g_{max}^{th}, mirror reflectivity and free-carrier absorption, as well as optical confinement factors, need to be taken into account. For estimation of the free-carrier absorption, available data on carrier mobility (Sizov et al., 1976) were used. The resulting J_{thr} are shown as a function of temperature by the dashed curves in Fig. 39, where the experimental and calculated J_{th} values are compared. The calculated J_{thr}, which correspond to

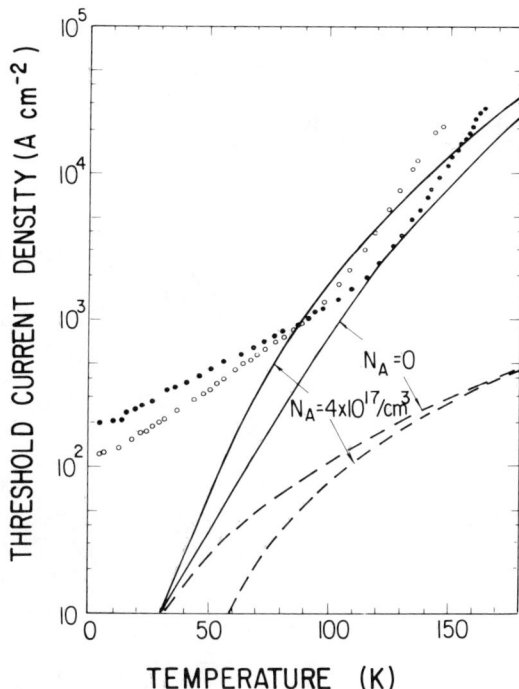

FIG. 39. Temperature dependence of J_{th} for $Pb_{0.85}Sn_{0.15}Se_{0.02}Te_{0.98}/PbSe_{0.08}Te_{0.92}$ DH lasers: (●) result for a nondoped active-region laser; (○) result for a Tl-doped (4×10^{17} cm^{-3}) active-region laser; (– – –) radiative component J_{thr} calculated by using the Anderson's model; (———) total J_{th} including the Auger component J_{tha} calculated with Emtage's expression. [From Horikoshi (1983).]

100% internal quantum efficiency, lie considerably below the experimental points.

As an intrinsic carrier-loss process, the band-to-band Auger effect, including collisions of carriers in different valleys, has been investigated for lead salt materials (Emtage, 1976; Ziep *et al.*, 1980; Rosman and Katzir, 1982). Emtage noticed the importance of the intervalley Auger process in determining the PbSnTe *p–n*-junction resistance and presented an analytical expression for lifetimes using a parabolic dispersion relation. This made it possible to evaluate the Auger lifetimes at n_{th} and the corresponding current component J_{tha}.

In Fig. 39, the calculated total threshold current $J_{th} = J_{thr} + J_{tha}$ is shown by solid curves. Although the validity of the Auger transition model mentioned earlier is restricted to a moderate temperature range, the calculated results show a similar temperature dependence of the experimental result at moderate to higher temperatures. This fact suggests that the band-to-band

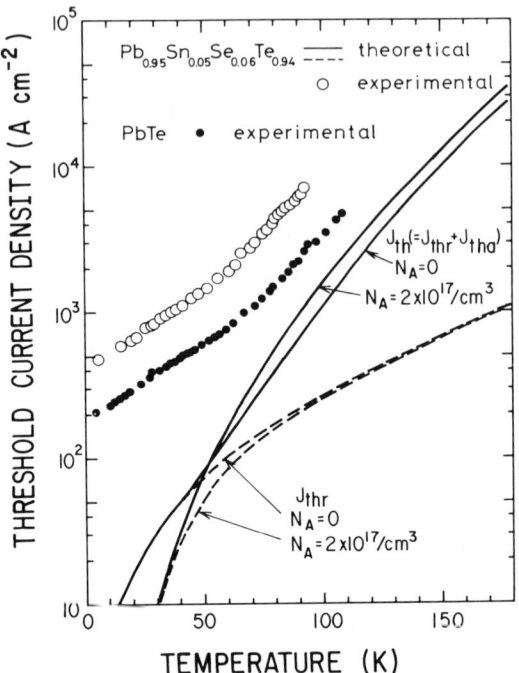

FIG. 40. Temperature dependence of J_{th} for $Pb_{0.95}Sn_{0.05}Se_{0.06}Te_{0.94}/PbSe_{0.08}Te_{0.92}$ DH lasers: experimental results for a nondoped active-region laser (○) and for a PbTe HJ laser (●); curves represent the calculated result with the models used in Fig. 36. [From Horikoshi 1983).]

Auger effect is very important in determining J_{th} at higher temperatures. A similar relationship can be seen in lasers of different wavelengths. Figure 40 shows a similar comparison for $Pb_{0.95}Sn_{0.05}Se_{0.06}Te_{0.94}/PbSe_{0.08}Te_{0.92}$ DH lasers with a 7-μm emission wavelength at 77 K. In Fig. 40, the experimental results for PbTe HJ lasers are also given, because the composition of the active region in this DH laser is very similar to that for PbTe. The importance of the Auger effect is again suggested in Fig. 40. However, Rosman and Katzir (1982) improved the Auger transition model by applying a more realistic dispersion relation (Kane, 1980) and concluded that the lifetimes predicted by Emtage's method, and therefore the Auger lifetimes shown in Fig. 34, are considerably underestimated. Therefore, the calculated J_{th} in Figs. 39 and 40 may be somewhat overestimated.

Low-temperature characteristics, on the other hand, necessitate an explanation in relation to some other loss processes. Zemel and Eger (1980) and Eger *et al.* (1981) respectively, studied forward-current–voltage ($I-V$) characteristics in PbSnTe heterojunction and HJ diodes. They found that the $I-V$ relationship can be expressed by $J = J_0(T) \exp[A(V - IR)]$, where R is the sum of the contact resistance and A is a temperature-independent parameter, rather than by $J = J_0(T) \exp[q(V - IR)/nkT]$ until the J_{th} values are reached. They interpreted this phenomenon in terms of a tunneling mechanism that involves localized states in the forbidden energy gap. They also showed that the current component due to this mechanism dominates the J_{th} values. Experimental results concerning $I-V$ measurement from Eger et al. (1981) are shown in Fig. 41, together with a plausible tunneling mechanism proposed by Riben and Feucht (1966). Similar characteristics, that is, relative temperature invariance of the exponential factor in an $I-V$ relationship, have been observed with PbSnSeTe/PbSeTe DH lasers having Tl- or In-doped active regions. However, for nondoped active-region DH lasers, the slope of the $I-V$ relationship changes with temperature. Therefore, the origins of the localized states responsible for the tunneling may be extrinsic defects, such as those due to impurities, or may have been generated during growth.

Figure 42 shows low-field current–voltage characteristics for the PbTe HJ laser having a lightly doped $p-n$ junction at various temperatures. In Fig. 42, the voltage drop across the series contact resistance was subtracted from the applied voltage. Except for the data for very low temperatures, experimental

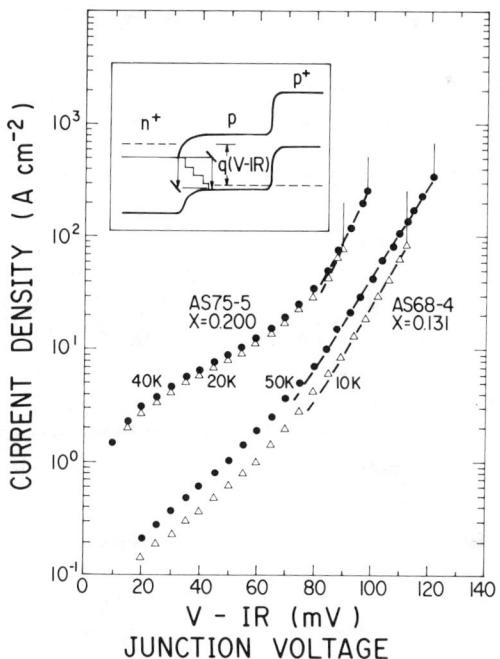

FIG. 41. $I-V$ Measurements on a stripe-geometry $Pb_{0.8}Sn_{0.2}Te$ HJ laser (AS 75-5) and a broad-area $Pb_{0.869}Sn_{0.131}Te$ HJ laser (AS 64-4); insert illustrates the tunneling model. [From Eger et al. (1981).]

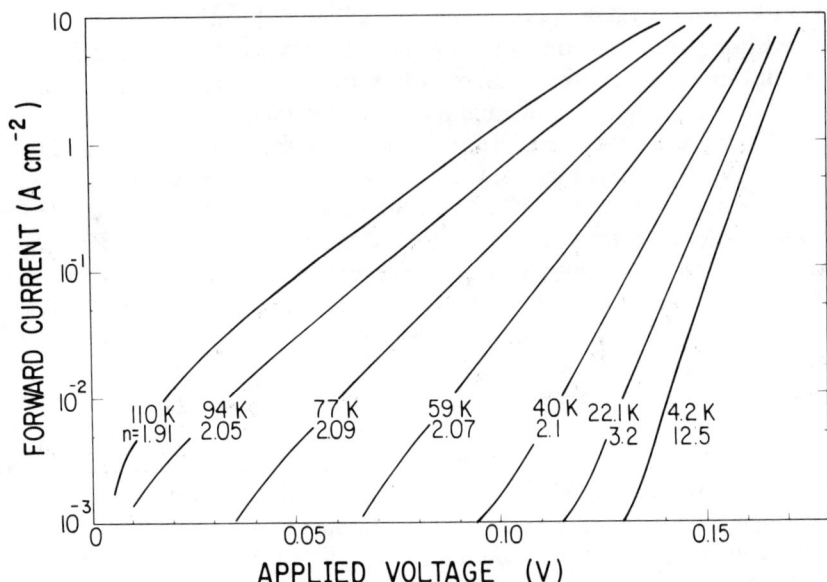

FIG. 42. Low-field I-V characteristics of a PbTe HJ laser at various temperatures; $2kT$ components dominate the characteristics except for very low temperatures. [From Horikoshi (1983).]

results are dominated by the $2kT$ component. To examine the importance of this component at J_{th}, the linear part of each characteristic was extrapolated to the voltage corresponding to the emission wavelength at each temperature. Figure 43 shows the comparison between the extrapolated current thus obtained and the measured J_{th}. The extrapolated currents (dashed curve) coincide with the J_{th} values below about 50 K and saturate above this temperature. Therefore, J_{th} values at low temperatures are determined by the recombination current, whereas high-temperature J_{th} values are dominated by a much more temperature-sensitive component, which is probably the diffusion component. A similar phenomenon was observed with PbSnSeTe/PbSeTe DH lasers having unintentionally doped active regions. However, in this case, the temperature range where J_{th} and the extrapolated current coincide showed an increase, and the saturation of the latter occurred at much higher temperatures.

9. Output Properties and High-Temperature Operation

a. Optical Output and Efficiency

As was described previously, laser diodes fabricated from narrow-energy-gap III–V compound semiconductors show high optical output and high

3. SEMICONDUCTOR LASERS WITH WAVELENGTHS >2 μm

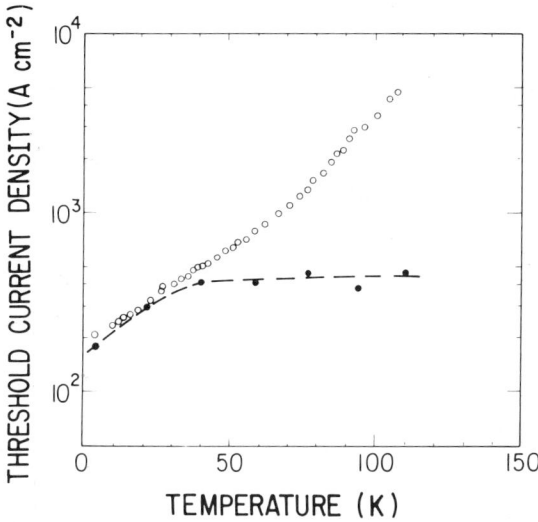

FIG. 43. Comparison between the currents extrapolated from low field values in Fig. 42 to the voltage corresponding to the emission wavelength (●, – – –) and J_{th} values of the same diode (○). [From Horikoshi (1983).]

external quantum efficiencies. Experimental results with InAs HJ lasers (Melngailis and Rediker, 1966) suggest the importance of the DH configuration in achieving high optical outputs and quantum efficiencies. In DH lasers, actually, optical output over several milliwatts is easily attained even when the active-layer thickness is very small. With the majority of the laser diodes made from IV–VI compound semiconductors, however, both optical output and efficiency are relatively low. Although higher output has been observed for PbSnTe HJ lasers with polished end faces rather than with cleaved faces (Walpole et al., 1973b; Lo et al., 1976), this method itself does not seem to be indispensable for obtaining high optical output lasers, since even with the lasers having cleaved end faces, fairly high output has been reported. Early work revealed that HJ and SH lasers show much better output characteristics than their DH counterparts. This is partly due to a more abrupt output saturation with the injected current in thinner active-region DH lasers. Indeed, high optical outputs, of over 10 mW, have been reported in relatively thick (~ 5 μm) active-region PbSnTe/PbTe DH lasers at 10 K (Yoshikawa et al., 1978). Lattice-matched heterostructures also improve the optical output and the external quantum efficiency (Kasemset and Fonstad, 1979b; Shinohara et al., 1982).

It has been found that the external differential quantum efficiency η_{ext} increases with current up to the maximum value and decreases with a further current increase (Walpole et al., 1973b; Oron and Zussman, 1980). Figure

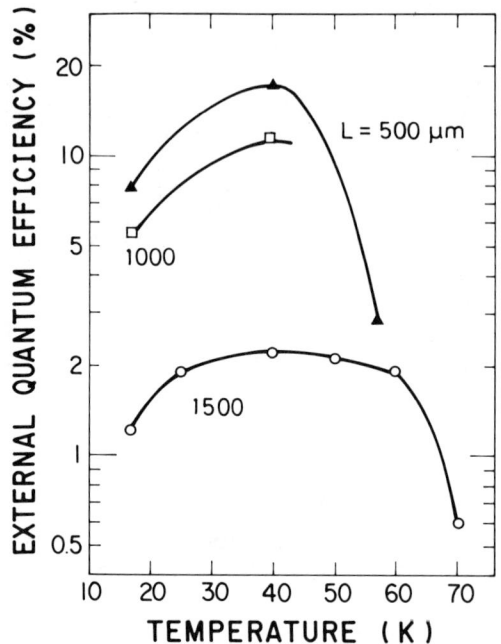

FIG. 44. Temperature dependence of the external quantum efficiency for LPE-grown n^+-p-p^+ $Pb_{0.85}Sn_{0.15}Te$ HJ laser diodes with various cavity lengths [From Oron and Zussman (1980).]

44 shows the maximum values of η_{ext} as a function of temperature for LPE-grown n^+-p-p^+ $Pb_{0.85}Sn_{0.15}Te$ HJ laser diodes of different cavity lengths (Oron and Zussman, 1980); η_{ext} increases with temperature, reaching a maximum at about 40 K, and then decreases. Similar behavior has been reported for MBE-grown p^+-p-n^+ PbTe HJ lasers (Partin and Lo, 1981). In general, η_{ext} can be expressed in terms of the internal quantum efficiency η_{int} as

$$\eta_{ext} = \eta_{int}\eta_c \frac{L^{-1} \ln R^{-1}}{\alpha_i + L^{-1} \ln R^{-1}}, \tag{5}$$

where η_c represents the carrier-injection efficiency to the gain region (Horikoshi, 1982) and α_i the optical loss due mainly to the free-carrier absorption.

Since the quantity $L^{-1} \ln R-1$ is relatively not very sensitive to temperature, the temperature dependence of η_{ext} may come from the temperature dependence of α_i, η_{int}, and η_c. The optical absorption loss increases monotonically with temperature (Kasemset and Fonstad, 1982), and the free-carrier absorption for $Pb_{0.85}Sn_{0.15}Te$ as calculated according to the expression presented by Anderson (1977) is about 10 cm^{-1} at 70 K, which is similar to the quantity $L^{-1} \ln R^{-1}$. Therefore, the temperature dependence of η_{ext} shown in Fig. 44 may not be dominated by the free-carrier absorption. On the other hand, η_{int} can be simply related to the carrier lifetime as

$$\eta_{\text{int}} = (\tau/\tau_r) = [1 + (\tau_r^{\text{st}}/\tau_n)]^{-1}, \quad (6)$$

where τ_r^{st} and τ_n represent the stimulated radiative lifetime and nonradiative lifetime, respectively. The stimulated radiative lifetime will be limited by the intraband relaxation time of injected carriers.

If we use the intraband relaxation time of 2×10^{-12} sec, which has been used by Kasemset and Fonstad (1982) to interpret the observed limit to the single longitudinal mode power in PbSnTe BH lasers, the η_{int} values will be very close to unity. This is because the lifetime predicted from the band-to-band Auger effect, probably the most dominant nonradiative process at relatively high temperatures, is in the 10^{-8}–10^{-10}-sec range for the threshold conditions as can be seen in Fig. 34. Therefore, the temperature dependence of η_{ext} has to be attributed to the change in the injection efficiency η_c. At low temperatures, both the tunneling current (Eger *et al.*, 1981) and the generation–recombination component discussed earlier may be responsible for the η_c degradation. At higher temperatures, on the other hand, carrier leakage from the gain region may result in low η_c values, since carrier confinement in HJ lasers is less effective.

The maximum available output power depends upon the emission wavelength. For example, output power reaching 200 mW has been reported (Ralston *et al.*, 1974) for PbS HJ lasers with a 4.3-μm emission wavelength. High output has also been reported for PbSSe HJ lasers (~ 70 mW) with ~ 4.5-μm emission wavelength (Linden *et al.*, 1977). This phenomenon may be partly attributed to the reduced free-carrier absorption. Single-mode output is limited to much lower levels than is total output power, although single longitudinal mode power as high as 6-mW has been reported in PbSnTe HJ lasers (Walpole *et al.*, 1973b).

Figure 45 shows the output characteristics of a PbSnTe/PbSeTe lattice-matched 2-μm-wide BH laser at 80 K, as reported by Kasemset and Fonstad (1982). The maximum single-mode power, which is 112 μW in this case, has been explained in terms of the gain saturation caused by intraband relaxation time, which is of the order of 2×10^{-12} sec. In Fig. 46, the gain saturation calculated for an output power level of 100 μW in one single mode is shown for various intraband relaxation lifetimes.

b. High-Temperature Laser Oscillation

In principle, laser oscillation is possible unless nonradiative lifetimes become close to or shorter than the stimulated radiative recombination life-

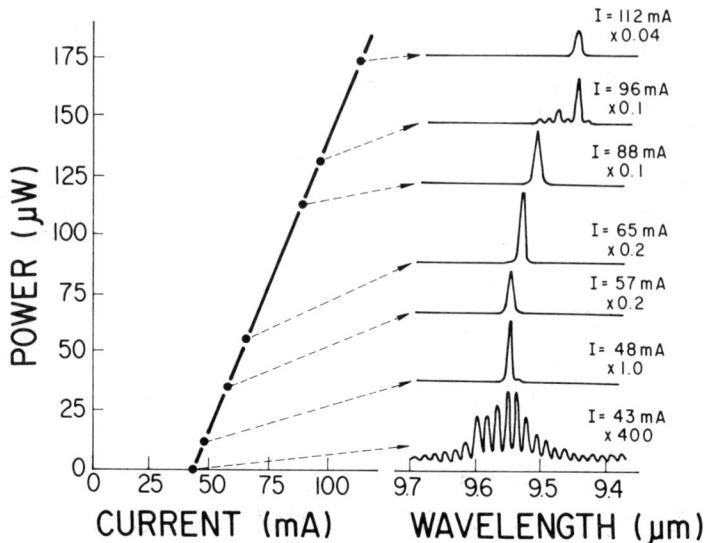

FIG. 45. The optical outputs and spectra of a PbSnTe/PbSeTe lattice-matched BH laser with a 2-μm stripe at 80 K. [From Kasemset and Fonstad (1982). © IEEE 1982.]

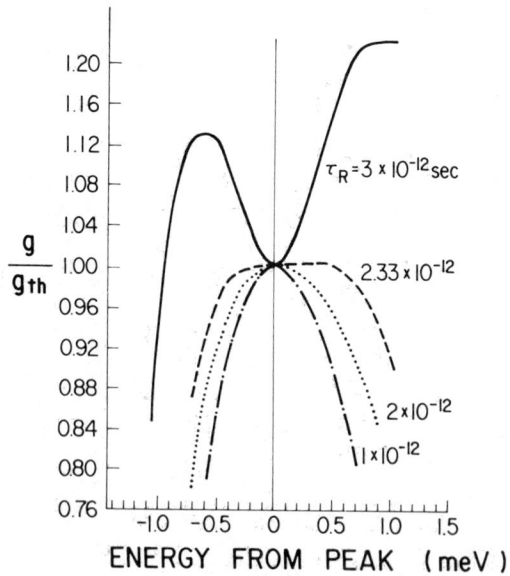

FIG. 46. The optical gain saturation calculated for an output power level of 100 μW at $T = 80$ K in one single mode for various intraband relaxation lifetimes. [From Kasemset and Fonstad (1982). © IEEE 1982.]

3. SEMICONDUCTOR LASERS WITH WAVELENGTHS >2 μm

time. The latter is dominated by the intraband relaxation time and is of the order of $10^{-12}-10^{-13}$ sec. Therefore, laser operation can be achieved up to fairly high temperatures, since nonradiative lifetimes are usually $10^{-8}-10^{-10}$ sec. Among the nonradiative processes, the band-to-band Auger effect shows the most dramatic temperature dependence. Therefore, this effect may limit the temperature range for laser oscillation. However, practical temperature limits are not simply determined by lifetimes but are affected by some additional effects, such as junction-temperature increase during operation, even when very short current pulses are employed.

Experimentally, both InAs HJ lasers (Melngailis and Rediker, 1966) and InAs/InAsPSb DH lasers with InAs active regions (Kobayashi and Horikoshi, 1980) cease laser action at about 150 K, although the J_{th} of these lasers are very different from each other. Similarly, both PbSnTe/PbSeTe (Kasemset and Fonstad, 1979b) and PbSnSeTe/PbSeTe DH lasers (Horikoshi et al., 1982a) show laser action up to about 165 K. This fact suggests the existence of a material limit for obtaining high-temperature laser operation.

Figure 47 shows the temperatures at which the calculated internal quantum efficiency falls to 2.5% for the materials discussed in this chapter. These

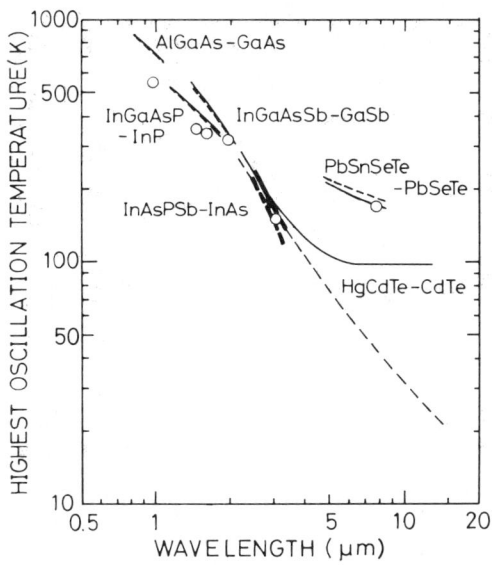

FIG. 47. High-temperature limits for laser oscillation in DH lasers with various materials under pulsed operation; curves represent the temperatures at which the calculated internal quantum efficiency falls to 2.5% for each DH material; results (– – –) for nondoped active-region lasers and (———) for lasers with p-type active regions ($p = 2 \times 10^{17}$ cm^{-3}). Experimentally observed highest lasing temperatures with 100-nsec-long current pulses are given for DH lasers made from various materials. [From Horikoshi (1983).]

are shown as a function of the emission wavelength, together with the experimentally observed maximum laser-operation temperatures. The value 2.5% corresponds to the calculated internal quantum efficiency of InAs/InAsPSb DH lasers at a maximum lasing temperature of 150 K, as was described before. This was tentatively assumed to be the minimum internal quantum efficiency required for laser oscillation in various materials. Here, the internal quantum efficiency was defined as $[1 + (\tau_r/\tau_a)]^{-1}$ or $J_{thr}/(J_{thr} + J_{tha})$. For III–V compound semiconductor lasers τ_r and J_{thr} were evaluated from the band-to-band k-conservation transition model, whereas τ_a and J_{tha} were calculated by the CCHC Auger process by considering the screening effect. The calculated values for InGaAsP/InP DH lasers may, therefore, be somewhat overestimated, since, in this system, the CHHS Auger process has been reported to be more probable than the CCHC process (Sugimura, 1981). The calculations for HgCdTe were performed using the same transition models, whereas for IV–VI semiconductors, the transition models used in Figs. 36 and 37 were applied. In each calculation, the DH configuration was assumed, and the active-region thickness d was chosen so as to minimize d/Γ. It should be emphasized that the laser diodes made from the various III–V semiconductors and from HgCdTe showed similar characteristics, whereas the results for IV–VI semiconductors showed much better temperature characteristics. This difference may be attributed to the difference in band structures and mass parameters.

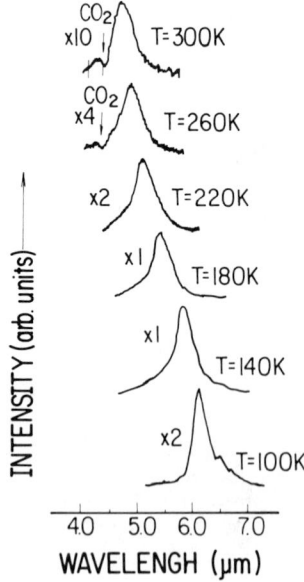

FIG. 48. Room-temperature spontaneous emission spectra from an $PbS_{0.1}Se_{0.9}$ HJ diode. [From Lo and Swets (1980).]

According to Rosman and Katzir (1982), Auger-lifetime calculation with Emtage's model leads to considerable underestimation. Therefore, much higher temperature operation may be possible with IV–VI semiconductor lasers. Since high-temperature operation can be achieved in shorter-wavelength lasers, as shown in Fig. 47, application of IV–VI-based materials to such shorter wavelengths as the 3–5-μm range seems very attractive. To date, however, high-temperature operation has not been reported for PbCdS lasers. It is noteworthy, though, that operating temperatures up to 230 K have been realized with PbSe/PbS DH lasers having a 5.5-μm emission wavelength (Preier et al., 1977). More recently, Partin (1984) has fabricated PbEuSeTe/PbTe single-quantum-well lasers with various active-region thicknesses, and demonstrated pulsed operation up to 241 K at 4.01-μm emission wavelength with a laser diode having an active-region thickness of 1200 Å. Observation of a 4.6-μm spontaneous emission at room temperature has been reported with PbSSe HJ diodes (Lo and Swets, 1980). The latter is also illustrated in Fig. 48. Although the power output for this diode is on the order of a few hundred nanowatts, this low-output nature has been attributed to the high refractive index for PbSSe, which produces considerable internal reflection at the diode surface. Therefore, output will be improved with application of a suitable diode structure.

Acknowledgments

The author would like to thank Drs. N. Kuroyanagi, Y. Furukawa, S. Shimada, and H. Okamoto for their encouragement through this work. He is also indebted to Dr. N. Kobayashi for discussions in connection with his work on narrow-gap III–V semiconductor lasers, and Messrs. M. Kawashima and H. Saito for IV–VI semiconductor-laser fabrication and measurement.

References

Abrikosov, N. Kh., Bankina, V. F., Poretskaya, L. V., and Shelimova, L. E. (1969). "Semiconducting II–VI, IV–VI, and V–VI Compounds." Plenum, New York.
Adler, M. S., Hewes, C. R., and Senturia, S. D. (1973). *Phys. Rev. B* **7**, 5186.
Anderson, W. W. (1977). *IEEE J. Quantum Electron.* **QE-13**, 532.
Andrews, A. M., Higgins, J. A., Longo, J. T., Gertner, E. R., and Pasko, J. G. (1972). *Appl. Phys. Lett.* **21**, 285.
Antcliffe, G. A., and Parker, S. G. (1973). *J. Appl. Phys.* **44**, 4145.
Antcliffe, G. A., and Wrobel, J. S. (1970). *Appl. Phys. Lett.* **17**, 290.
Bauer, G. (1978). *Int. Conf. Appl. High Magn. Fields Semicond. Phys., 1978*, Oxford, p. 153.
Bellavance, D. W., and Johnson, M. R. (1976). *J. Electron. Mater.* **5**, 363.
Bis, R. F., Dixon, J. R., and Lowney, J. R. (1972). *J. Vac. Sci. Technol.* **9**, 226.
Bleicher, M., Wurzinger, H.-D., Maier, H., and Preier, H. (1977). *J. Mater. Sci.* **12**, 317.
Brebrick, R. F., and Allgaier, R. S. (1960). *J. Chem. Phys.* **32**, 1826.
Butler, J. F., Calawa, A. R., Phelan, R. J., Harman, T. C., Strauss, A. J., and Rediker, R. H. (1964). *Appl. Phys. Lett.* **5**, 75.

Calawa, A. R. (1972). *Proc. Cent. Res. Appl. Adv. Technol. Symp.*
Casey, H. C., and Panish, M. B. (1978). "Heterostructure Lasers," Part B, p. 105. Academic Press, New York.
Chu, M., Vanderwyck, A. H. B., and Cheung, D. T. (1980). *Appl. Phys. Lett.* **37,** 486.
Dalven, R. (1969). *Infrared Phys.* **9,** 141.
Daniel, D. R., Maier, H., and Preier, H. (1977). *J. Cryst. Growth* **38,** 145.
Dimmock, J. O., Melngailis, I., and Strauss, A. J. (1966). *Phys. Rev. Lett.* **16,** 1193.
Dionne, C., and Wooley, J. C. (1972). *Phys. Rev. B* **6,** 3898.
Dolginov, L. M., Druzhinina, L. V., Eliseev, P. G., Kryukova, I. V., Leskovitch, V. I., Milvidskii, M. G., Sverdlov, B. N., and Shevchenko, E. G. (1977). *IEEE J. Quantum Electron.* **QE-12,** 609.
Dolginov, L. M., Eliseev, P. G., Lapshin, A. N. and Milvidskii, M. G. (1978a). *Krist. Tech.* **13,** 631.
Dolginov, L. M., Druzhinina, L. V., Eliseev, P. G., Lapshin, A. N., Mil'vidskii, M. G., and Sverdlov, B. N. (1978b). *Sov. J. Quantum Electron. (Engl. Transl.)* **8,** 416.
Dolginov, L. M., Korchagin, Yu. N., Kryukova, I. V., Leskovich, V. I., Matveenko, E. V., Mil'vidskii, M. G., and Stepanov, V. M. (1978c). *Sov. Tech. Phys. Lett. (Engl. Transl.)* **4,** 580.
Dolginov, L. M., Drakin, A. E., Druzhinina, L. V., Eliseev, P. G., Milvidsky, M. G., Skripken, V. A., and Sverdlov, B. N. (1981). *IEEE J. Quantum Electron.* **QE-17,** 593.
Duh, K., and Preier, H. (1975). *J. Mater. Sci.* **10,** 1360.
Eger, D., Oron, M., Zemel, A., and Zussman, A. (1981). *Appl. Phys. Lett.* **39,** 471.
Emtage, P. R. (1976). *J. Appl. Phys.* **47,** 2565.
Faurie, J. P., and Million, A. (1981). *J. Cryst. Growth* **54,** 582.
Faurie, J. P., and Million, A. (1982). *Appl. Phys. Lett.* **41,** 264.
Fukui, T., and Horikoshi, Y. (1981). *Jpn. J. Appl. Phys.* **20,** 587.
Gertner, E. R., Cheung, D. T., Andrews, A. M., and Longo, J. T. (1977). *J. Electron. Mater.* **6,** 163.
Glisson, T. H., Hauser, J. R., Littlejohn, M. A., and Williams, C. K. (1978). *J. Electron. Mater.* **7,** 1.
Gorbylev, V. A., Pak, G. T., Petrov, A. I., Chernousov, N. P., Shveikin, V. I., and Yashumov, I. V. (1972). *Sov. J. Quantum Electron. (Engl. Transl.)* **1,** 505.
Grisar, R., Heime, A., Pfeiffer, H., and Preier, H. (1979). *IEEE Device Res. Conf., Boulder, Colorado, 1979.*
Grisar, R., Riedel, W. J., and Preier, H. M. (1981). *IEEE J. Quantum Electron.* **QE-17,** 586.
Groves, S. H. (1977). *J. Electron. Mater.* **6,** 195.
Groves, S. H., Nill, K. W., and Strauss, A. J. (1974). *Appl. Phys. Lett.* **25,** 331.
Guillaume, C. B., and Lavallard, P. (1963). *Solid State Commun.* **1,** 148.
Harman, T. C. (1971). In "Physics of Semimetals and Narrow Gap Semiconductors" (D. L. Carter and R. T. Bate, eds.), p. 363. Pergamon, Oxford.
Harman, T. C. (1973). *J. Nonmet.* **1,** 183.
Harman, T. C. (1979). *J. Electron. Mater.* **8,** 191.
Harman, T. C., and Melngailis, I. (1974). *Appl. Solid State Sci.* **4,** 1.
Harris, J. S., Longo, J. T., Gertner, E. R., and Clarke, J. E. (1975). *J. Cryst. Growth* **28,** 334.
Hinkley, E. D., and Freed, C. (1969). *Phys. Rev. Lett.* **23,** 277.
Hohnke, D. K., Holloway, H., Young, K. F., and Hurley, M. (1976). *Appl. Phys. Lett.* **29,** 98.
Horikoshi, Y. (1982). In "GaInAsP Alloy Semiconductors" (T. P. Pearsall, ed.), Chapter 15. Wiley, Chichester.
Horikoshi, Y. (1983). *Proc. Soc. Photo. Opt. Instrum. Eng.*

3. SEMICONDUCTOR LASERS WITH WAVELENGTHS $>2~\mu$m

Horikoshi, Y., Kawashima, M., and Saito, H. (1981). *Jpn. J. Appl. Phys.* **20**, L897.
Horikoshi, Y., Kawashima, M., and Saito, H. (1982a). *Jpn. J. Appl. Phys.* **21**, 77.
Horikoshi, Y., Kawashima, M., and Saito, H. (1982b). *Jpn. J. Appl. Phys.* **21**, 198.
Hsieh, H. H., and Fonstad, C. G. (1979). *Tech. Dig.—Int. Electron Devices Meet.*, p. 126.
Hunter, A. T., Smith, D. L., and McGill, T. C. (1980). *Appl. Phys. Lett.* **37**, 200.
Hwang, C. J., Patel, N. B., Sacilotti, M. A., Prince, F. C., and Bull, D. J. (1978). *J. Appl. Phys.* **49**, 29.
Ilegems, M., and Panish, M. B. (1974). *J. Phys. Chem. Solids* **35**, 409.
Ilegems, M., and Pearson, G. L. (1975). *Ann. Rev. Mater. Sci.* **5**, 345.
Ivanov-Omskii, V. I., Maltseva, V. A., Britov, A. D., and Sivachenko, S. D. (1978). *Phys. Status Solidi A* **46**, 77.
Jakobus, T., and Hornung, J. (1978). *J. Cryst. Growth* **45**, 224.
Kane, E. O. (1980). *In* "Narrow Gap Semiconductors" (W. Zawadzki, ed.), p. 13. Springer-Verlag, Berlin and New York.
Kasai, I., and Bassett, D. W. (1974). *J. Cryst. Growth* **27**, 215.
Kasai, I., Hornung, J., and Baars, J. (1975). *J. Electron. Mater.* **4**, 299.
Kasemset, D., and Fonstad, C. G. (1979a). *Appl. Phys. Lett.* **34**, 432.
Kasemset, D., and Fonstad, C. G. (1979b). *Tech. Dig.—Int. Electron Devices Meet.*, p. 130.
Kasemset, D., and Fonstad, C. G. (1979c). *IEEE J. Quantum Electron.* **QE-15**, 1266.
Kasemset, D., and Fonstad, C. G. (1982). *IEEE J. Quantum Electron.* **QE-18**, 1078.
Kasemset, D., Rotter, S., and Fonstad, C. G. (1981). *J. Electron. Mater.* **10**, 863.
Kennedy, C. A., and Linden, K. J. (1970). *J. Appl. Phys.* **41**, 252.
Kobayashi, N. (1982). Private communication.
Kobayashi, N., and Horikoshi, Y. (1980). *Jpn. J. Appl. Phys.* **19**, L641.
Kobayashi, N., and Horikoshi, Y. (1981). *Jpn. J. Appl. Phys.* **20**, 2301.
Kobayashi, N., and Horikoshi, Y. (1982). *Jpn. J. Appl. Phys.* **21**, 201.
Kobayashi, N., Horikoshi, Y., and Uemura, C. (1979). *Jpn. J. Appl. Phys.* **18**, 2169.
Kobayashi, N., Horikoshi, Y., and Uemura, C. (1980). *Jpn. J. Appl. Phys.* **19**, L30.
Konnerth, K., and Lanza, C. (1964). *Appl. Phys. Lett.* **4**, 120.
Kucherenko, I. V., Taktakishvili, M. S., and Shotov, A. P. (1976). *Sov. Phys.—Semicond. (Engl. Transl.)* **10**, 807.
Kucherenko, I. V., Mityagin, Yu. A., Vodoplyanov, L. K., and Shotov, A. P. (1977). *Sov. Phys.—Semicond. (Engl. Transl.)* **11**, 282.
Lanir, M., Wang, C. C., and van der Wyck, A. H. B. (1979). *Appl. Phys. Lett.* **34**, 50.
Laugier, A., Cadon, J., Faure, M., and Moulin, M. (1974). *J. Cryst. Growth* **21**, 235.
Linden, K. J., Nill, K. W., and Butler, J. F. (1977). *IEEE J. Quantum Electron.* **QE-13**, 720.
Lo, W. (1976). *Appl. Phys. Lett.* **28**, 154.
Lo, W. (1977). *IEEE J. Quantum Electron.* **QE-13**, 591.
Lo, W., and Gifford, F. E. (1980). *J. Electrochem. Soc. (Solid-State Sci. Technol.)* **127**, 1372.
Lo, W., and Swets, D. E. (1978). *Appl. Phys. Lett.* **33**, 938.
Lo, W., and Swets, D. E. (1980). *Appl. Phys. Lett.* **36**, 450.
Lo, W., Montgomery, G. P., Jr., and Swets, D. E. (1976). *J. Appl. Phys.* **47**, 267.
Longo, J. T., Gertner, E. R., and Joseph, A. S. (1971). *Appl. Phys. Lett.* **19**, 202.
Longo, J. T., Harris, J. S., Jr., Gertner, E. R., and Chu, J. C. (1972). *J. Cryst. Growth* **15**, 107.
Lopez-Otero, A. (1978). *Thin Solid Films* **49**, 1.
Lowney, J. R., and Cammack, D. A. (1974). *Mater. Res. Bull.* **9**, 1639.
McLane, G. F., and Sleger, K. J. (1975). *J. Electron. Mater.* **4**, 465.
Maier, H., Daniel, D. R., and Preier, H. (1976). *J. Cryst. Growth* **35**, 121.
Melngailis, I. (1963). *Appl. Phys. Lett.* **2**, 176.

Melngailis, I. (1968). *J. Phys., Colloq. (Orsay, Fr.)* **C4**, Suppl., No. 11-12.
Melngailis, I., and Harman, T. C. (1970). In "Semiconductors and Semimetals" (R. K. Willardson and A. C. Beer, eds.), Vol. 5, p. 111. Academic Press, New York.
Melngailis, I., and Harman, T. C. (1973). Solid State Res. Rep., p. 6. Lincoln Lab., MIT, Cambridge, Massachusetts.
Melngailis, I., and Rediker, R. H. (1966). *J. Appl. Phys.* **37**, 899.
Melngailis, I., and Strauss, A. J. (1966). *Appl. Phys. Lett.* **8**, 179.
Moon, R. L., Antypas, G. A., and James, L. W. (1974). *J. Electron. Mater.* **3**, 635.
Muszynski, Z., Davarashvili, O. I., Riabtsev, N. G., and Shotov, A. P. (1979). *J. Cryst. Growth* **46**, 487.
Nakajima, K., Osamura, K., Yasuda, K., and Murakami, Y. (1977). *J. Cryst. Growth* **41**, 87.
Nathan, M. I., Dumke, W. P., Burns, G., Dill, F. H., and Lasher, C. (1962). *Appl. Phys. Lett.* **3**, 62.
Nill, K. W., Strauss, A. J., and Blum, F. A. (1973). *Appl. Phys. Lett.* **22**, 677.
Novoselova, A. V., Zlomanov, V. P., Karbanov, S. G., Matveyev, O. V., and Gas'kov, A. M. (1972). *Prog. Solid State Chem.* **7**, 85.
Onabe, K. (1982). *Jpn. J. Appl. Phys.* **21**, 964.
Oron, M., and Zussman, A. (1980). *Appl. Phys. Lett.* **37**, 7.
Panish, M. B., and Hayashi, I. (1974) *Appl. Solid State Sci.* **4**, 235.
Panish, M. B., and Ilegems, M. (1972). *Prog. Solid State Chem.* **7**, 39.
Partin, D. L. (1983). *Appl. Phys. Lett.* **43**, 996.
Partin, D. L. (1984). *Appl. Phys. Lett.* **45**, 487.
Partin, D. L., and Lo, W. (1981). *J. Appl. Phys.* **52**, 1579.
Partin, D. L., and Thrush, C. M. (1984). *Appl. Phys. Lett.* **45**, 193.
Phelan, J., Calawa, A. R., Rediker, R. H., Keyes, R. J., and Lax, B. (1963). *Appl. Phys. Lett.* **3**, 143.
Pinnow, D. A., Gentile, A. L., Standlee, A. G., Timper, A. J., and Hobrock, L. M. (1978). *Appl. Phys. Lett.* **33**, 28.
Preier, H. (1979). *Appl. Phys.* **20**, 189.
Preier, H., Bleicher, M., Riedel, W., and Maier, H. (1976). *J. Appl. Phys.* **47**, 5476.
Preier, H., Bleicher, M., Riedel, W., Pfeiffer, H., and Maier, H. (1977). *Appl. Phys.* **12**, 277.
Ralston, R. W., Walpole, J. N., Calawa, A. R., Harman, T. C., and McVittie, J. P. (1974). *J. Appl. Phys.* **45**, 1323.
Ralston, R. W., Walpole, J. N., Harman, T. C., and Melngailis, I. (1975). *Appl. Phys. Lett.* **26**, 64.
Riben, A. R., and Feucht, D. L. (1966). *Int. J. Electron.* **20**, 583.
Rolls, W., Lee, R., and Eddington, R. J. (1970). *Solid-State Electron.* **13**, 75.
Rosman, R., and Katzir, A. (1982). *IEEE J. Quantum Electron.* **QE-18**, 814.
Sankaran, R., and Antypas, G. A. (1976). *J. Cryst. Growth* **36**, 198.
Schmit, J. L., and Bowers, J. E. (1979). *Appl. Phys. Lett.* **35**, 457.
Schoolar, R. B., and Jensen, J. D. (1977). *Appl. Phys. Lett.* **31**, 536.
Schoolar, R. B., Jensen, J. D., and Black, G. M. (1977). *Appl. Phys. Lett.* **31**, 620.
Shinohara, K., Nishijima, Y., and Akamatsu, T. (1982). *IEEE Conf. Laser Electro-Opt. Syst., Phoenix, Arizona, 1982.*
Sizov, F. F., Lashkarev, G. V., Radchenko, M. V., and Orletskii, V. B. (1976). *Sov. Phys.— Semicond. (Engl. Transl.)* **10**, 236.
Smith, D. L., and Pickhardt, V. Y. (1976). *J. Electron. Mater.* **5**, 247.
Smith, D. L., and Pickhardt, V. Y. (1978). *J. Electrochem. Soc. (Solid-State Sci. Technol.)* **125**, 2042.

Starik, P. M., Britov, A. D., Luchitskii, R. M., Lototskii, V. B., Mikityuk, V. I., and Karavaev, S. M. (1978). *Sov. Phys.—Semicond. (Engl. Transl.)* **12**, 1353.
Stern, F. (1973). *IEEE J. Quantum Electron.* **QE-9**, 290.
Stringfellow, G. B. (1972). *J. Appl. Phys.* **43**, 3455.
Stringfellow, G. B. (1982). *J. Cryst. Growth* **58**, 194.
Sugimura, A. (1981). *IEEE J. Quantum Electron.* **QE-17**, 627.
Sugimura, A. (1982). *IEEE J. Quantum Electron.* **QE-18**, 352.
Tamari, N., and Shtrikman, H. (1979). *J. Appl. Phys.* **50**, 5736.
Tomasetta, L. R., and Fonstad, C. G. (1974a). *Appl. Phys. Lett.* **24**, 567.
Tomasetta, L. R., and Fonstad, C. G. (1974b). *Appl. Phys. Lett.* **25**, 440.
Tomasetta, L. R., and Fonstad, C. G. (1974c). *Mater. Res. Bull.* **9**, 799.
Tomasetta, L. R., and Fonstad, C. G. (1975). *IEEE J. Quantum Electron.* **QE-11**, 384.
Verie, C., and Granger, R. (1965). *C. R.* **261**, 3349.
Wagner, J. W., and Thompson, A. G. (1970). *J. Electrochem. Soc.* **117**, 936.
Walpole, J. N., Calawa, A. R., Ralston, R. W., Harman, T. C., and McVittie, J. P. (1973a). *Appl. Phys. Lett.* **23**, 620.
Walpole, J. N., Calawa, A. R., Ralston, R. W., and Harman, T. C. (1973b). *J. Appl. Phys.* **44**, 2905.
Walpole, J. N., Calawa, A. R., Harman, T. C., and Groves, S. H. (1976a). *Appl. Phys. Lett.* **28**, 552.
Walpole, J. N., Calawa, A. R., Chinn, S. R., Groves, S. H., and Harman, T. C. (1976b). *Appl. Phys. Lett.* **29**, 1976.
Walpole, J. N., Groves, S. H., and Harman, T. C. (1977). *IEEE Electron Devices* **ED-24**, 1214.
Wang, C. C., and Lorenzo, J. S. (1977). *Infrared Phys.* **17**, 83.
Wang, C. C., Shin, S. H., Chu, M., Lanir, M., and Vanderwyck, A. H. B. (1980). *J. Electrochem. Soc. (Solid-State Sci. Technol.)* **127**, 175.
Wemple, S. H., and DiDomenico, M. (1971). *Phys. Rev. B* **3**, 1338.
Williams, C. K., Glisson, T. H., Hauser, J. R., and Littlejohn, M. A. (1978). *J. Electron. Mater.* **7**, 639.
Wrobel, J. S., Antcliffe, G. A., and Bate, R. T. (1973). *J. Nonmet.* **1**, 217.
Wu, B. Y., and Pearson, G. L. (1972). *J. Phys. Chem. Solids* **33**, 409.
Yoshikawa, M., Shinohara, K., and Ueda, R. (1977). *Appl. Phys. Lett.* **31**, 699.
Yoshikawa, M., Koseto, M., and Ueda, R. (1978). *IEEE/OSA Conf. Laser Electro-Opt. Syst. 1978, San Diego.*
Yoshikawa, M., Ito, M., Shinohara, K., and Ueda, R. (1979). *J. Cryst. Growth* **47**, 230.
Zayachuk, D. M., and Starik, P. M. (1980). *Sov. Phys.—Semicond. (Engl. Transl.)* **12**, 1353.
Zemel, A., and Eger, D. (1980). *Solid-State Electron.* **23**, 1123.
Ziep, O., Genzow, D., Mocker, M., and Herrmann, K. H. (1980). *Phys. Status Solidi B* **99**, 129.

CHAPTER 4

The Functional Reliability of Semiconductor Lasers as Optical Transmitters

B. A. Dean and M. Dixon

AT&T BELL LABORATORIES
ALLENTOWN, PENNSYLVANIA

I.	INTRODUCTION	153
II.	EXPERIMENTAL CONDITIONS.	156
	1. *General Considerations*	156
	2. *A 45-Mbit/sec AlGaAs Transmitter Test Set*	158
III.	LIFE-TEST RESULTS FOR A 45-Mbit/sec, AlGaAs SYSTEM . .	159
	3. *Average Power Drift*	163
	4. *Extinction-Ratio Degradation*	166
	5. *Laser-Degradation Modes*.	168
IV.	FUNCTIONAL LIFE TESTING FOR IMPROVED DESIGN.	170
	6. *Effect of Bias Conditions on Aging Rates*	170
	7. *Temperature Dependence of Transmitter Aging*.	175
	8. *Preliminary Results of Functional Reliability Test for Tap-Controlled Transmitters*.	179
V.	FUNCTIONAL RELIABILITY OF AN AlGaAs LASER TRANSMITTER FOR SPACE COMMUNICATION	184
VI.	SUMMARY.	187
	REFERENCES	189

I. Introduction

Lightwave transmission systems are rapidly being integrated into the existing hierarchy of telecommunication networks (Pellegrini, 1983; DeWitt, 1983; Jacobs, 1983). Semiconductor lasers from the ternary system AlGaAs and from the quaternary system InGaAsP have been developed during the past decade for transmission windows at the wavelengths of 0.8–0.9 and 1.1–1.6 µm (Kressel *et al.*, 1982). Significant progress has been made in material growth techniques and in understanding the physics of operation and the degradation mechanism encountered in such devices (Botez and Herskowitz, 1980). In fact, the median lifetime of AlGaAs multimoded lasers operated under cw conditions has been demonstrated to be well in excess of 10^5 hr at room temperature (Hartman *et al.*, 1977; Goodwin *et al.*, 1979; Kressel *et al.*, 1978; Imai *et al.*, 1981; Wolf *et al.*, 1981). Single-mode lasers of InGaAsP have quickly approached and surpassed this quality (Nakamura, 1983; Rosiewicz *et al.*, 1983; Ettenburg *et al.*, 1984; Mizuishi *et al.*, 1984; Hakki *et al.*, 1985).

The cw aging studies invariably test stud-mounted laser diodes for their ability to emit a prescribed amount of light as a function of time. The lifetime is measured generally by the two criteria of changes in the lasing threshold and lasing efficiency. The optical power is usually measured from a single lasing facet into a large-area detector placed in the laser far-field pattern. Often the measurement is performed at elevated temperatures and the results extrapolated by an Arrhenius relationship to room temperature (Hartman *et al.*, 1977; Ettenberg and Kressel, 1980). The results of such laser-diode life testing have been invaluable to the rapid advances in semiconductor laser technology and are a necessary precursor to the design of an optical transmitter. However, we and others have found that such results cannot be translated directly into a transmitter lifetime (Dean and Dixon, 1982; Garrett and Midwinter, 1980).

The difficulty in using the previously described laser reliability is illustrated in Fig. 1. This block diagram indicates a complex relationship between the laser structure and the final objective of reliable system design. In the past, only those interactions to the left of the dotted line have been considered in discussions of laser reliability. For example, the influence of thermal

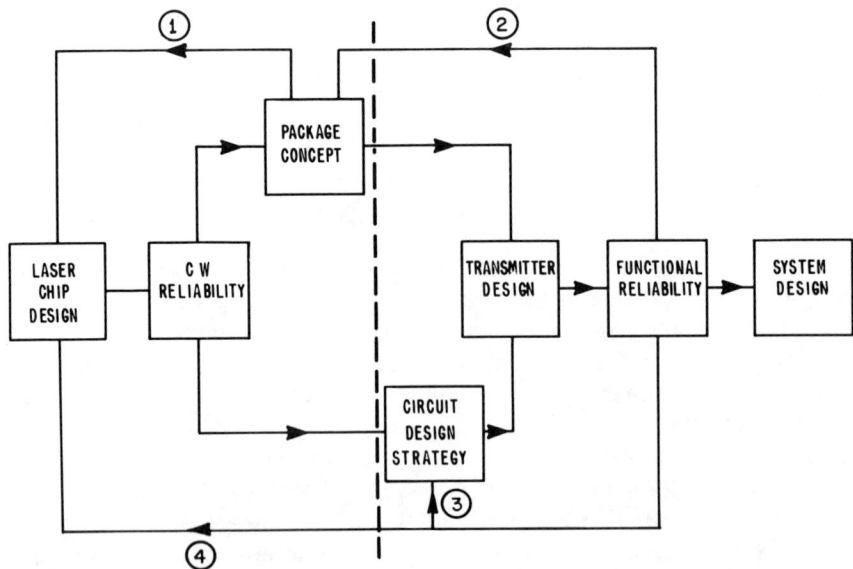

FIG. 1. Schematic diagram showing feedback loops leading to a reliable design of a laser transmitter subsystem.

4. SEMICONDUCTOR LASERS AS OPTICAL TRANSMITTERS 155

heat sinking (Ritchie *et al.*, 1978) and the influence of thermal bond metallurgy (Fujiwara *et al.*, 1979) on laser design and reliability (represented by loop 1) have been investigated in some laboratories. However, in systems transmitting information, criteria on the *quality* of light transmitted along an optical fiber must be applied and the performance judged in terms of system parameters. Thus, we have found it necessary to define a *functional lifetime* for laser diodes, which is obtained by exercising them in a transmitter circuit and judging their performance by system criteria. It can be seen that this functional lifetime depends not only on the cw aging characteristics of the laser diode, but also on its packaging, which includes mounting of control photodiodes and output optical fibers, as well as on the circuit strategy, the design of the modulation and bias circuits, and the system operating conditions.

The advantages of such a functional testing procedure are that it can lead (1) to improvements in both laser packaging concepts and circuit-design strategy (as represented by loops 2 and 3 in Fig. 1) as well as to improvements in the choice of laser design (loop 4), and (2) to a more realistic evaluation of the trade-offs and reliability available to the system designer in any system configuration adopted. This second advantage does not imply that we shall be evaluating system reliability. This is decided ultimately by the appropriate allocation of "operating-power margins," including safety margins, for each component of the system (Garrett and Midwinter, 1980). The functional reliability we shall discuss is just a more realistic evaluation of the margin to be allocated to a laser transmitter.

The disadvantage of functional testing, of course, is the degree of specificity of the results of such a functional life test. Indeed, few discussions of laser-diode reliability in terms of transmitter performance have appeared in the literature (Dixon and Dean, 1981; Thompson *et al.*, 1981; Dean and Dixon, 1982); other studies have been based on field work and not on controlled laboratory experiments (Schwartz *et al.*, 1978; Horiguchi *et al.*, 1981; Miskovic, 1982).

In our laboratory, systems on test have included short-wavelength multimode laser transmitters at 45 and 90 Mbits/sec and long-wavelength single-mode laser transmitters at 274 and 432 Mbits/sec. Since each functional reliability test is specific to a system, writing a general review is difficult. Our discussion is based mainly on results we have obtained since 1978 on multimode AlGaAs lasers exercised in a 45-Mbit/sec transmitter test set, simply because the data here are most complete. Although these results are not of universal application, they illustrate the type of aging mechanisms that should be considered for lasers used in transmitters for optical telecommunication systems. They also illustrate the general principle shown in Fig. 1 by which such functional reliability data can produce a feedback effect on laser design, laser packaging, and circuit strategy and lead to improved overall system performance.

Thus in Part II, there will be a description of the various components of our functional test set, a discussion of the parameters it can record, and the type of analyses applied to the raw results. Part III presents data illustrating our main conclusion that average power drift and extinction-ratio degradation limit the functional lifetime of our packaged multimode AlGaAs lasers to about 10^4 hr when using a single average power feedback loop with back-facet monitoring. In Part IV, we discuss three "second-generation" experiments: an investigation of the effects of different biasing conditions in a transmitter circuit, a study of the temperature dependence of transmitter reliability, and a life-test study of transmitters with a redesigned optical feedback scheme. These experiments are examples of how functional reliability results provide feedback for transmitter-design decisions. In Part V, we review data by other workers (Barry *et al.*, 1985) on a single-mode AlGaAs laser transmitter for space communication. Finally, in Part VI, we summarize our conclusions of functional life testing.

II. Experimental Conditions

1. General Considerations

Test-set design for functional life testing can be quite simple or very elaborate, depending on the number of units and parameters to be measured. However, to ensure flexibility in the design of experiments and to minimize ambiguity during data analyses, temperature control and provisions for measuring both static and dynamic properties are basic requirements.

A typical transmitter functional life-test set should contain the following components:

(1) power supplies and a pulsed data source, which may be common to all units;

(2) a temperature-controlling device, either a common environmental chamber or individual thermoelectric coolers;

(3) multimeters for measuring such static transmitter properties as the bias current, the modulation current, and the average power; and

(4) devices such as photodetectors and an oscilloscope for measuring dynamic parameters, or a bit-error-rate system, so that changes in the optical "eye" margin can be recorded.

Data might be recorded hourly, daily, monthly, or after a prescribed change in transmitter parameters has occurred. It is soon apparent that, even in an experiment involving only ten transmitters, the volume of data is such that a computer-controlled test set is a necessity. Figure 2 shows a test set designed for evaluating the end-of-life criteria for a transmitter using a 1.3-μm single-mode laser as the source. The individual environmental chambers use thermoelectric coolers to provide temperature controls for the entire transmitter.

The introduction of the computer simplifies the routine gathering and storing of digital data. Because measurements can be made more frequently than is possible with a manual test set and at times when there is no operator present, a computer-controlled reliability test set makes it possible to verify to 99% certainty the very low degradation rates ($\sim 0.2\%$/khr) imposed by some system applications on the current aging of the semiconductor laser (Hakki *et al.*, 1985). There is a serious problem, however, with the use of unattended automated test equipment. Because a typical life test operates for at least a year, the experimenter is in effect life testing not only the optical subsystem but also the test equipment. We have found that the data files of about 20% of our transmitters contain spurious data introduced by unprotected malfunctions in our complex test equipment. Thus, it is not always possible to reduce the data without carefully studying the raw results, identifying the causes of error, and possibly eliminating data points. This continu-

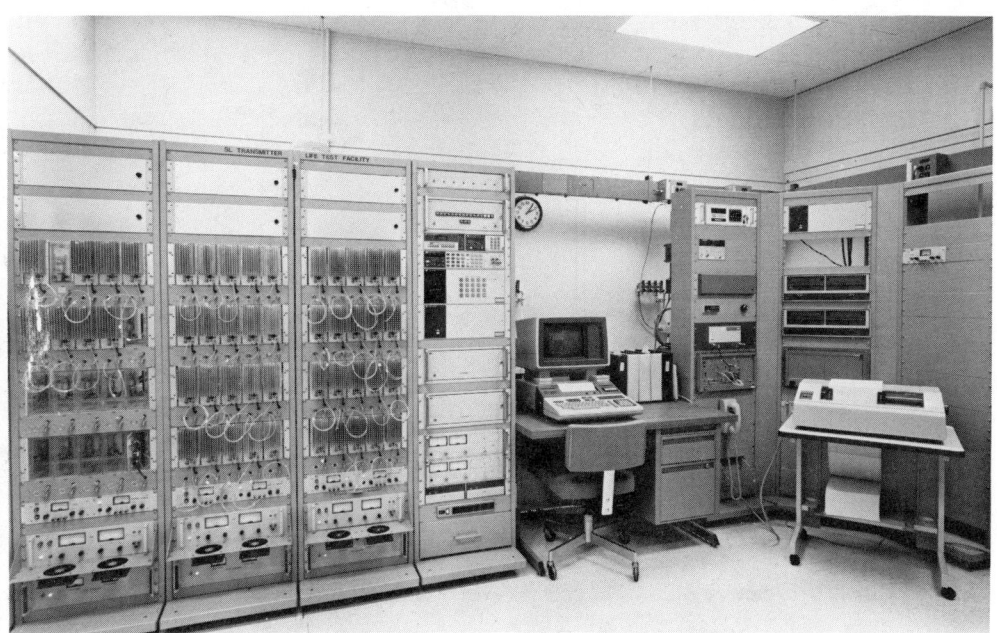

FIG. 2. Photograph of a functional life-test set for 274-Mbit/sec single-mode laser transmitters (similar to the test set described in the text). (Courtesy of AT&T Bell Laboratories.)

ous monitoring of both the software and hardware used for data collection often becomes a major component of the time expended in obtaining aging results.

2. A 45-Mbit/sec AlGaAs Transmitter Test Set

There are two points that our test was designed to address. First, it was not at all clear when these tests were initiated whether cw aging above threshold was a "worst case" or if pulsing from below threshold during aging introduced new or accelerated existing degradation mechanisms. Second, as we have described earlier, it is necessary to evaluate the performance of the packaged laser in transmitter circuits in terms of system requirements. Our experiments address these points by aging lasers in transmitter circuits under pulsed conditions and then applying stringent failure criteria to the average optical power, extinction ratio, self-pulsation frequency, optical noise level, and pattern dependence.

The lasers in these studies are stripe-geometry, double-heterostructure $Al_xGa_{1-x}As/Al_yGa_{1-y}As$ lasers, with 5-μm-wide stripes delineated by shallow proton bombardment (Dyment et al., 1972; Dixon and Joyce, 1980). A schematic of the structure is shown in Fig. 3. Most of the lasers were grown by liquid-phase epitaxy (LPE); five were grown by molecular-beam epitaxy (MBE). The lasers are hermetically packaged with a 50-μm core, 0.23 NA fiber, which collects light from the front facet, and a Si $P-I-N$ photodiode, which collects light from the back facet.

Before subjecting lasers to functional life testing, the performance of the

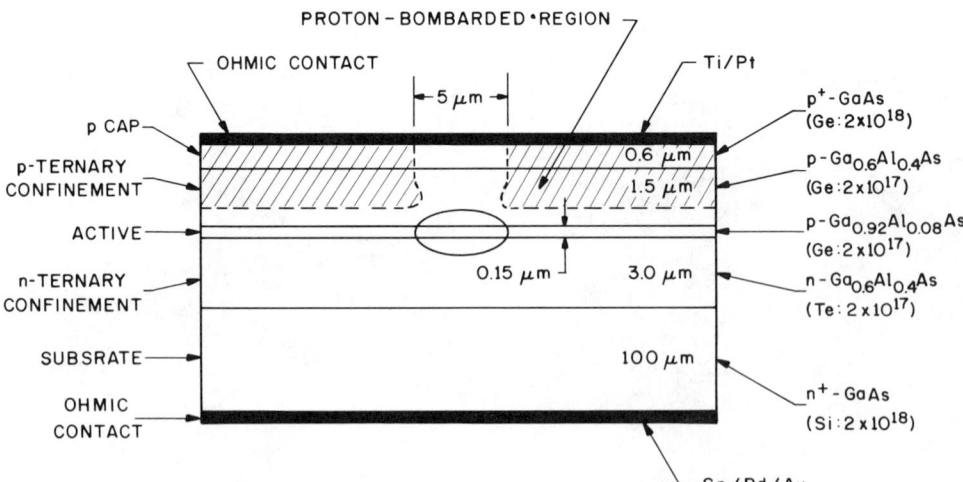

FIG. 3. Schematic diagram of the stripe-geometry proton-bombarded laser structure.

packaged laser is evaluated and must meet several screening criteria. Laser diodes with initial rapid aging of the threshold current and the external quantum efficiency, as determined by cw burn-in, are eliminated from the population of lasers to be used in transmitters (Tsang *et al.*, 1981). Similarly, on the transmitters in our experiment, screens are applied to other laser characteristics that influence transmitter performance, such as the power level of nonlinearities (Dixon *et al.*, 1976); the frequency of self-pulsations (Paoli, 1977); the optical noise level (Biesterbos and den Boef, 1981; Dixon, 1985); and the characteristic temperature T_0.

The screened, packaged lasers are built into transmitter circuits and subsequently mounted on "mother boards," which provide heat sinking and an interface to the computer-controlled test set. This life-test set is similar to that in Fig. 2. The test facility provides the transmitter with a system environment, with the exception that the laser package in our experiment is mounted on a thermoelectric cooler. The transmitters usually are aged at 30°C, pulsed at 45 Mbits/sec with a pseudorandom word sequence of 50% duty cycle. The bias current I_b, average power P_{av}, and temperature are measured hourly and recorded daily. On a monthly basis, the input-data stream is disabled, and a cw light–current ($L-I$) characteristic is scanned on the light from both facets; cw scans are also automatically taken if certain preestablished changes in the monitored parameters are exceeded. We also have the capability of measuring such dynamic characteristics of the transmitter as self-pulsation frequency, optical rise and fall times, and pattern dependence (Dixon, 1985). Since the degradation of these properties were not observed to have a significant impact on the functional lifetime of these transmitters, they are not discussed further here. This statement does not imply that these properties should be universally ignored, because they have the potential for being a limiting factor in the functional lifetime of a transmitter (Biesterbos and Salemink, 1981).

III. Life-Test Results for a 45-Mbit/sec AlGaAs System

The experience gained from our functional life test has been that cw laser aging and the failure criteria adopted for end of life are inadequate for estimating the system lifetime of optical transmitters. Our system lifetime is limited in fact by the degradation of the average light power and the extinction ratio.

We illustrate this major point of our investigation by presenting the life-test results on ten transmitters that have been aged at 30°C for over 10,000 hr (Dean and Dixon, 1982). These transmitters aim at controlling the average optical power launched into the front fiber by feedback control (Shumate *et al.*, 1978). The feedback signal is derived from the light from the back laser

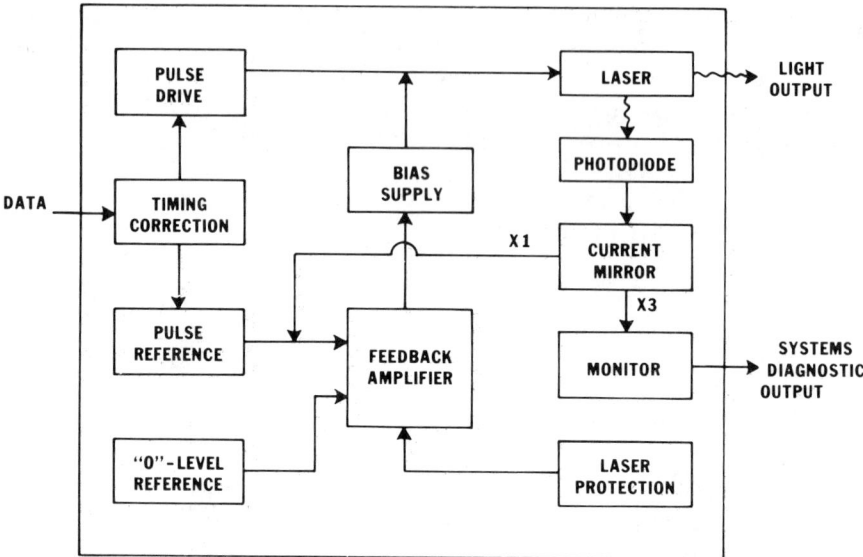

FIG. 4. Schematic diagram of the single-feedback-loop laser-transmitter circuit.

facet, which is detected by a photodiode mounted in the laser package. A schematic of the circuit is shown in Fig. 4. The transmitter is tuned initially at a dc bias current I_b, which is 8–25 mA below lasing threshold. A fixed modulation current I_{mod} is superimposed on the dc bias so that the laser is driven from a spontaneous emission power of P_0 to a stimulated emission power of P_1. The values of I_b and I_{mod} are set so that an average power $(P_0 + P_1)/2$ of -2.2 dBm and an extinction ratio P_1/P_0 of at least 15 are detected out of the front-facet fiber.

The data are derived from the cw L–I characteristics, which are scanned monthly. A typical example of the aging at 30°C of the L–I characteristic is given in Fig. 5, where we observe changes in both the threshold current I_{th} and the external quantum efficiency η_f (as detected out of the fiber). In Fig. 6, the rates of change of both I_{th} and η_f for the ten lasers are shown on a log-normal probability plot, as is customary (Joyce *et al.*, 1976). The median and standard deviations for the rate of change of the threshold current are %$\Delta I_{th} = 1.0$% khr^{-1} and $\sigma = 1.8$. A summary appears in Table I. The medians for these production-grown lasers are comparable to values reported for research-type lasers aged under cw conditions and the values found for other populations of laser transmitters that we have studied (Dixon and Dean, 1981).

Presented in Table II are the median times to failure based on these cw laser-aging criteria as they apply to the transmitter design. From the rate of

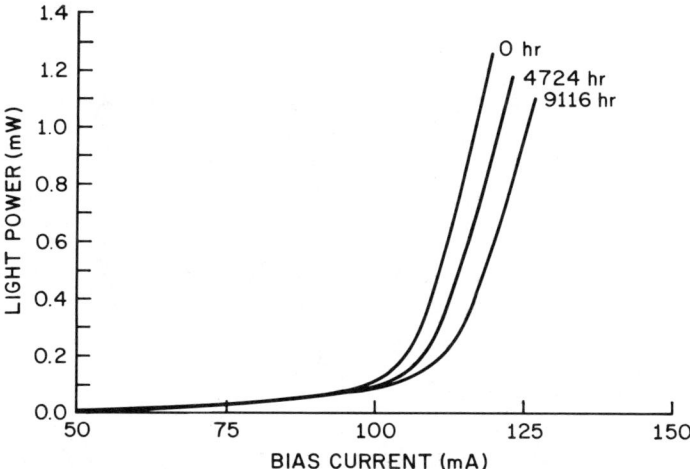

FIG. 5. The aging of the $L-I$ characteristic at 30°C of a laser aged in a pulsed circuit.

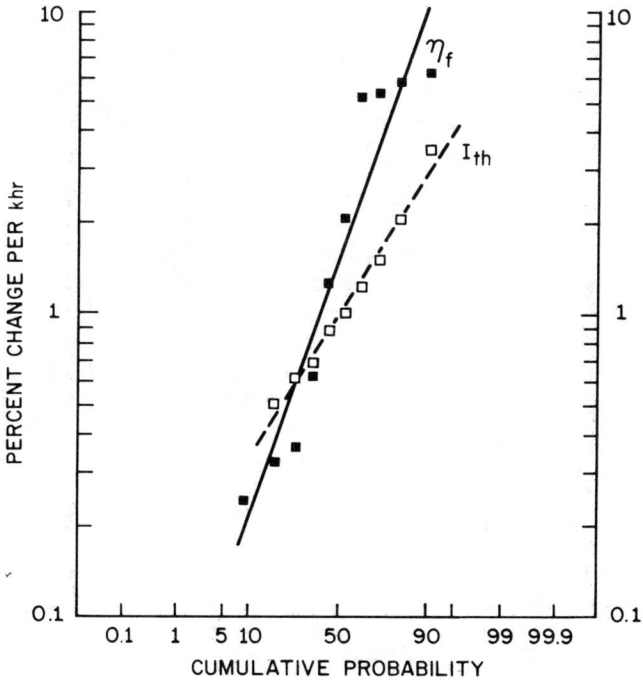

FIG. 6. The log-normal distribution of the percent change in threshold current I_{th} and the external quantum efficiency out of the fiber η_f per kilohour for the transmitters aged at 30°C.

TABLE I

STATISTICS FOR TEN TRANSMITTERS DERIVED
FROM A LOG-NORMAL DISTRIBUTION

Parameter	Median Δ (%Δ khr^{-1})	σ
I_{th} (mA)	+1.01	1.8
η_f (mW/mA)	−1.62	3.5
P_{av} (mW)	1.59	2.0
ε	−2.45	1.8
P_0 (mW)	+1.90	4.9

increase of I_{th}, and using the criterion that transmitter failure occurs when I_{th} increases to 2.5 times its initial value, we extrapolate a median time to failure at 30°C of 150 khr with $\sigma = 1.8$. The median failure time calculated from the rate of decrease of η_f and the failure criterion that η_f decreases to half its initial value is substantially smaller, 31 khr with $\sigma = 3.5$.

The much shorter failure time due to the decrease in η_f is expected despite the comparable median aging rates of both η_f and I_{th}. The major reason for this is the three-times-greater dynamic range available in a transmitter for degradation of I_{th}, compared to η_f. The criterion we used for η_f is the one commonly found in the literature for evaluating cw laser failure data.

We also observed, however, that the time to failure based on the degradation of η_f is also somewhat shorter than that found for cw-aged lasers (Kressel et al., 1978). There are two reasons for this result. First, it has been noted previously that the narrower (5 μm versus 8 or 10 μm) stripe devices have a larger rate of decrease of η_f and hence a shorter lifetime based on this quantity (Hartman and Dixon, 1980). Second, the change in the apparent quantum efficiency of the light emitted from the laser-package fiber can be due to

TABLE II

STATISTICS FOR TEN TRANSMITTERS DERIVED FROM A LOG-NORMAL DISTRIBUTION[a]

	Median projected failure times (khr)	σ	Criterion
I_{th}	149	1.8	I_{th} increased to 2.5 times its initial value
η_f	30.9	3.5	η_f decreased to one half its initial value
P_{av}	20.0	2.0	$P_{av} = -4$ dBm
ε	19.7	1.8	$\varepsilon = 10:1$

[a] Based on laser and transmitter criteria.

factors other than a change in the external quantum efficiency of the light emitted from the laser facet, such as changes in coupling efficiency.

Since the degradation rates of I_{th} and η_f in pulsed operation are similar to those found in cw laser testing, the design of this *simple feedback loop* transmitter, with average power control from a back-facet photodiode, would perform well if those degradations were solely responsible for the degradation of transmitter performance. However, we have observed two other aging characteristics as the limiting factors in the functional lifetime using this transmitter strategy. These aging effects are the following:

(1) *Average power drift,* which is a result of the asymmetric aging of the light coupled into the fiber compared to the light coupled to the back-facet photodiode (i.e., mistracking).

(2) *Extinction-ratio degradation* in which an increase in the spontaneous light at the bias current P_0 and the drift in the average power P_{av} translate into a decrease in the extinction ratio, $\varepsilon = (2P_{av} - P_0)/P_0$.

We now turn to a discussion of the mechanisms that have been proposed to account for these aging effects and the redesigns of the transmitter circuit that can minimize their effect on functional lifetime.

3. Average Power Drift

We have found that average power drift due to mistracking is a very serious problem when using back-facet feedback control. An example of mistracking with time is shown in the $L-I$ curves in Fig. 7. The circuit adjusts the bias current to the laser to maintain a constant back-face photodiode current. As the transmitter ages, this constant back-face current corresponds to a different light power out of the fiber. The power drift is shown in Fig. 8a for five transmitters, including the best and worst examples.

The absolute rates of change of P_{av} for all ten transmitters are shown in Fig. 9. The moduli of the percent change are used since both the upward and downward drift of P_{av} are caused by the same mechanism. The median and standard deviations for the rate of change of the average power are $|\%\Delta P_{av}| = 1.6\%$ khr^{-1} and $\sigma = 2.0$. From Table II, it is seen that the median failure time for the transmitters with a downward power drift is 20.0 khr with a $\sigma = 2.0$. In this case, failure is defined as an average light output power of -4 dBm, i.e., a decrease in power of 1.8 dB.

An *increase* in average power (which occurred in one transmitter, #86 of Fig. 8a) can also have serious ramifications for some system situations, but this situation is not reflected in this definition of failure. There are two consequences of increased average power. First, the laser will be operating at a higher power level, which could cause it to age more rapidly, and second,

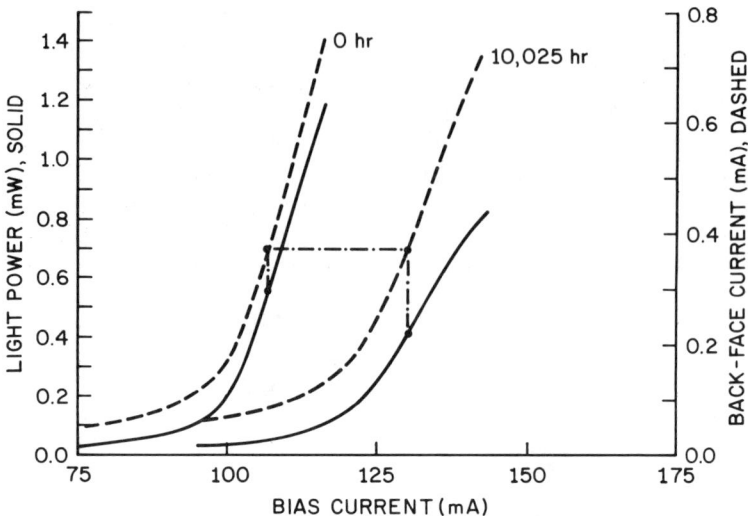

FIG. 7. The cw L–I characteristics at 30°C of a laser showing "mistracking." The circuit adjusts the bias current to the laser to maintain a constant back-diode current. As the transmitter ages, this constant back-diode current corresponds to a different light power out of the fiber.

for system applications that use avalanche photodetection, the higher power level can cause saturation of the optical receiver.

The mistracking we observed in Fig. 7 is superficially similar to the asymmetric aging previously reported (Paoli, 1977). However, unlike Paoli who observed asymmetry directly from the laser facets, we are monitoring the light outside the package. In our case, the asymmetry could be caused by the laser or by several of the package components. From our work on transmitters containing lasers grown by MBE (Tsang et al., 1983), the contribution of package relaxation, or photodiode aging, to average power drift appeared to be minimal (see also Part IV).

It is possible to eliminate the effect of mistracking by using an optical tap in the path of the front fiber (Karr et al., 1978, 1979). This solution avoids changes in P_{av} with time, provided that the tap is insensitive to any changes in fiber-mode content that might accompany the aging of the laser. Any aging of the photodiode should be lessened because of the decreased light intensity impinging on it. The drawbacks of a tap are that it introduces added loss to the transmission path, and it introduces the problem of the reliability of the tap and the laser–tap interface. A front-fiber tap version of the transmitter has been designed, and functional life-test results are presented in Part V.

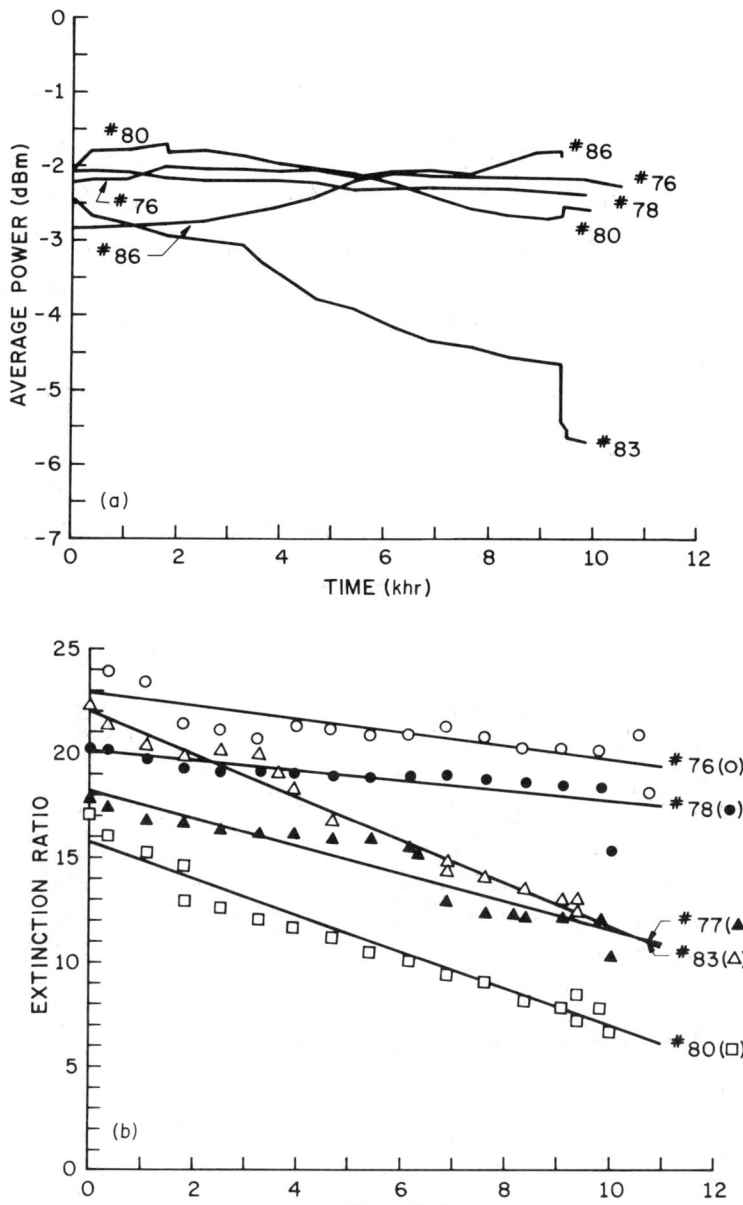

FIG. 8. (a) The drift in average power out of the fiber for five transmitters, including the best (#76) and the worst (#83) examples. (b) The extinction-ratio degradation for five transmitters, including the best (#78) and the worst (#80) examples.

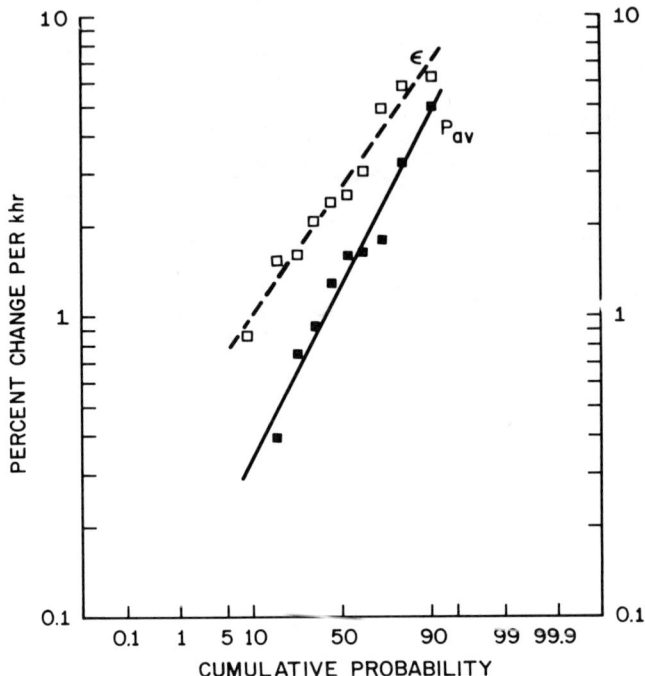

FIG. 9. The log-normal distribution of the percent change in average power P_{av} and extinction ratio ε per kilohour for the transmitters aged at 30°C.

4. Extinction-Ratio Degradation

The degradation in extinction ratio for five transmitters is shown in Fig. 8b. The moduli of the percent change of ε are plotted as a log-normal distribution in Fig. 9, which shows that the median change is $|\%\Delta\varepsilon| = 2.5\%$ khr^{-1} and $\sigma = 1.8$.

The median time to failure by extinction-ratio degradation for these ten transmitters is listed in Table II as 19.7 khr with $\sigma = 1.8$. Failure is defined as $\varepsilon = 10:1$. Degradation of ε to this value produces a degradation of eye margin in systems with avalanche photodiode (APD) receivers that is comparable to that produced by the 1.8-dB decrease in optical power we previously established on the average power (Smith and Personick, 1982). The failure time is similar to that observed for the average power drift and is an order of magnitude less than that expected by considering the cw criteria.

The degradation of the extinction ratio can arise by three mechanisms for the transmitter strategy used: (1) softening of the lasing threshold, which causes an increase in the spontaneous light at I_b; (2) a decrease in the external quantum efficiency, which moves I_b closer to the lasing threshold and increases P_0 independently of any softening; and (3) a decrease in P_{av} due to mistracking, which leads to a greater decrease in P_1 than P_0. All three mechanisms give rise to extinction-ratio degradation in our single-feedback-loop average power-control strategy, because I_{mod} is fixed, and there are no constraints on the individual values of P_0 and P_1. The influence of these changes in P_{av} and P_0 on the changes of ε with time can be obtained by partial differentiation of the extinction-ratio equation, which yields

$$\frac{d\varepsilon}{dt} = \frac{2}{P_0}\left[\frac{dP_{av}}{dt} - \left(\frac{P_{av}}{P_0}\right)\frac{dP_0}{dt}\right], \quad (1)$$

provided the mechanisms which affect P_{av} and P_0 can be considered independent.

The coefficient of dP_0/dt in the two terms on the right-hand side of Eq. (1) is negative and significantly larger than that of dP_{av}/dt. However, it is possible for ε to either increase or decrease with time if P_{av} changes dramatically by mistracking. In fact, ε has been observed to increase in some transmitters in the short term (less than ~ 3 khr). However, for longer times, the increase in P_0 tends to dominate any change in P_{av}.

Figure 10 presents a typical example of extinction-ratio degradation. The dotted line represents the degradation expected if only the decrease in exter-

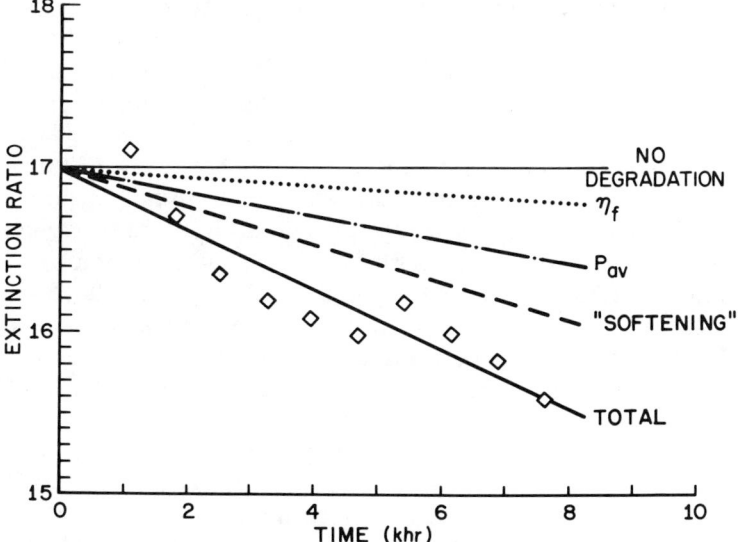

FIG. 10. The individual contributions to the extinction-ratio degradation: ($\cdots$) contribution due to a decrease in the external quantum efficiency out of the fiber; ($-\cdot-$) contribution due to a decrease in average power; ($---$) contribution due to the "softening" of the lasing transition.

nal quantum efficiency contributed; the dot–dash line is the degradation expected if only an average power decrease contributed; and the dashed line is the exclusive contribution of the "softening" of the lasing transition to the decrease in extinction ratio. Of the two mechanisms leading to the increase in P_0, the softening of the lasing transition is seen to be the more important. It should be noted that the value of $\%\Delta P_0$ khr^{-1} in Table I has been corrected for changes in quantum efficiency η_f and so represents the softening contribution only.

The problem of extinction-ratio degradation could be dealt with in the transmitter by providing a second feedback loop to the circuit which controls P_1 or $P_1 - P_0$ by changing the amplitude of I_{mod} to the laser (Gruber et al., 1978; Chen, 1980). This strategy will work effectively unless the aging of the lasing transition is accompanied by such a large broadening of the source wavelength that chromatic dispersion begins to degrade the performance of the system (Chen, 1981) or if an increasing modulation current adversely affects the degradation rate of the laser. This latter consideration is discussed in more detail in Section 6.

5. Laser-Degradation Modes

The conclusions of these experimental results have raised concern about degradation modes in gain-guided lasers which lead to softening of the lasing transition (giving extinction-ratio degradation) and to asymmetric aging of the coupling of light to the front-facet fiber and back-facet photodiode (giving the average power drift). Mechanisms that increase optical loss and thus result in increased softening include, for example, the growth of internal defects or faults, which scatter light out of the guiding region (Nash et al., 1976), and changes in mirror reflectivity (Nash et al., 1979). Although they cause an increase in I_{th}, such losses do not cause asymmetric aging.

To explain asymmetric aging of the light *emitted* from the two laser facets, we turn to a rate-equation model which predicts facet asymmetry in gain-guided injection lasers (Marcuse and Nash, 1982). Experimental results such as those shown in Fig. 7 can be obtained by introducing an isotropic optical loss *asymmetrically* located between identical facets. The critical length for this local disturbance has been shown to be on the order of 50–100 μm (Agrawal et al., 1983).

This situation can lead to the different optical power levels emitted from the two facets, a result that has been confirmed experimentally using a cleaved-coupled-cavity laser (W. T. Tsang, unpublished). The loss must be power dependent and increase monotonically, i.e., nonsaturable. Marcuse and Nash (1982) suggest that the physical origin of this loss is a *local* "kinking" of the lasing mode away from the gain peak at the stripe center toward the lossier edges of the stripe (Dixon, 1980; Thompson et al., 1978), as shown schematically in Fig. 11. It is proposed further that changes in asymmetry with time can occur if there is: (1) facet erosion on uncoated or inadequately coated devices (Nash and Hartman, 1980); (2) optical annealing of the proton-bombardment damage along the edge of the stripe (Schorr and Tsang, 1980; Dyment and Smith, 1981; Dixon and Dean, 1981); or (3) aging of the thermal bond (Plumb et al., 1979; Kobayashi and Iwane, 1977).

The variation with time of the local kinking has important implications in our case. Any asymmetry that is present initially is accommodated in the manufacture of the laser package and the tuning of the transmitter. If, however, the asymmetry changes with time, then not only will the beam of light from one facet be shifted with respect to its initial position, but the intensity of light from that facet will also be decreased since the mode is propagating in a lossier region, as shown in Fig. 11. These events will be more serious in coupling to the fiber than to the photodiode since the fiber has the smaller numerical aperture. The average power drift in the transmitter population will be biased downward, which is, in fact, our experience.

An alternative proposal (Kardontchik, 1982) suggests that asymmetric optical loss can be due to variations in the absorption coefficient along the stripe. This loss would be saturable at sufficiently high injection currents. The proposed origin of the absorption coefficient variation is a variation in the depth of proton bombardment caused by nonuniformity in thickness of the p ternary layer. Changes in asymmetry with time in this model require some form of annealing of the proton bombardment adjacent to the stripe.

It is interesting to note that both this and the rate-equation model suggest that proton-bombarded MBE-grown material should exhibit less asymmetric aging behavior than LPE material due to the uniformity of composition and thickness of MBE-grown epilayers. Our aging studies on MBE lasers (Tsang et al., 1983) support this conclusion.

We must also note that the mechanisms proposed to account for average power drift also lead to softening of the lasing transition and thus to extinc-

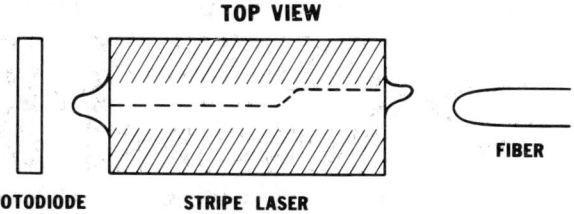

FIG. 11. Schematic diagram of the effect of a local kink on the far-field properties of the emitted light and the development of asymmetry between the two laser facets. The dashed line indicates the center of the mode; shading represents the insulating region.

tion-ratio degradation; but, as mentioned, there are also mechanisms that produce softening with no asymmetry and average power drift.

IV. Functional Life Testing for Improved Design

In addition to determining the functional reliability of lasers in transmitters, functional life testing has also been used to investigate specific questions of transmitter aging and to test the viability of transmitter redesigns. We discuss these aspects in Part IV.

6. Effect of Bias Conditions on Aging Rates

One of the variables in setting the initial operating points of a transmitter is the position of the bias current in relation to the threshold current. Given that the transmitter design suffers from a slow degradation of extinction ratio, it is important to investigate the effect of the ratio of I_b to I_{th} or, conversely, the magnitude of I_{mod} on the aging rates of laser and transmitter properties. The results of these studies will have an impact on future transmitter specifications and designs.

a. Early Results

In an early study (Dixon and Dean, 1981), we analyzed a group of laser transmitters that exhibited little front-to-back-facet mistracking and hence small changes in average power (all less than 1% khr^{-1}). For these transmitters, it was found that a strong correlation existed between the rate of extinction-ratio degradation and the difference between the initial bias and threshold currents ($I_{th} - I_b$). With one exception, the transmitters reported on had negligible degradation of their external quantum efficiency as detected from the fiber. Thus, having eliminated changes in average power and external quantum efficiency from the extinction-ratio degradation, the analysis established that the correlation existed between the percent increase of the spontaneous light emitted at the bias current and the position of that bias current with respect to the lasing threshold. The data are replotted in Fig. 12 as %ΔP_0 khr^{-1} versus I_b/I_{th}; the open diamond represents the one transmitter with a large change in η_f (> 5% khr^{-1}).

As we pointed out in Section 5, there are mechanisms of laser degradation, such as increased bulk optical loss and mirror darkening, which result in an increase in the power of spontaneous emission. In these cases, as the lasing threshold moves to a higher current due to the increased loss, the subthreshold light undergoes additional amplification (Thompson, 1980a). One expects, then, a correlation between the increases in the spontaneous light and the threshold current. However, we found no correlation between %ΔP_0

4. SEMICONDUCTOR LASERS AS OPTICAL TRANSMITTERS

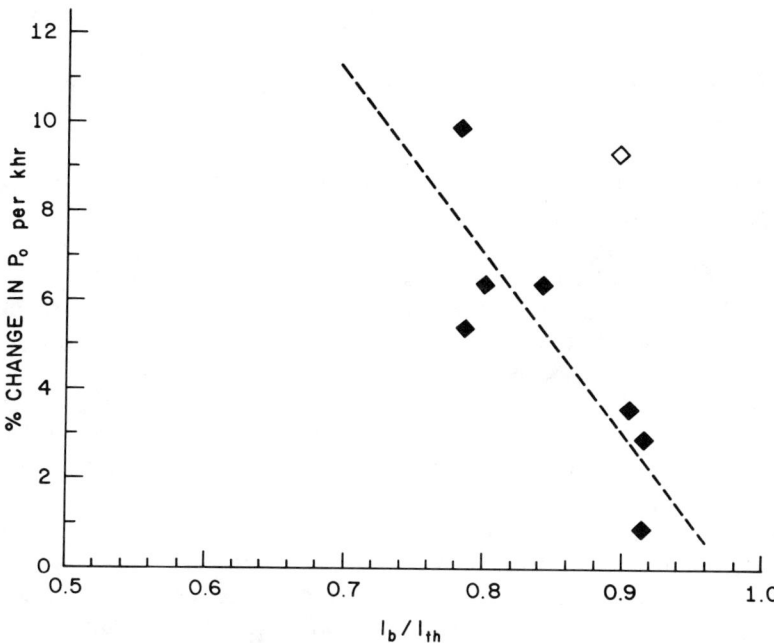

FIG. 12. The percent change in the spontaneous light power at the bias current as a function of the initial biasing conditions with respect to threshold.

khr^{-1} and the change in lasing threshold %ΔI_{th} khr^{-1} for these 5-μm, shallow-proton-bombarded lasers.

The origin of these two observations, the correlation between %ΔP_0 khr^{-1} and the bias position and the lack of correlation between %ΔP_0 and %ΔI_{th}, is consistent with a model of optical annealing of the proton damage in and near the active region. The driving force for the anneal would be nonradiative absorption in regions where the optical field overlaps the proton damage. Support for such a model can be drawn from experimental results (Dyment and Smith, 1981), which show that, for proton-bombarded stripe lasers, characteristics such as the threshold current, external quantum efficiency (from the laser facet), and the amount of spontaneous emission are determined principally by the proton damage at the edge of the lasing stripe and *not* by the peak proton damage.

The presence of this small proton damage along the edges of the *active stripe* introduces an optical loss mechanism that has a more pronounced effect on the amount of spontaneous light emitted from the facet than on the stimulated emission, since the spontaneous light is nondirectional inside the cavity. Annealing of this proton damage will decrease the loss, causing an increase in the amount of spontaneous light emitted. At the same time, since the spontaneous light overlaps the proton-damaged region more than the stimulated light (Kirby et al., 1977), it is also more effective at producing optical annealing. Thus, for pulse-modulated lasers, the rate of annealing should increase as the laser is biased further into the spontaneous region. Hence, we have a mechanism that correlates the rate of increase of the spontaneous light with a decreasing ratio I_b/I_{th}, as observed.

In contrast, the lack of correlation during aging of the spontaneous emission with threshold current arises from the complex dependence of the threshold current and spontaneous emission on proton-bombardment depth. Experimental results (Dyment and Smith, 1981) showed that the threshold current decreases with increasing proton-bombardment energy and reaches a minimum around 250 keV. This energy corresponds to a peak damage position, which was approximately 1 μm above the active region. At higher proton energies, as the peak damage approaches the active region, the threshold increases again. On the contrary, the amount of spontaneous emission at a *fixed bias current* remains constant as the proton energy increases until the cross-over value of 250 keV, observed in threshold current, is reached. At that point, the spontaneous light power abruptly decreases to approximately one third of its former value and then again remains constant. Thus, in this model, the behavior of the threshold current and the amount of spontaneous light *need not be correlated* on annealing of the proton damage if the processing of the laser resulted in a device operating near the cross-over point. In fact, bombarding the laser to the cross-over point is desirable, since it results in the minimum threshold current. As evidence that some lasers in our experiment operate near this regime, we present, in Fig. 13, data from a transmitter that exhibited an increase in spontaneous emission while the threshold current decreased.

b. Further Examination of the Effects of Bias Conditions

The difficulty of confirming the proton-damage annealing mechanism, then, is that the appearance of the effect is very dependent upon such laser-processing parameters as the thickness of epilayers and the energy of proton bombardment. In the previous example, the transmitters were selected from a larger group on the basis of their small average power drift. They contained lasers cleaved from three wafers and processed in a two-week period. (One of the wafers, grown by MBE, contributed five of the eight lasers.) During this time, the processing parameters were not changed.

Realizing this restriction, we have attempted to extend the range of I_b/I_{th} of Fig. 12 to a value below 0.7. In this experiment, we selected seven transmitters containing lasers from three different wafers and tuned the transmitters within each wafer subset to cover as large a range of I_b/I_{th} as possible.

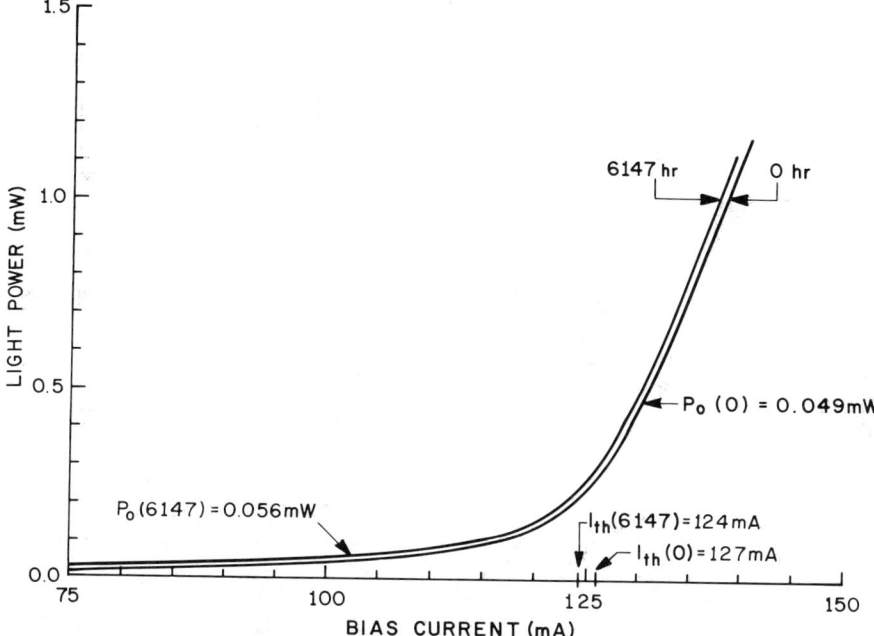

FIG. 13. The aging of the cw $L-I$ characteristics for a laser transmitter for which the bias current decreases and the power emitted at the bias current increases with time.

These data are shown in Fig. 14, where the data for three transmitters containing lasers from two sequentially grown wafers are treated as one group, and the data for the four transmitters containing lasers from another wafer are treated as a second group. For these transmitters, $\%\Delta\eta_f$ khr^{-1} was not always insignificant (in one case, it was as high as 20%), so the data plotted have been corrected to remove the contribution of the change in slope to the increase in the spontaneous emitted power. The results of this experiment again show an inverse relationship between $\%\Delta P_0$ khr^{-1} and I_b/I_{th}. A dependence on wafer processing is evident, since the data points from different wafer groups do not lie in the same area.

It appears from the data in Fig. 14, especially that from wafer RA068, that the degradation becomes markedly worse for I_b/I_{th} less than 0.4. In fact, every measure of degradation, including the degradation of the threshold current, *increases* rapidly in this regime as I_b/I_{th} decreases.

c. Comparison with Other Workers

This new result, that a small value of I_b/I_{th} leads to a large degradation of threshold current, is in sharp contrast to the results of a pulsed-laser life-test

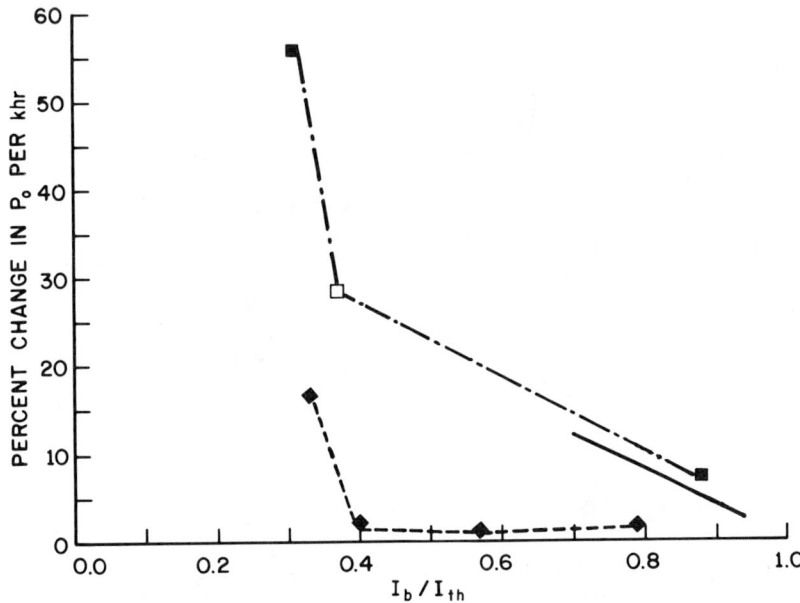

FIG. 14. The percent change in the spontaneous light power at the bias current (corrected for changes in the external quantum efficiency) as a function of the initial biasing conditions for three groups of transmitters. Solid diamond is for wafer RA068; solid square is for wafer RB054; open square is for wafer RB053. Solid line is taken from Fig. 11.

study on Zn-diffused, planar-stripe AlGaAs lasers (Yoshida et al., 1982). In that case, an increase in threshold current was observed which was proportional to the square of the total injected current density; i.e., the rate of increase of I_{th} correlated with an increasing I_b/I_{th}. However, only the values of I_b/I_{th} of 0.5 and 0.9 were investigated.

The initial performance characteristics of the two laser structures, Zn-diffused and shallow-proton implanted, are comparable (Thompson, 1980b). In the case of the Zn-diffused laser, the threshold current increase was thought to be induced by an increase in nonradiative recombination centers (Yoshida et al., 1982). We do not yet have the data to fully understand which of the several mechanisms discussed here are responsible for the rapid degradation seen in proton-bombarded lasers when I_b/I_{th} is less than 0.4.

On the other hand, work on 20-μm oxide-isolated-stripe AlGaAs lasers (Robertson, 1983) yields no statistical difference in the degradation of cw and 140-Mbit/sec pulsed devices aged at 55°C. However, the pulsed-circuit strategy adopted in this experiment keeps $I_b/I_{th} = 1$ at all times by using a lasing threshold-detector circuit (Smith, 1978).

In the *limit of* $I_b/I_{th} = 1$, our results are consistent with two major conclusions of this latter work. First, there are similar rates of degradation of I_{th}

when cw and pulsed operation are used. Second, only a small rate of growth of the spontaneous light at threshold is observed (Fig. 14). Because of the choice of circuit strategy, the influence of I_b/I_{th} was not investigated for the oxide-isolated lasers. Thus it is unknown at present what the influence of pulsed biasing conditions below I_{th} have on either threshold or extinction-ratio degradation for these devices.

Thus, from the *laser* lifetime viewpoint, the mechanisms of laser degradation under small-bias-current, large-pulse-current conditions are still an open question. However, for our *functional* lifetime, it is clear that the proton-bombarded lasers should be biased as close to threshold as possible without compromising the initial value of the extinction ratio or its margin for aging.

In the case of the Zn-diffused planar-stripe lasers, improved reliability occurred when operated at small bias currents well below I_{th}. For the oxide-stripe devices, it appears that pulsed aging can be predicted from cw data when biasing at threshold, but it is unknown whether this is the optimum pulsed biasing condition.

7. Temperature Dependence of Transmitter Aging

The room-temperature aging studies revealed that the functional lifetime of our transmitters was not limited by the same degradation modes as reported earlier for the high-temperature "pseudoconstant power" cw aging of the stud-mounted lasers (Hartman and Dixon, 1975).

Our purpose here is to extend this study and evaluate the temperature dependence of the functional lifetime of our transmitters in order to establish whether the aging mechanisms influencing system performance are also thermally activated. In the case of the pseudoconstant-power cw-aging study on 10-μm stripe lasers, it was found that thermal stress increased the rate of degradation of the threshold current. A thermal activation energy of 0.7 eV was calculated from data at 50, 70, and 90°C (Hartman and Dixon, 1975). However, little change in the quantum efficiency was found at any temperature.

The experimental procedures were essentially those already described. Twenty transmitters, which had similar initial laser and transmitter properties, were selected. Half were aged at 50°C and half at 70°C laser-stud temperatures, corresponding to laser-junction temperatures of 52 and 73°C, respectively. (Initial characteristics of the ten transmitters aged at 30°C, reported in Part III, are statistically different from this group; hence, those results are not incorporated here.) All the transmitters passed initial specifications at their aging temperatures at the beginning of the test.

The results of the aging are given in Table III and Fig. 15a–d. The data are presented as degradation rates rather than as percent changes in order to

TABLE III

DEGRADATION RATES OF LASER AND TRANSMITTER PROPERTIES FOR TRANSMITTERS AGED AT 50°C AND 70°C

	dI_{th}/dt (mA khr^{-1})	σ	$d\eta_f/dt$ (mW mA^{-1} khr^{-1})	σ	dP_{av}/dt (mW khr^{-1})	σ	$d\varepsilon/dt$ (khr^{-1})	σ
50°C	+1.36	1.5	−0.00143	1.7	0.00942	2.5	−0.503	1.6
70°C	+3.57	1.6	−0.00238	1.5	0.0156	1.8	−0.432	1.6

eliminate any weighting due to very different initial values of the measured parameters at the two temperatures. The degradation rates of external quantum efficiency, average power, and extinction ratio showed no thermal dependence, indicating that the mechanisms responsible for their degradations are not thermally activated. The only characteristic for which there was a statistically significant difference at the two temperatures, at the 75%

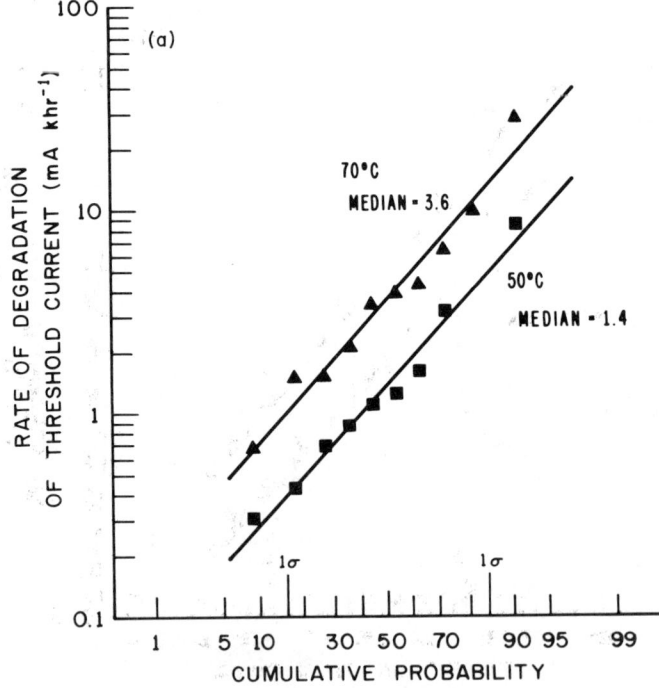

FIG. 15. The log-normal distributions for: (a) the rate of degradation of the threshold current; (b) the rate of degradation of the external quantum efficiency; (c) the rate of degradation of the average power; and (d) the rate of degradation of the extinction ratio, for transmitters aged at 50 and at 70°C.

4. SEMICONDUCTOR LASERS AS OPTICAL TRANSMITTERS 177

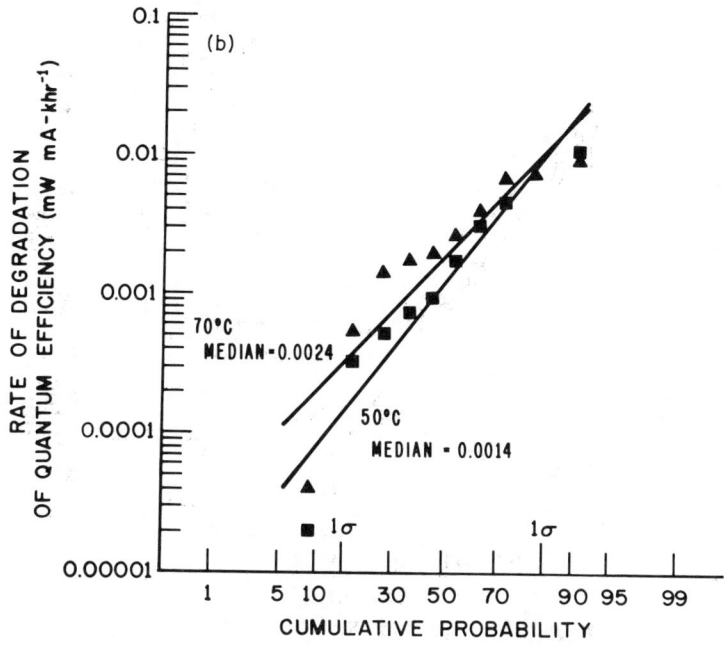

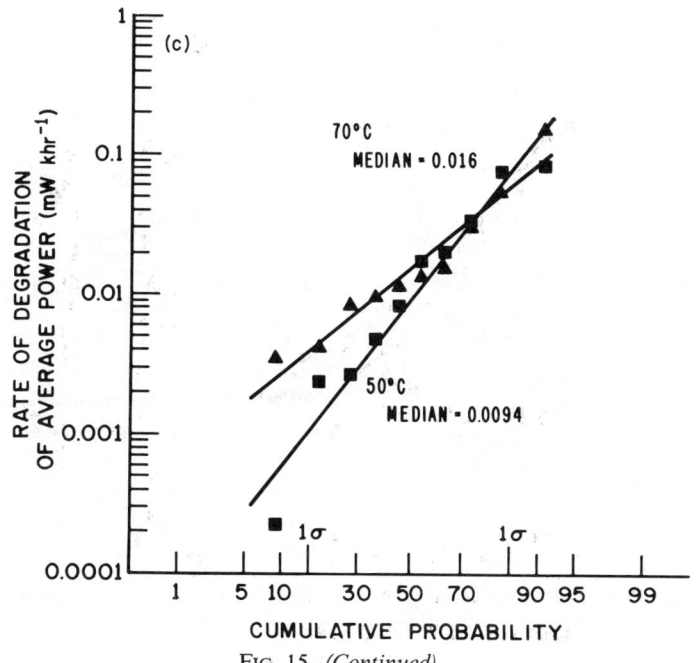

FIG. 15. *(Continued)*

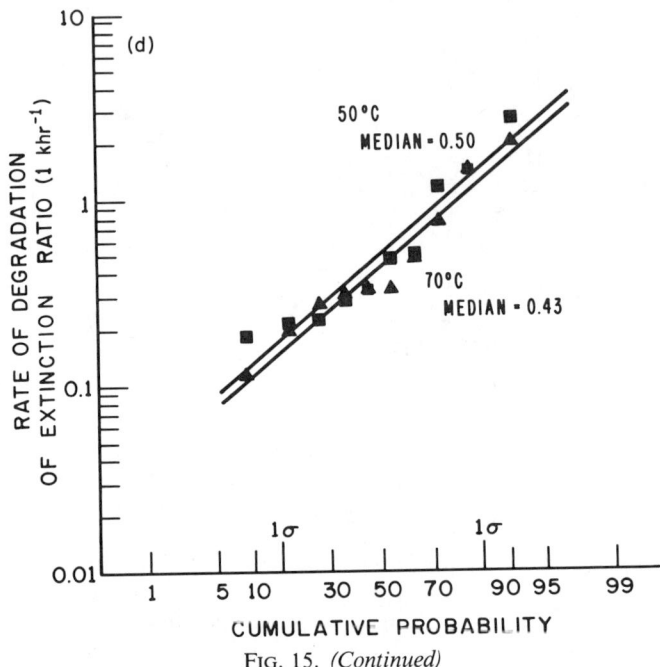

FIG. 15. *(Continued)*

confidence level, was the threshold current. The cumulative probability of the aging rates of the threshold current, external quantum efficiency, average power, and extinction ratio are shown in Fig. 15a–d.

A Fisher's F test confirms that the standard deviations of the rate of degradation of the threshold current for the two populations are equal within the experimental error. This result in turn implies that an Arrhenius extrapolation is valid for this experiment. By using the standard Arrhenius equation, we calculate an "activation" energy of 0.45 eV. This value is lower than the value of 0.7 eV found earlier, but it should be remembered that the device structure here is slightly different, the aging modes are different (cw versus pulsed), and the basis for extrapolation is different. We have used a *degradation rate* as the parameter to which the Arrhenius equation was applied. Earlier workers have used *failure times*. As has been pointed out (Joyce *et al.*, 1976), the activation energy that is calculated under those circumstances is dependent on the failure criterion that is used. In addition, each degradation mode has its own activation energy, and in a proper analysis these must be singled out and calculated individually (Ritchie *et al.*, 1978). It is likely that the relative importance of the several different degradation modes is very different for the earlier uncoated 10-μm stripe deep-bombarded lasers and our facet-coated 5-μm stripe shallow-bombarded lasers.

In a subsystem test such as this one, it is impossible to apply an increased temperature stress without also applying an increased current stress. [For the power levels here, light power is not a factor in the degradation of I_{th} (Dixon and Hartman, 1977).] By analyzing the covariances (Dixon and Massey, 1969) of the rate of threshold degradation versus the average operating current $I_{op} = \langle I_b + 1/2 I_{mod} \rangle$, it was found that the significant difference in the degradation of the threshold current between the two populations was due entirely to the different operating currents of the two populations (again, at the 75% confidence level). The same result is obtained by fitting the data at each temperature to a current modifying factor of the Arrhenius equation (Mizuishi *et al.*, 1979):

$$dI_{th}/dt = \Lambda_0(T) \exp(\gamma I_{op}), \tag{2}$$

where $\Lambda_0(T)$ is the standard Arrhenius factor. For our data, γ is 0.2 mA^{-1} (compared to 0.25 found by Mizuishi *et al.*, 1979), and all the "temperature" dependence of the degradation can be accounted for in the current-dependent factor.

In summary, we have attempted to discover if thermal stress accelerates the degradation of transmitter properties in a way not predicted by the cw high-temperature aging of the laser diode. We have not found that the change in external quantum efficiency, average power drift, extinction-ratio degradation, and hence the laser-degradation modes which cause them, are thermally activated. Only the degradation of the threshold current was significantly affected by the application of thermal stress, and the degree of acceleration was less than that found for cw lasers. Finally, a more careful analysis of the data revealed that all the differences in degradation rates of the threshold current between the two populations at two temperatures was accounted for by considering the effects of the different operating currents for the two populations.

As a result, we must conclude that caution must be exercised when trying to obtain a transmitter lifetime by performing high-temperature "accelerated" testing and then extrapolating to room temperature by using the 0.7-eV activation energy commonly quoted for AlGaAs lasers.

8. Preliminary Results of Functional Reliability Test for Tap-Controlled Transmitters

In Part III, we demonstrated that the functional lifetime of our single-feedback-control transmitters was limited by the average power drift. This drift was caused by "mistracking," i.e., the difference in aging of the light from the front and back facets of the laser. Since the coupling of the light from the front-laser facet to the fiber is more sensitive to misalignment than the coupling of the light from the back-laser facet to the photodiode, the optical power drift was predominantly downward.

Further, we proposed that average power drift should be controllable by providing the bias current feedback signal from the front laser facet by tapping off light launched into the system fiber (Karr *et al.*, 1979). We have designed such a "tap-controlled" transmitter using a graded-index (GRIN) lens with a multilayer dielectric coating on its end surface to tap off light from the fiber (Tomlinson, 1980). The tap is shown schematically in Fig. 16. The quarter-wavelength pitch GRIN lens is used as an image-transfer device between the input and output fibers. Light from the input fiber is converted into a parallel beam by the lens before it strikes the partially transparent dielectric mirror. The transmitted, or "tapped," component of the beam is detected by a photodiode and provides the feedback signal to the bias control circuit. The percentage of tapped light is determined by the composition and number of dielectric layers, and in our case is approximately 6%. The resulting photodiode signal is small and must be amplified before it is used as the feedback signal for the bias current. The reflected component of the beam is focused by the same GRIN lens into an output fiber, adjacent to the input fiber. The input fiber of the tap is connected directly to the output fiber of the laser package.

The tap has an intrinsic loss of 0.6 ± 0.1 dB when using light that has an equilibrium-mode distribution within the fiber. This loss arises from the GRIN lens aberration, fiber-to-lens misalignment, fiber microbending, scattering and absorption in the dielectric films, and the 6% coating transmission. To this loss must be added any loss due to the connection between the laser package and the tap input fiber.

The insertion of this new component into the transmitter raises at least two questions. First, is the functional reliability of the transmitter adversely affected by the insertion loss of the tap, which necessitates the use of higher

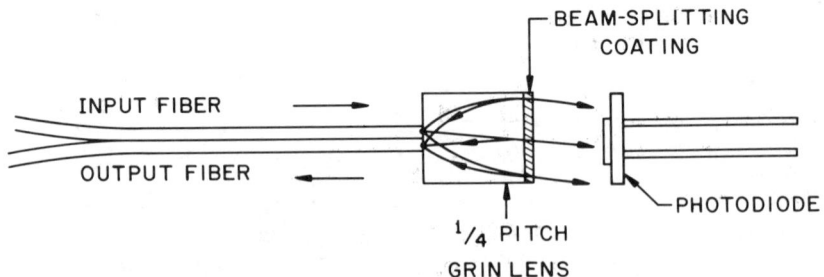

FIG. 16. Schematic diagram of the GRIN lens optical tap.

laser facet powers to achieve the same output power? Second, is the tap stable with time and temperature? These questions are best answered through functional reliability tests.

A preliminary investigation of the functional reliability of these transmitters has been conducted. Ten transmitters, tuned to −2.2 dBm output power, were placed on test at 30°C. Improvements in the coupling efficiency between the lasing facet and the output fiber since our earlier experiments compensated for the tap-insertion loss. Thus, the average facet powers of the lasers in this population were approximately the same as in previous populations. The laser-biasing conditions were also comparable. Since the "back-facet" photodiode was still in the laser package, although not providing the feedback signal, it was monitored during the experiment in order to evaluate the magnitude of "mistracking" in this laser population.

The results of over 7000 hr of aging are summarized in Table IV. The laser- and transmitter-degradation rates $\%\Delta I_{th}$, $\%\Delta \eta_f$, and $\%\Delta \varepsilon$ khr^{-1} are similar to those obtained with other transmitter populations using "back-diode" control. The average power drift $\%\Delta P_{av}$ khr^{-1} was an order of magnitude smaller. A typical example of the behavior of the average power and the back-diode current is given in Fig. 17. The change in back-diode current $\%\Delta I_{bd}$ khr^{-1} in Table IV provides a lower limit on the average power drift, which would have occurred if these transmitters had been under back-diode control. The distribution of average power drift found in this experiment is shown in Fig. 18, in which it can be contrasted to the distribution of average power drift of the earlier back-diode-controlled transmitters, taken from Fig. 9. That the improvement in the amount of average power drift is due to tap control can also be seen from the third distribution in Fig. 18. This distribution of $\%\Delta I_{bd}$ khr^{-1} has a median value approximately equal to that of the average power drift observed in the earlier back-diode-controlled transmit-

TABLE IV

STATISTICS FOR TEN TAP-CONTROLLED TRANSMITTERS DERIVED FROM A LOG-NORMAL DISTRIBUTION[a]

Parameter	(% Median Δ (khr^{-1}))	σ
I_{th} (mA)	+0.568	1.7
η_f (mW/mA)	−1.22	1.3
P_{av} (mW)	0.301	1.4
ε	−1.42	1.9
P_0 (mW)	+1.26	2.1
I_{bd} (mA)	1.77	2.1

[a] Compare with Table I

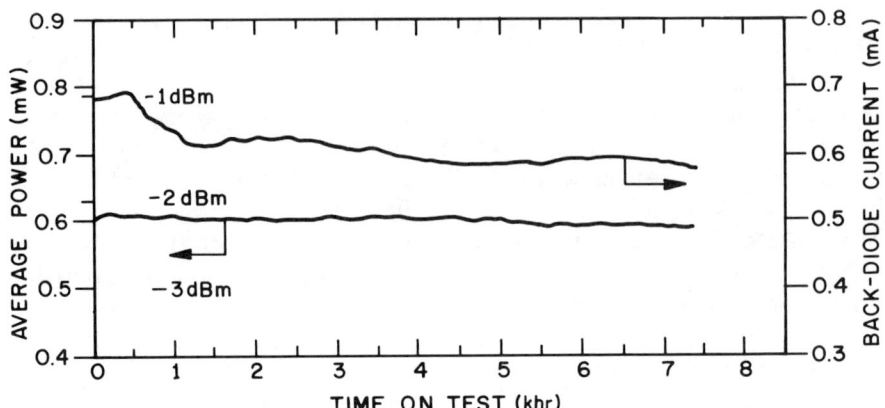

FIG. 17. The degradation of the average power out of the fiber and of the back-diode current for a tap-controlled transmitter.

ters. From these results, we can infer that the lasers used in the back-diode-controlled and tap-controlled transmitters had similar mistracking problems.

There are several mechanisms that could have led to the small residual average power drift in the tap-controlled transmitters. The list includes mechanical instability of the tap, chemical or physical changes in the dielectric coating and other tap components, and changes in the wavelength of the light and the fiber-mode content. An analysis of the taps at the conclusion of this experiment eliminated the first mechanism from consideration. Transmission changes due to chemical or physical changes brought about by variations in temperature or humidity can be estimated by using the results of the initial tap performance specification. The maximum change in transmission through the dielectric coating observed in humidity/temperature cycling (85% RH/85°C) is 4% (H. P. Hsu, private communication), which means a maximum 0.3-dB change in average power. The dielectric coating on the GRIN lens is designed to produce a minimum in transmissivity at the nominal system wavelength of 0.825 μm. Based on initial performance, the total transmission change in the wavelength range 0.81–0.84 μm is also 4%. A drift in the lasing wavelength with time could contribute at most another 0.3 dB to the average power drift.

The GRIN lens is an effective mode stripper, removing approximately half of the nonpropagating cladding modes from the fiber-mode spectrum of the input fiber when measured at the output fiber. At the same time, the active area of the photodiode is about three times as large as the end surface area of the lens, so that it captures all the light, including light in the cladding modes,

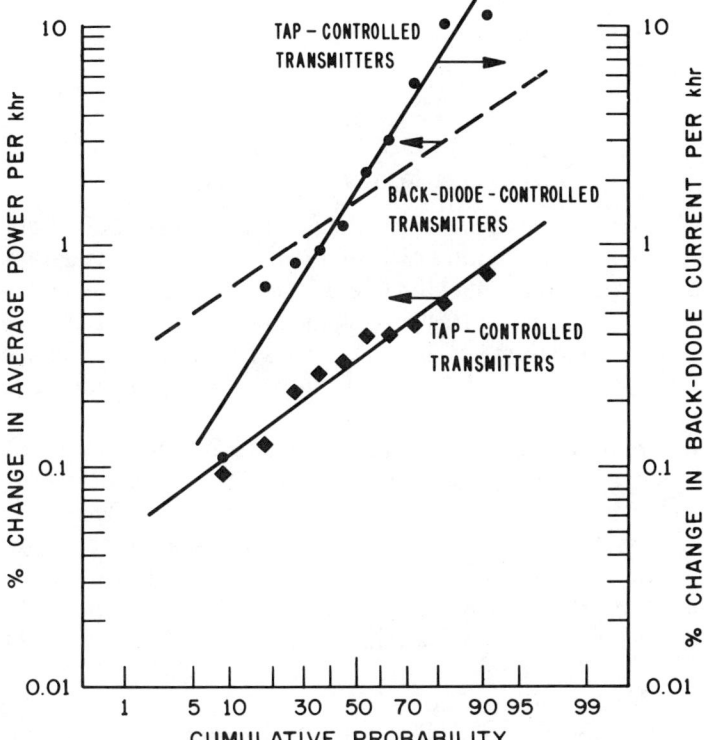

FIG. 18. The log-normal distributions of the percent change in average power and the percent change in back-diode current per kilohour for a tap-controlled transmitter. The dotted line represents the percent change in average power per kilohour for earlier back-diode-controlled transmitters (Fig. 9).

that is transmitted through the dielectric coating. There is the potential for mistracking if the amount of light launched into the cladding changes with time. Initially, the amount of light found in the cladding at the output fiber of the laser package is less than 5%, typically 3%. But, how the mode filling of the fiber changes as the laser ages is not known and needs to be investigated, since this mechanism could be the most serious cause of average power drift in the tap-controlled transmitters.

For these tap-controlled transmitters, we find that the average power drifts upward and downward with equal probability. By using the data from the five transmitters in which the average power drifted downward, we calculate a median time to failure based on $P_{av} = -4$ dBm of 174 khr. This figure represents an improvement factor of 5 over back-diode-controlled transmitters. We conclude that the tap-controlled transmitters have achieved the goal of stabilizing the average output power without deleteriously affecting laser-aging characteristics.

V. Functional Reliability of an AlGaAs Laser Transmitter for Space Communication

We have repeatedly stressed that the results of functional reliability tests are specific to the system for which they were designed. The data presented thus far have been for a 45-Mbit/sec digital system for voice transmission over optical fibers. In contrast, we shall review here a functional life test designed to evaluate semiconductor lasers as optical sources for space communication (Thompson et al., 1981; Einhorn et al., 1982; Barry et al., 1983, 1985).

The system under test uses AlGaAs lasers in a 2-Mbit/sec pulsed communication system for free-space applications. The laser is pulsed to 50 mW peak power with a 25-nsec pulsewidth at 1% duty cycle. The constraints of the application require that the laser have a single spatial mode, small longitudinal mode content, and a useful system lifetime of 10 years. In this case, it is obvious from the requirement of a peak power of 50 mW that the results of cw testing (usually at 5 mW) are inadequate for estimating the functional lifetime of the laser. In addition, the inaccessibility of the laser once it is in service and the requirement of a useful lifetime of up to 10 years make it desirable to apply an accelerating stress to the laser diodes. The functional life test was designed (1) to determine if temperature is a stress for accelerating the degradation of the lasers under pulsed operation and, if so, to determine the activation energy for pulsed degradation using the Arrhenius equation; and (2) to determine the failure statistics.

Failure is defined in terms of parameters that are important to system performance, such as average laser power, wavelength, pulse width, and single spatial mode quality. The lasers used in the test were commercially available AlGaAs, single-mode lasers, nominally emitting at 0.83 μm. The device had a double-heterostructure channel-substrate planar (CSP) laser structure and was grown by LPE. The laser facets were coated with SiO_2. The lasers were unencapsulated and were not subject to any screening other than by the manufacturers before they were used for the life test.

In the life test, the laser diodes were mounted on boards with individual pulse-forming networks (PFN). These PFN boards were attached to thermoelectrically cooled heat sinks. The temperature of the heat sink was continuously monitored and automatically adjusted to maintain the test temperatures of 40, 55, and 70°C. The lasers with their PFN boards were enclosed in individual chambers into which flowed dry, inert N_2. The individual assem-

4. SEMICONDUCTOR LASERS AS OPTICAL TRANSMITTERS

TABLE V

STATISTICS FOR LASERS TESTED FOR SPACE COMMUNICATION SYSTEM[a]

Nominal temperature (°C)	Number on test	Number failed	Maximum accumulated hours
40	15	5	9,500
55	15	7	9,400
70	20	13	14,700

[a] After Barry et al., 1985. © IEEE 1985.

blies were then mounted in such a way that the measuring equipment could be brought to the lasers and the measurements made without disturbing them. A summary of the lasers and test conditions is given in Table V.

The PFN boards were all tuned at 27°C to deliver a peak power of approximately 50 mW per facet with a pulse width of 25 nsec. The pulse format used pulse-position modulation (M-ary with $M = 32$) at a 1% duty cycle. A typical optical pulse is shown schematically in Fig. 19. The lasers were then aged at the higher temperatures of 40, 55, and 70°C. However, at these higher temperatures, some lasers did not necessarily reach threshold.

Measurements were taken weekly at 27°C but to avoid stress, the lasers

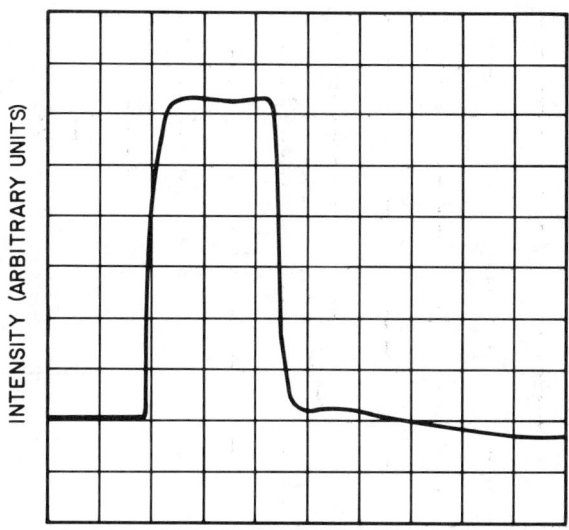

FIG. 19. Schematic drawing of the shape of the optical pulse emitted from a CSP laser driven by the PFN board. [After Barry et al. (1985). © IEEE 1985.]

were cooled from their aging temperatures to 27°C at a rate of less than 1°C min^{-1}. The measurements included recordings of (1) the pulse width and shape; (2) the longitudinal mode spectrum to determine the peak wavelength and amount of energy in the side modes; (3) the absolute power at the facet; (4) the near- and far-field patterns; (5) the polarization of the light; and (6) the $L-I$ curve. Functional failure was defined as occurring when (1) the output power decreased by 50%; (2) the peak wavelength changed by ± 1 nm or the width of the emission spectra increased to 2 nm; (3) the pulse width increased by 25%; or (4) the single spatial mode became multilobed.

The results of over 9000 hr of testing are summarized in Table VI. All lasers experienced a decrease in average power during the aging. Typically, the laser power would drop by 15% within the first 2000 hr of testing followed by a gradual decrease in power on the order of 0.5% khr^{-1}. Failure of the diode was characterized by a sudden, sharp decrease in lasing power.

In addition, during the aging time, most lasers showed an increase in the peak lasing wavelength by about 1 nm. Of these, a significant fraction (~30%) also experienced an increase in the spectral width to values over 10 nm. The occurrence of spectral broadening was strongly correlated with a degradation of the single spatial mode to a multimoded pattern. This last failure mode invariably occurred within the first 2000 hr of testing.

The failure times for all mechanisms were found to follow a log-normal distribution, as is shown in Fig. 20 for the 70°C data. Since failures due to changes in the far-field pattern all occur during the early part of testing and can presumably be screened from a population to be used in space, the authors have analyzed their data in two ways: first, by considering failures due to average power drift for all lasers, and, second, by analyzing all failure modes for only those lasers that retained a single spatial mode throughout

TABLE VI

FAILURE STATISTICS FOR CSP LASERS PULSED TO 50 mW FROM A LOG-NORMAL DISTRIBUTION[a]

	Temperature (°C)	Number in set	Median time to failure (khr)	σ
All lasers: power failure mechanism	40	15	16.8[b]	—
	55	15	9.88	1.5
	70	20	6.32	1.2
Single-mode lasers: all failure mechanisms	40	9	17.4[b]	—
	55	12	9.14	1.0
	70	15	5.65	1.0

[a] From Barry et al., 1985. © IEEE 1985.
[b] Projected time.

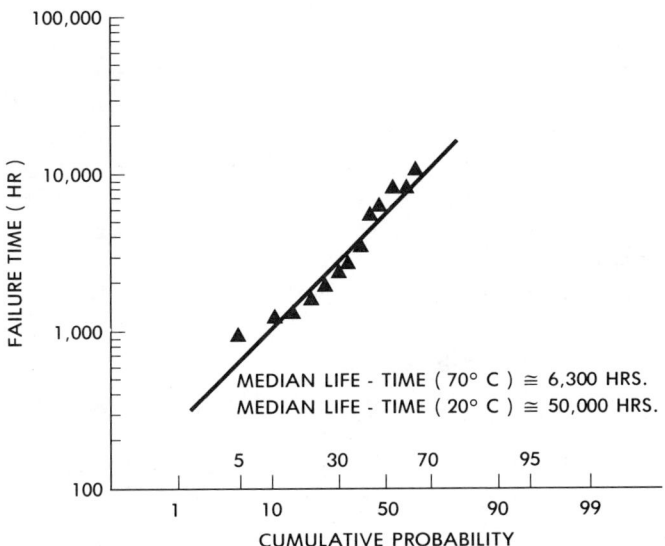

FIG. 20. The log-normal distribution of the failure times of CSP lasers aged pulsed at 70°C. The median life-time at 70°C was about 6300 hr and at 20°C about 0.5×10^5 hr. [After Barry et al. (1985). © IEEE 1985.]

the test. The median times to failure at the three temperatures were used in the Arrhenius equation to calculate an activation energy of 0.34 eV for the power failure of all laser diodes and 0.39 eV for the system failure due to the loss of single-spatial-mode characteristics of the laser diodes. By using the latter value, the extrapolated median time to failure at 20°C is found to be 50 khr. This *pulsed* time to failure is about an order of magnitude lower than the best (cw) lifetimes published. This result is not totally unexpected given the extreme conditions of this system test, particularly the peak power of 50 mW.

In summary, the results of the functional life test of single-mode AlGaAs lasers aged pulsed to 50 mW peak power have shown that the degradation processes leading to system failure are thermally activated and that there is an early system failure mode due to degradation of the spatial and longitudinal mode patterns. The authors are continuing the functional life test to find the dependence of functional lifetime on the peak optical power.

VI. Summary

Optical transmitters using semiconductor lasers as sources have been successfully introduced into commercial lightwave systems. These first-generation transmitters for multimode fiber applications have used for the most part, gain-guided AlGaAs devices with projected lifetimes greater than 10^5 hr. Initial system performance has been excellent (Jacobs, 1983). As a result, system requirements have tightened, and second-generation transmitters currently using index-guided InGaAsP lasers are being introduced into both multimode and monomode fiber systems. However, the long-term reliability of even the first-generation systems, and the advisibility of using cw laser-aging results for estimates of the laser's contribution to the total system lifetime are still unanswered questions.

We have reviewed our own work on the functional reliability of optical transmitters operated at 45 Mbits/sec using proton-bombarded lasers. Our general conclusions are as follows:

(1) cw laser-aging criteria are inadequate for estimating the system lifetime of optical transmitters.

(2) The functional lifetime of transmitters is very specific to the system and dependent on the circuit strategy adopted.

(3) The laser degradation mechanisms limiting system lifetime are different from those limiting the cw light-emitting capability.

(4) Extrapolation to room-temperature performance from high-temperature data, using acceleration parameters derived from the cw results, is not valid.

Points 1 and 4 have been confirmed by other workers (Barry *et al.*, 1985).

The results of some other workers on the pulsed aging of lasers prepared by other fabrication techniques are not in complete agreement with these conclusions, in particular, points 1 and 3 (Yoshida *et al.*, 1982; Robertson, 1983). Their results further emphasize the need for careful functional reliability testing of new laser and transmitter designs.

We wish to reiterate that second-generation laser transmitters must be vigorously evaluated by functional life testing. Demands on characteristics such as laser beam stability and mechanical package stability are orders of magnitude more severe in monomode fiber systems. The trend toward long-wavelength systems using InGaAsP introduces potential problems not found in AlGaAs devices, such as carrier-current leakage (Ikegami *et al.*, 1983). These degradation mechanisms have an unknown effect on the quality of the light emitted from aging lasers.

In conjunction with increasing the types of transmitters on functional life testing, the test sets themselves must become more sophisticated. Although we attempted in our studies to establish objective criteria for judging the system performance of lasers, we feel that this test was only a useful first step in evaluating the reliability of packaged lasers in an optical transmission system. A more stringent and valid method of evaluation is continuous bit-error-rate testing of pulse-code-modulated transmitters. This technique reveals any short-duration instabilities that would seriously affect system performance but which are not found by our present testing. In short, then, as the complexity of optical transmission systems increases, the need for functional reliability testing becomes more crucial.

Acknowledgments

A long-running and complicated experiment such as the one we have described in our laboratory is not possible without the contributions of many people. It is our pleasure, first and foremost, to acknowledge the initial and continued support of C. E. Barnes to the area of optical transmitter functional life-testing. The existence and scope of our testing program are due to him. We have also enjoyed the full support of our management, and wish to thank L. F. Moose and J. M. Goldey for their encouragement, and R. G. Smith and L. K. Anderson for their continuing interest and the time to write this report.

We also wish to acknowledge the past and continuing contributions of R. W. Folajtar to the areas of hardware and software design. B. Owen and G. M. Palmer also contributed to the design of the original test set.

We wish to thank P. D. Yeates, D. E. Durgin, C. A. Young, R. M. Neifert, M. J. Gabala and P. M. Moatz for their able technical support.

We must also thank W. T. Tsang for the invitation to write this report and for editorial support. Finally, we thank J. D. Barry and A. J. Einhorn for sending us the results of their testing before publication.

One of us (B.A.D.) dedicates this chapter to Adam Mantz.

References

Agrawal, G. P., Joyce, W. B., Dixon, R. W., and Lax, M. (1983). *Appl. Phys. Lett.* **43,** 11.
Barry, J. D., Einhorn, A. J., Dye, R. A., and Nussmeier, T. A. (1983). *Tech. Dig. IEEE/OSA Conf. Lasers Electroopt., 3rd, 1983,* p. 152.
Barry, J. D., Einhorn, A. J., Mecherle, G. S., Nelson, P., Dye, R., and Archambeault, W. J. (1985). *IEEE J. Quantum Electron.* **QE-21** (submitted for publication).
Biesterbos, J. W. M., and den Boef, J. A. (1981). *IEEE J. Quantum Electron* **QE-17,** 701.
Biesterbos, J. W. M., and Salemink, H. W. M. (1981). *Proc. Eur. Conf. Opt. Commun., 7th, 1981,* p. 10.2.
Botez, D., and Herskowitz, G. J. (1980). *Proc. IEEE* **68,** 689.
Chen, F. S. (1980). *Electron. Lett.* **16,** 7.
Chen, Y. C. (1981). *IEEE J. Quantum Electron* **QE-17,** 2262.
Dean, B. A., and Dixon, M. (1982). *Proc. Soc. Photo-Opt. Instrum. Eng.* **328,** 35.
DeWitt, R. G. (1983). *Tech. Dig. IEEE/OSA Top. Meet. Opt. Fiber Commun., 6th, 1983,* p. 79.
Dixon, R. W. (1980). *Bell Syst. Tech. J.* **59,** 669.
Dixon, M. (1985). *IEEE J. Quantum Electron.* (to be published).
Dixon, M., and Dean, B. A. (1981). *Tech. Dig. Int. Conf. Integr. Opt. Opt. Commun., 3rd, 1981,* p. 36.
Dixon, R. W., and Hartman, R. L. (1977). *J. Appl. Phys.* **48,** 3225.
Dixon, R. W., and Joyce, W. B. (1980). *Bell Syst. Tech. J.* **59,** 975.
Dixon, W. J., and Massey, F. J., Jr. (1969). *In* "Introduction to Statistical Analysis," 3rd ed., Chapter 12. McGraw-Hill, New York.

Dixon, R. W., Nash, F. R., Hartman, R. L., and Hepplewhite, R. T. (1976). *Appl. Phys. Lett.* **29**, 372.
Dyment, J. C., and Smith, G. (1981). *IEEE J. Quantum Electron* **QE-17**, 750.
Dyment, J. C., D'Asaro, L. A., North, J. C., Miller, B. I., and Ripper, J. E. (1972). *Proc. IEEE* **60**, 726.
Einhorn, A. J., Hall, D. B., and Nelson, P. R. (1982). *Proc. Soc. Photo-Opt. Instrum. Eng.* **328**, 28.
Ettenberg, M. (1981). *J. Appl. Phys.* **52**, 3845.
Ettenberg, M., and Kressel, H. (1980). *IEEE J. Quantum Electron.* **QE-16**, 186.
Ettenburg, M., Olsen, G. H., Ladany, I., Webb, P., DiGiuseppe, N. J., Zamerowski, T. J., and Appert, J. (1984). *Tech. Dig., IEEE Int. Semicond. Laser. Conf. 9th, 1984,* p. 170.
Fujiwara, K., Fujiwara, T., Hori, K., and Takusagawa, M. (1979). *Appl. Phys. Lett.* **34**, 668.
Garrett, I., and Midwinter, J. E. (1980). *In* "Optical Fiber Communications" (M. J. Howes and D. V. Morgan, eds.), Chapter 6. Wiley, New York.
Goodwin, A. R., Kirby, R. A., Davies, I. G. A., and Baulcomb, R. S. (1979). *Appl. Phys. Lett.* **34**, 647.
Gruber, J., Marten, P., Petschacher, R., and Russer, F. (1978). *IEEE Trans. Commun.* **COM-26**, 1088.
Hakki, B. W., Fraley, P. E., and Eltringham, T. F. (1985). *AT&T Bell Lab. Tech. J.* **64** (to be published).
Hartman, R. L., and Dixon, R. W. (1975). *Appl. Phys. Lett.* **26**, 239.
Hartman, R. L., and Dixon, R. W. (1980). *J. Appl. Phys.* **51**, 4014.
Hartman, R. L., Schumaker, N. E., and Dixon, R. W. (1977). *Appl. Phys. Lett.* **31**, 756.
Horiguchi, T., Miki, T., and Ishio, H. (1981). *Jpn. Telecommun. Rev.* **23**, 125.
Ikegami, T., Takahei, K., Fukuda, M., and Kuroiwa, K. (1983). *Electron. Lett.* **10**, 282.
Imai, H., Hori, K., Takusagawa, M., and Wakita, K. (1981). *J. Appl. Phys.* **52**, 3167.
Jacobs, I. (1983). *Tech. Dig. IEEE/OSA Top. Meet. Opt. Fiber Commun., 6th, 1983,* p. 80.
Joyce, W. B., Dixon, R. W., and Hartman, R. L. (1976). *Appl. Phys. Lett.* **28**, 684.
Kardontchik, J. E. (1982). *IEEE J. Quantum Electron.* **QE-18**, 1287.
Karr, M. A., Rich, T. C., and DiDomenico, M., Jr. (1978). *Appl. Opt.* **17**, 2215.
Karr, M. A., Chen, F. S., and Shumate, P. W. (1979). *Appl. Opt.* **18**, 1262.
Kirby, P. A., Goodwin, A. R., Thompson, G. H. B., and Selway, P. R. (1977). *IEEE J. Quantum Electron.* **QE-13**, 705.
Kobayashi, T., and Iwane, G. (1977). *Jpn. J. Appl. Phys.* **16**, 1403.
Kressel, H., Ettenberg, M., and Ladany, I. (1978). *Appl. Phys. Lett.* **32**, 305.
Kressel, H., Ettenberg, M., Wittke, J. P., and Ladany, I. (1982). *In* "Semiconductor Devices for Optical Communications," 2nd ed., p. 21. Springer-Verlag, New York.
Marcuse, D., and Nash, F. R. (1982). *IEEE J. Quantum Electron.* **QE-18**, 30.
Miskovic, E. (1982). *Proc. Soc. Photo-Opt. Instrum. Eng.* **328**, 51.
Mizuishi, K., Chinone, N., Sato, H., and Ito, R. (1979) *J. Appl. Phys.* **50**, 6668.
Mizuishi, K., Sawai, M., Hirao, M., Kajimura, T., and Nakamura, N. (1984). *Tech. Dig., IEEE Int. Semicond. Laser Conf., 9th, 1984,* p. 172.
Nakamura, M. (1983). *Tech. Dig., IEEE/OSA Top. Meet. Opt. Fiber Commun., 6th, 1983,* p. 96.
Nash, F. R., and Hartman, R. L. (1980). *IEEE J. Quantum Electron.* **QE-16**, 1022.
Nash, F. R., Wagner, W. R., and Brown, R. C. (1976). *J. Appl. Phys.* **47**, 3992.
Nash, F. R., Hartman, R. L., Denkin, N. M., and Dixon, R. W. (1979). *J. Appl. Phys.* **50**, 3122.
Paoli, T. L. (1977). *IEEE J. Quantum Electron.* **QE-13**, 351.
Pellegrini, G. (1983). *Tech. Dig. IEEE/OSA Top. Meet. Opt. Fiber Commun., 6th, 1983,* p. 78.

Plumb, R. G., Goodwin, A. R., and Baulcomb, R. S. (1979). *Solid State Electron. Devices* **3**, 206.
Ritchie, S., Godfrey, R. F., Wakefield, B., and Newman, D. H. (1978). *J. Appl. Phys.* **49**, 3127.
Robertson, M. J. (1983). *Electron. Lett.* **19**, 88.
Rosiewicz, A., Kirby, P. A., and Janssen, A. P. (1983). *Proc. Eur. Conf. Opt. Commun., 9th, 1983,* post-deadline papers.
Schorr, A. J., and Tsang, W. T. (1980). *IEEE J. Quantum Electron.* **QE-16**, 898.
Schwartz, M. I., Reenstra, W. A., Mullins, J. H., and Cook, J. S. (1978). *Bell Syst. Tech. J.* **54**, 1981.
Shumate, P. W., Jr., Chen, F. S., and Dorman, P. W. (1978). *Bell Syst. Tech. J.* **57**, 1823.
Smith, D. W. (1978). *Electron. Lett.* **14**, 775.
Smith, R. G., and Personick, S. D. (1982). *In* "Semiconductor Devices for Optical Fiber Communication Systems," 2nd ed., p. 45. Springer-Verlag, New York.
Thompson, G. H. B. (1980a). "Physics of Semiconductor Laser Devices," Sect. 2.7.3. Wiley, New York.
Thompson, G. H. B. (1980b). "Physics of Semiconductor Laser Devices," p. 390. Wiley, New York.
Thompson, G. H. B., Lovelace, D. F., and Turley, S. E. (1978). *Solid State Electron. Devices* **2**, 12.
Thompson, M. D., Dye, R. A., Archambeault, W. J., Einhorn, A. J., and Barry, J. D. (1981). *Natl. Telecommun. Rec.* **1**, B10.3.
Tomlinson, W. J. (1980). *Appl. Opt.* **19**, 1127.
Tsang, W. T., Fraley, P. E., and Holbrook, W. R. (1981). *Appl. Phys. Lett.* **38**, 7.
Tsang, W. T., Dixon M., and Dean, B. A. (1983). *IEEE J. Quantum Electron.* **QE-19**, 59.
Wolf, H. D., Metler, K., and Zachaver, K. H. (1981). *Jpn. J. Appl. Phys.* **20**, L693.
Yoshida, J.-I., Chino, K.-I., and Wakita, K. (1982). *IEEE J. Quantum Electron.* **QE-18**, 879.

CHAPTER 5

Light-Emitting-Diode Device Design

R. H. Saul

AT&T BELL LABORATORIES
MURRAY HILL, NEW JERSEY

Tien Pei Lee* and C. A. Burrus

AT&T BELL LABORATORIES
CRAWFORD HILL LABORATORY
HOLMDEL, NEW JERSEY

I.	INTRODUCTION	193
II.	LED STRUCTURES	194
III.	LED POWER AND MODULATION BANDWIDTH	198
	1. *Interfacial Recombination and Self-Absorption*	199
	2. *Carrier Lifetime Due to Radiative Recombination*	204
	3. *Modulation Bandwidth*	206
	4. *Carrier Leakage*	209
IV.	LED DEVICE CHARACTERISTICS	211
	5. *Current–Voltage Characteristics*	211
	6. *Light Output versus Current and Temperature*	213
	7. *Spectral Characteristics*	216
	8. *Output Power and Modulation Bandwidth*	220
V.	COUPLING OF LEDs TO OPTICAL FIBERS	221
	9. *Direct Coupling to Fibers*	222
	10. *Coupling with Lenses*	225
	11. *Sensitivity of Coupling Efficiency to Mechanical Alignment*	229
	12. *Effect of Coupling on Speed*	230
VI.	DUAL-WAVELENGTH LEDs	232
VII.	LEDs MADE BY MBE AND VPE	233
VIII.	CONCLUSIONS	234
	REFERENCES	235

I. Introduction

Optical fiber systems using light-emitting diode (LED) sources can provide reliable, inexpensive alternatives to laser-based systems. LEDs utilize

---

* Present address: Bell Communications Research, Murray Hill, New Jersey.

relatively simple driving circuits without the need for feedback to control output power, and they are capable of operating over a wide range of temperatures with projected device lifetimes one to two orders of magnitude longer than those of laser diodes made of the same material. First-generation LED systems used GaAlAs sources emitting in the 0.85–0.9-μm-wavelength region, but because of the relatively high attenuation and chromatic dispersion of silica-based fibers in this wavelength region, transmission distances are limited to ≤ 4 km, and data rates are moderate (≤ 50 Mbits/sec). Thus, such systems are primarily useful for inexpensive data links. Second-generation systems utilize long-wavelength InGaAsP LEDs emitting near 1.3-μm wavelength, where silica fibers exhibit low attenuation and minimum dispersion. Repeater spacings of tens of kilometers and data rates of a few hundred megabits per second are feasible (Gloge et al., 1980).

For LEDs to be used as light sources in optical fiber systems, the ultimate performance requirements are (1) high output power, (2) large modulation bandwidth, and (3) efficient coupling to optical fibers — requirements that are quite different from those of the more familiar indicator-lamp LEDs. The chapter opens with a brief description of LED structures suitable for communications use at both 0.9- and 1.3-μm wavelengths and then discusses in some detail the formulation of design principles to achieve the first two of these requirements. Next, the characteristics and performance of practical devices, as well as practical coupling to fibers, are described. Finally, the more recent extension of LED technologies to the production of multiple wavelengths from single devices is discussed. Several earlier review papers on small-area high-radiance LEDs (Lee, 1980, 1982; Saul, 1983) may be used as supplementary reading material.

II. LED Structures

Two basic structures for LEDs suitable for use as sources in optical fiber systems have evolved. The first to be designed specifically for coupling to fibers is a small-area surface emitter, shown in Fig. 1a configured as a GaAlAs/GaAs double-heterostructure (DH) (Burrus and Miller, 1971), which is an adaptation of an earlier GaAs homostructure etched-well construction (Burrus and Dawson, 1970). Briefly, it is a large $p-n$ junction with a small emitting area, typically 20–50 μm in diameter, made by oxide masking of the p-side contact. The GaAs substrate material above the emitting area is removed by chemical etching, forming a well, to reduce absorption loss in the GaAs substrate and, simultaneously, to improve the coupling of the emitted light to fibers. Alternatively, planar devices have been fabricated by totally removing the substrate (Berkstresser et al., 1980; Abe et al., 1977), and Schottky barriers have been used in place of an oxide-defined p

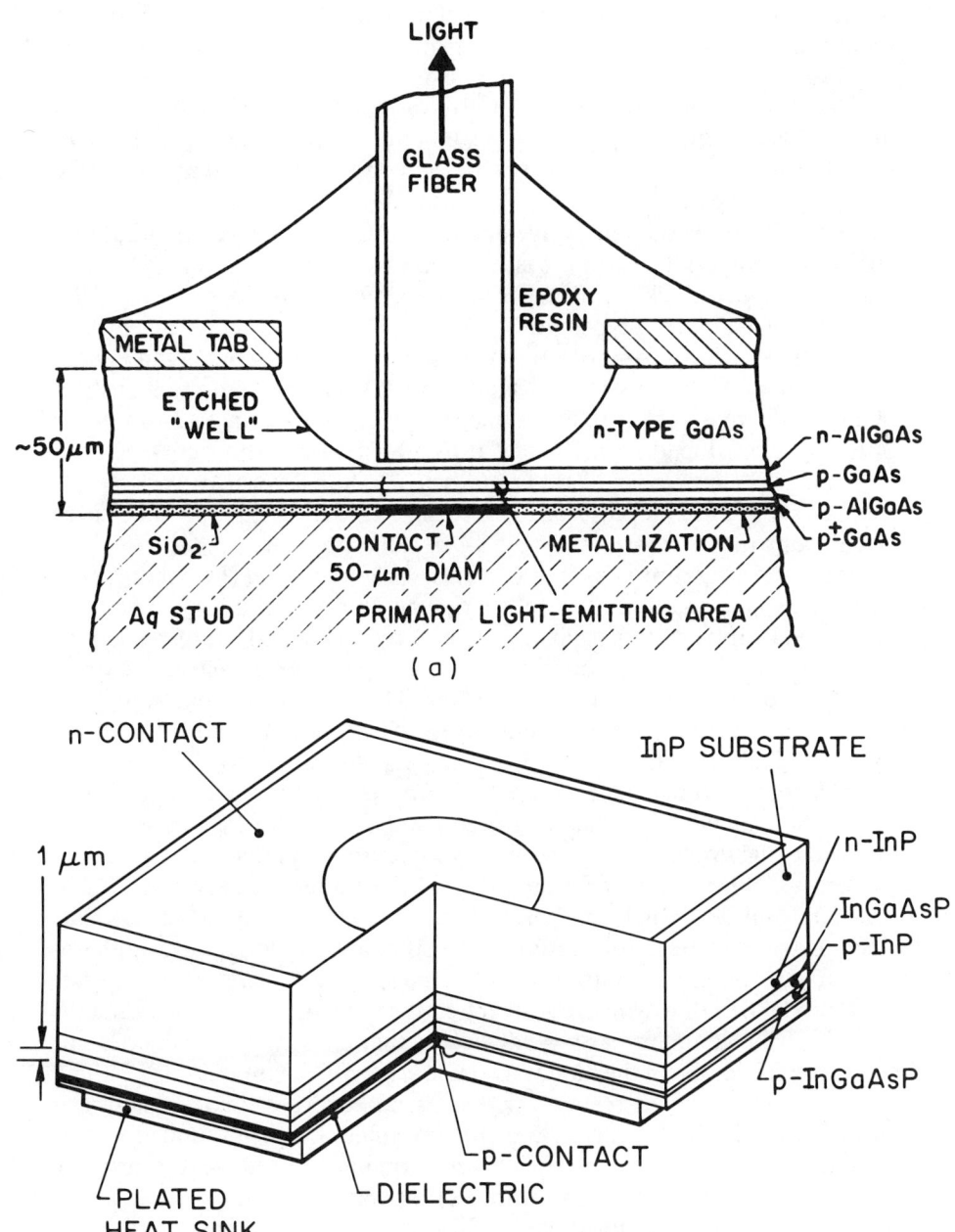

FIG. 1. Surface-emitting LED structures: (a) GaAlAs/GaAs; (b) InGaAsP/InP.

contact (Chin et al., 1981). The DH wafer, usually grown by liquid-phase epitaxy (LPE), consists of four layers (from the substrate): an n-type (10^{18} cm^{-3}) Ga$_{1-x}$Al$_x$As confining layer; a p-type ($10^{17}-10^{19}$ cm^{-3}) GaAs or Ga$_{1-y}$Al$_y$As ($y < x$) active layer; another p-type (10^{19} cm^{-3}) Ga$_{1-x}$Al$_x$As confining layer; and a top p-type ($> 10^{19}$ cm^{-3}) GaAs contacting layer to assure low p-contact resistance; Sn or Te is used for the n-type dopant and Ge as the p-type dopant.

The structure for the long-wavelength (1.3-μm) LED is very similar, except that the material system is based on InGaAsP/InP, as shown in Fig. 1b. The first 2-5-μm-thick n-InP buffer layer is generally Sn doped ($\sim 2 \times 10^{18}$ cm^{-3}), followed by an n-type ($\sim 5 \times 10^{17}$ cm^{-3}) InGaAsP active layer (0.5-1.5 μm thick) whose composition is chosen to provide a band-gap energy corresponding to an emission wavelength of 1.3 μm and, simultaneously, a crystal-lattice match to InP. Subsequent layers consist of a 1-2-μm p-InP confining layer doped with Zn or Cd ($0.5 \times 10^{18}-5 \times 10^{18}$ cm^{-3}) and a 0.5-μm thick "cap" layer of InGaAsP heavily doped with Zn ($\sim 1 \times 10^{19}$ cm^{-3}) to reduce contact resistance. Etching of a well may be omitted since the InP substrate is transparent to 1.3-μm radiation.

The second structure, shown in Fig. 2a, is an edge emitter (Ettenberg et al., 1976), a configuration similar to a stripe-geometry diode laser. Lasing is suppressed by introducing cavity losses, e.g., by roughened or antireflection-coated facets. Self-absorption along the light path in the active region is minimized by using a very thin active layer (0.05-0.1 μm) so that the optical field is spread out into the confining layers, which have less absorption loss (due to the higher band-gap energy of Ga$_{0.6}$Al$_{0.4}$As relative to Ga$_{0.9}$Al$_{0.1}$As). This structure also provides a waveguide, in the direction perpendicular to the junction plane, which serves to reduce the beam divergence and thus provides a higher efficiency in coupling power to fibers than does a surface emitter. A similar structure for the InGaAsP/InP system is shown in Fig. 2b.

A third emitter, which in operation does not reach the lasing state but which provides some amplification of the spontaneous emission (and therefore a relatively narrow output beam) is the superluminescent (or superradiant) diode (SLD). A typical construction is shown in Fig. 3a. It is similar to that of a stripe-geometry laser diode except that the stripe contact does not extend the full length of the diode chip (Lee et al., 1973; Ure et al., 1982). The resulting unpumped rear region serves as an absorber for the backward waves, and only the forward waves are amplified. Other methods for suppressing Fabrey-Perot resonance utilize an oblique output facet (Kurbatov et al., 1971) or, more recently, antireflection coatings on the cleaved facets (Wang et al., 1982; Kaminow et al., 1983). An example is shown in Fig. 3b. At low currents, the device operates as an edge-emitting LED; at high currents, the output power increases superlinearly, and the spectral width narrows considerably as a result of the onset of the optical gain. To achieve the

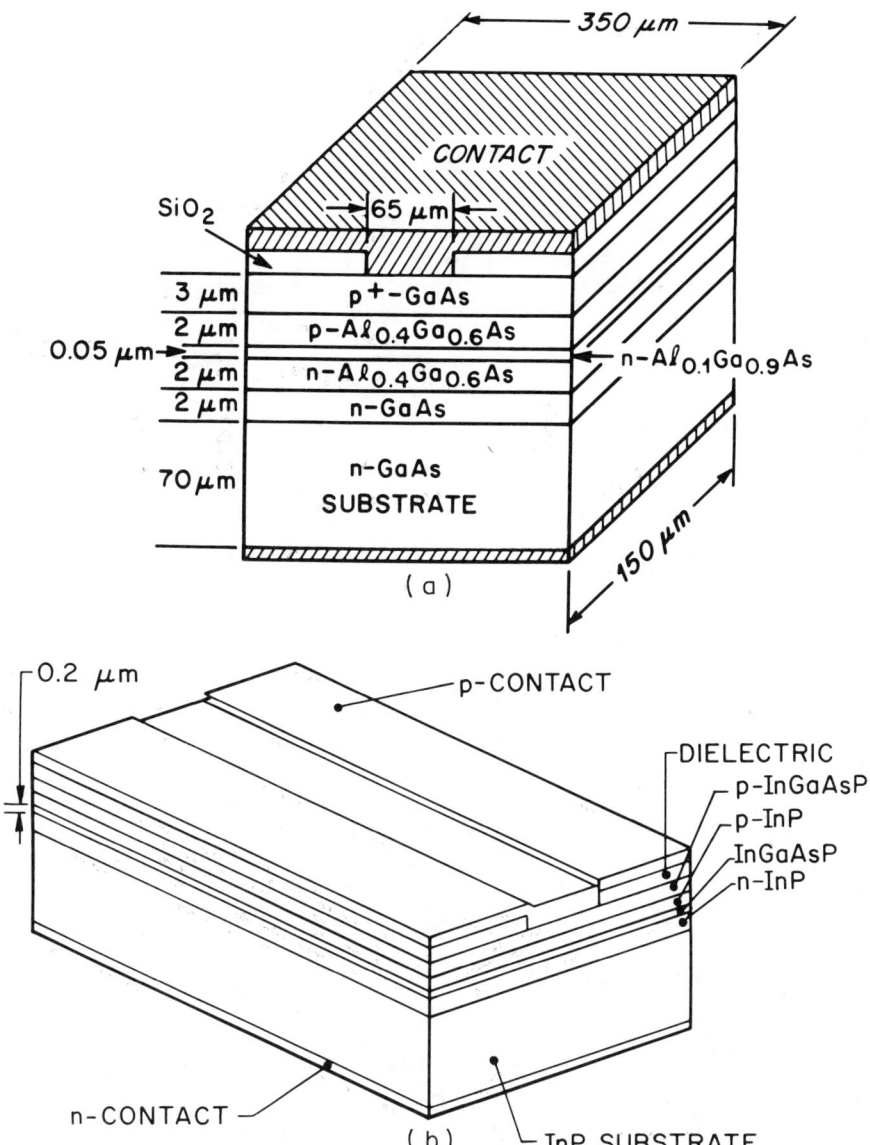

FIG. 2. (a) GaAlAs stripe-geometry edge emitter; (b) InGaAs stripe-geometry edge emitter.

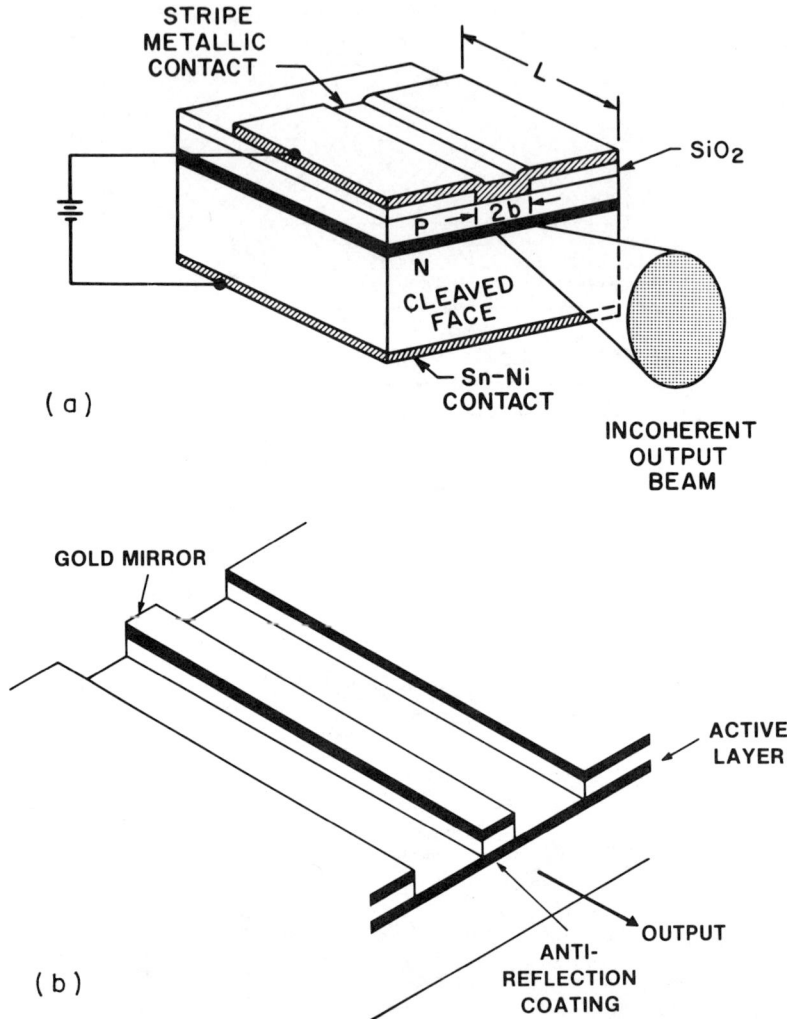

FIG. 3. (a) SLD with absorbing region; (b) InGaAsP ridge-waveguide SLD with front-facet AR coating.

same output, SLDs operate at higher current densities than do lasers; thus recent improvements in laser diodes, resulting in lowered thresholds and higher output per watt input, have made the SLD more practical.

III. LED Power and Modulation Bandwidth

Factors affecting the output power (or quantum efficiency) and modulation bandwidth of an LED include nonradiative interfacial recombination, radiative bimolecular recombination, self-absorption, and carrier confinement. These factors are in turn related to device parameters, such as doping concentrations, diffusion length of minority carriers, absorption coefficient, active-layer thickness, differential band-gap energies between the active layers and the confining layers (in DHs), and injection-current density. These parameters, however, are not independent but interrelated. The relationships among them give rise to a power × bandwidth product for LEDs, and maximization of this product dictates the choice of device parameters. In Part III, we illustrate the interplay among these parameters with a simple model, and we note those obtained in practice for both GaAlAs LEDs emitting at 0.9-μm wavelength and InGaAsP LEDs emitting at 1.3-μm wavelength. Deviations from this simple model are also discussed.

1. INTERFACIAL RECOMBINATION AND SELF-ABSORPTION

We now consider surface-emitting LEDs, using GaAlAs devices for illustration. Figure 4a is a schematic diagram of various regions surrounding the active layer. We assume that the active layer is p type and that light is extracted from the n-type substrate side, as is usual in actual devices. Electrons are injected by the forward-biased p–n heterojunction. The electron-density distribution (determined later) is depicted in Fig. 4c. The injected electrons recombine with holes in the active region via both radiative and nonradiative processes. The total carrier lifetime τ is

$$\tau^{-1} = \tau_r^{-1} + \tau_{nr}^{-1}, \tag{1}$$

where τ_r is the radiative recombination time and τ_{nr} the nonradiative recombination time of carriers in the bulk region. The internal quantum efficiency is defined as

$$\eta_{int} = \tau/\tau_r. \tag{2}$$

However, for thin active layers, carriers also can recombine nonradiatively at the heterointerfaces, thereby shortening the total carrier lifetime and reducing the net quantum efficiency. We now examine this carrier lifetime due to interfacial recombination and its effect on net efficiency.

Assuming that the electron diffusion current dominates the junction current, the steady-state continuity equation is

$$D\, d^2n/dx^2 - n/\tau = 0, \tag{3}$$

where $n = n(x)$ is the electron-density distribution function, D the diffusion constant (cm^2 sec^{-1}), and τ the carrier recombination time in seconds in the bulk region (away from the surface). The boundary conditions, assuming equal surface recombination velocity s (cm sec^{-1}) at both heterointerfaces,

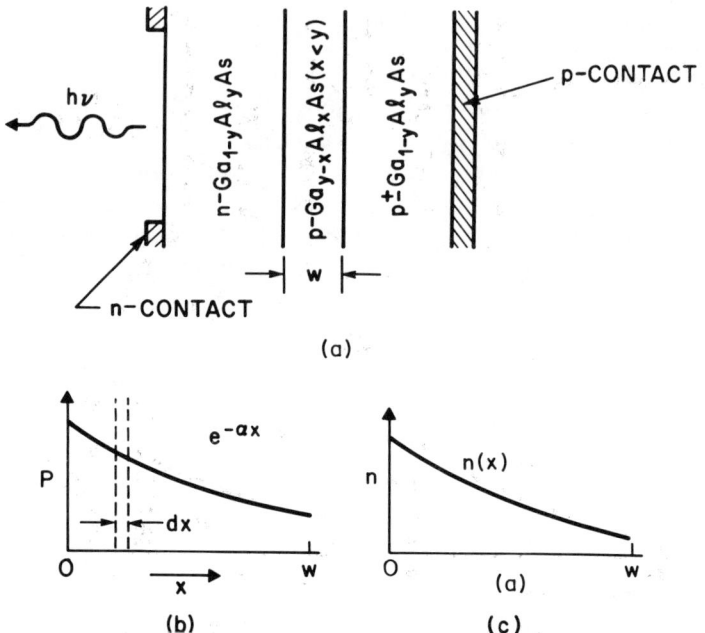

FIG. 4. (a) A one-dimensional schematic diagram of the light-generation region in a surface-emitting LED. (b) Emitted power as a function of x. (c) Electron density distribution.

are

$$-dn/dx|_{x=0} = J/eD - sn(0)/D, \quad (4)$$

$$dn/dx|_{x=w} = -sn(w)/D, \quad (5)$$

where J is the injection-current density (A cm^{-2}) at the p–n heterojunction and e the electronic charge (1.6×10^{-19} C). The function $n(x)$ can be obtained readily by Eqs. (3)–(5), as shown later.

$$n(x) = \frac{JL}{eD} \left\{ \frac{\cosh[(w-x)/L] + (sL/D)\sinh[(w-x)/L]}{[(sL/D)^2 + 1]\sinh(w/L) + (2sL/D)\cosh(w/L)} \right\}, \quad (6)$$

where $L = \sqrt{D\tau}$ is the electron diffusion length in centimeters. The average electron density over the active region of width w is

$$\bar{n} = \frac{1}{w} \int_0^w n(x)\, dx = \frac{J\tau_{\text{eff}}}{ew}, \quad (7)$$

where the effective carrier lifetime τ_{eff} is given by

$$\tau_{\text{eff}} = \tau \left\{ \frac{\sinh(w/L) + (sL/D)[\cosh(w/L) - 1]}{[(sL/D)^2 + 1]\sinh(w/L) + (2sL/D)\cosh(w/L)} \right\}. \quad (8)$$

The bracketed term is less than unity if $s > 0$, which implies a reduction in the carrier lifetime due to interfacial recombination, as we wished to show. When the active-layer thickness is smaller than the diffusion length, and the surface recombination velocities are small, i.e., $(w/L) < 1$ and $(sL/D)^2 \ll 1$, Eq. (8) reduces to a simpler form, namely,

$$\tau_{\text{eff}}^{-1} = \tau^{-1} + 2s/w \quad (9)$$

Now, we consider the light output power. Because the active region absorbs some of its own emitted light, the power output cannot be increased indefinitely by increasing the thickness of the active region. In addition, the photon intensity in regions away from the $p-n$ junction decreases because of the reduced electron density, as given by Eq. (6). To calculate the LED power output, we approximate the photon intensity of spontaneous emission to be a Gaussian function of wavelength given by (Marcuse, 1977).

$$W(\lambda) = W_0 \exp[-4(\lambda - \lambda_0)^2/(\Delta\lambda)^2], \quad (10)$$

where W is the number of photons generated per unit volume per unit wavelength range; $\Delta\lambda$ is the intrinsic emission linewidth, which is not modified by the self-absorption in the active region; and the emission is near the band edge, where the absorption is asymmetrical. Typical emission and absorption spectra are shown in Fig. 5.

The photon generation rate per unit volume W_0 in Eq. (10) is a function of distance x from the $p-n$ junction,

$$W_0(x) = n(x)/\tau_r. \quad (11)$$

Thus the power emitted into air by a surface-emitting LED is

$$P = \eta_c A(hc) \int_0^\infty \frac{d\lambda}{\lambda} \int_0^w \frac{n(x)e^{-\alpha(\lambda)x}}{\tau_r} dx \quad (12)$$

where A is the light-emitting area, h is Planck's constant, c the velocity of light in vacuum, α the absorption coefficient, and η_c is the light escape probability. For a flat emitting surface, η_c can be written as

$$\eta_c = (1 - R)\sin^2\theta_c, \quad (13)$$

where

$$R = [(n_r - 1)/(n_r + 1)], \quad (14)$$

$$\theta_c = \sin^{-1}(1/n_r), \quad (15)$$

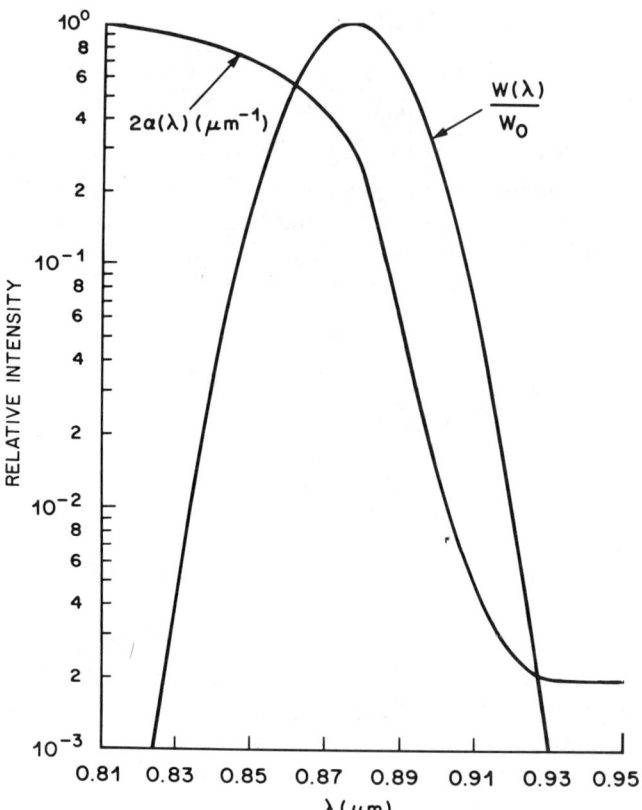

FIG. 5. Typical emission and absorption curves of the active material of a GaAlAs LED. [From Marcuse (1977). © 1977 IEEE.]

and R is the reflectivity of the semiconductor substrate, θ_c the critical angle, and n_r the refractive index of the semiconductor. Reflection at the heterointerface is neglected. For GaAs, $n_r = 3.6$ and Eq. (13) yields $\eta_c = 0.0525$. Thus, only 5% of the generated light can be extracted.

The power P_0 at the emission peak wavelength λ_0 can be obtained from Eq. (12) by letting $\lambda = \lambda_0$ and $\alpha(\lambda_0) = \alpha_0$; it is given by

$$P_0 = \frac{\eta_c A_c(hc/\lambda_0)}{\tau_r} \int_0^w n(x)e^{-\alpha_0 x}\, dx. \tag{16}$$

Using Eq. (6) for $n(x)$ in Eq. (16), we obtain

$$P_0 = \eta_c \eta_s (\tau/\tau_r)(hc/e\lambda_0)I, \tag{17}$$

where $I = AJ$ is the injection current, assuming no current spreading, and η_s

is a factor which reduces the quantum efficiency due to surface recombination and self-absorption. The parameter η_s is given by (Lee and Dentai, 1978)

$$\eta_s = \{2[(S^2 + 1)\sinh(w/L) + 2S\cosh(w/L)]\}^{-1}$$
$$\times \{[(1 + S)/(1 + \alpha L)](1 - e^{w(1+\alpha L)/L})e^{w/L}$$
$$- [(1 - S)/(1 - \alpha L)](1 - e^{w(1-\alpha L)/L})e^{-w/L}\}, \qquad (18)$$

where $S = sL/D$.

Figure 6 is a plot of Eq. (18) for two values of s to demonstrate the importance of low interfacial recombination in achieving high efficiency. For heterostructures, the interfacial recombination velocity can be related

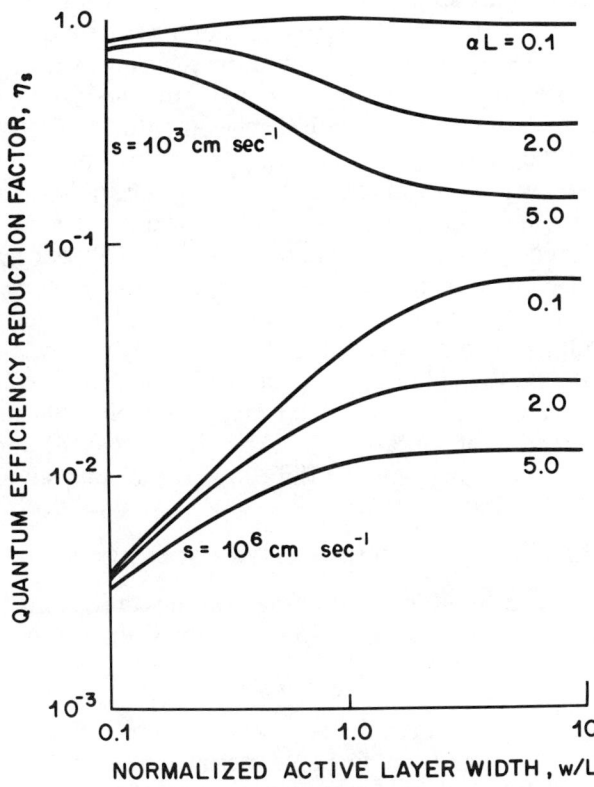

FIG. 6. Reduction factor of the quantum efficiency as a function of the ratio of active-layer width w to the diffusion length L; s is the interfacial recombination velocity and α the absorption coefficient.

approximately to the strain due to lattice mismatch by (Kressel et al., 1980):

$$s \simeq (2 \times 10^7)(\Delta a/a), \qquad (19)$$

where a is the lattice constant and Δa is the mismatch in lattice constants. Thus, in order to have internal quantum efficiency higher than 50% in the absence of bulk nonradiative centers, $(\Delta a/a) < 10^{-3}$ is required. For GaAlAs/GaAs DH structures, an interfacial recombination velocity of 2×10^3 cm sec^{-1} can be achieved routinely (Lee and Dentai, 1978) and a value as low as 500 cm sec^{-1} has been reported (Nelson and Sobers, 1978). A higher value of 10^4 cm sec^{-1} was measured for InGaAsP/InP (Sakai et al., 1980).

In actual devices, the output power can be maximized by varying the thickness and doping concentration of the active layer. The optimum values depend on the interfacial recombination velocity and the diffusion length of minority carriers. For GaAlAs/GaAs diodes at 0.9-μm wavelength, maximum output power has been obtained experimentally with w between 2 and 2.5 μm and with 2×10^{17} cm^{-3} p-type (Ge) doping (Lee and Dentai, 1978). For InGaAsP/InP 1.3-μm LEDs, maximum power was achieved with w between 1 and 1.5 μm and an undoped ($\sim 10^{17}$ cm^{-3}) n-type layer (Wada et al., 1981). In the latter case, junction displacement in the active layer was found to be an additional factor affecting the output power. The output power decreases as the $p-n$ junction moves into the InGaAsP active layer from the InP/InGaAsP heterointerface and as the doping in the active layer changes from predominantly n type to predominantly p type (Temkin et al., 1982). This suggests that the radiative recombination efficiency of p-type InGaAsP must be lower than that of n-type material. The junction displacement can result from contamination of the active-layer growth solution by Zn vapor from the neighboring p confining-layer growth solution. This contamination can be eliminated by covering the melt wells and adjusting the flow and direction of the carrier gas or by use of Cd as the dopant (Wada et al., 1981), since Cd has a smaller diffusion coefficient than Zn (Kundukhov et al., 1967).

2. CARRIER LIFETIME DUE TO RADIATIVE RECOMBINATION

The band-to-band spontaneous recombination rate R_{sp} under conditions not requiring momentum conservation takes the simple form

$$R_{sp} = Bnp, \qquad (20)$$

where B (cm^3 sec^{-1}) is the radiative recombination probability, a characteristic of the band structure depending on material. Under the nonequilibrium condition in which excess electrons Δn are generated (for example, by the absorption of incident photons) in the conduction band and an equal number of holes $\Delta p = \Delta n$ are created in the valence band, the excess carriers recombine spontaneously at a rate given by

$$R_{sp} = B(n_0 + \Delta n)(p_0 + \Delta p), \tag{21}$$

where n_0 and p_0 are electron and hole concentrations at thermal equilibrium, and $n_0 p_0 = n_i^2$, n_i being the intrinsic carrier density of the material. For n-type material, $n_0 \simeq N_D$, the donor concentration, and for p-type material $p_0 \simeq N_A$, the acceptor concentration. Using $\Delta p = \Delta n$, Eq. (21) becomes

$$R_{sp} = B n_0 p_0 + B \Delta n (n_0 + p_0 + \Delta n). \tag{22}$$

The first term, $B n_0 p_0$, represents the thermal equilibrium radiative recombination rate, which is usually negligible. Thus, the radiative recombination rate can be written as

$$R_{sp} = B \Delta n (n_0 + p_0 + \Delta n). \tag{23}$$

Let us define a carrier lifetime due to radiative recombination by

$$\tau_r = \Delta n / R_{sp}. \tag{24}$$

Then, with the help of Eq. (23), we obtain

$$\tau_r = [B(n_0 + p_0 + \Delta n)]^{-1}. \tag{25}$$

For p-type material, $p_0 \simeq N_A$, $n_0 \ll p_0$, and if the excess carrier density is less than the doping (acceptor) concentration, i.e., $\Delta n \ll N_A$ (the case at low injection levels), the carrier lifetime is inversely proportional to the doping concentration. That is,

$$\tau_r = (B N_A)^{-1}. \tag{26}$$

Similarly, for n-type material N_A in Eq. (26) is replaced by N_D, the donor concentration. Thus, the value of B can be determined by the measurement of τ_r versus N_A. For GaAs, B is $0.64-1.3 \times 10^{-10}$ cm^3 sec^{-1} (Casey and Stern, 1976; Acket et al., 1974). The value of B ranges from 0.35×10^{-10} cm^3 sec^{-1} (Wada et al., 1979) to 0.98×10^{-10} cm^3 sec^{-1} (Wada et al., 1981) for InGaAsP.

At high injection levels such that Δn is much larger than the background doping concentration, i.e., $\Delta n > N_A$ (or $\Delta n > N_D$), Eq. (25) becomes

$$\tau_r = (B \Delta n)^{-1}. \tag{27}$$

Since Δn must equal Δp to satisfy the charge neutrality condition, an equally large number of holes must be injected into the active region, and this can be achieved in a DH only by means of hole injection at the P-p junction (P represents wider band-gap material). This condition is called bimolecular recombination. The injected-carrier density in a 50-μm-diameter LED usually ranges from 1×10^{18} cm^{-3} to 5×10^{18} cm^{-3} at ordinary operating conditions, and bimolecular recombination is dominant only for active layers lightly doped (below $\sim 1 \times 10^{18}$ cm^{-3}). However, shortening the radiative lifetime through heavy doping, according to Eq. (26), does not necessarily improve the internal quantum efficiency because heavy doping may also introduce nonradiative recombination through an Auger process. In fact, heavy doping is used to increase the speed of an LED at the expense of output power, as we shall describe in detail later.

3. MODULATION BANDWIDTH

a. Modulation Bandwidth Limited by Carrier Lifetime

Intensity modulation of LED light output can be accomplished by direct modulation of the injection current. Parasitic elements such as space-charge capacitance cause a delay of carrier injection into the junction and consequently a delay in the light output (Lee, 1975). The ultimate limit on the modulation rate is determined by the carrier lifetime. If the current is modulated at an angular frequency ω, the intensity of the light output will vary with ω as (Joyce *et al.*, 1974; Namizaki *et al.*, 1974; Liu and Smith, 1975)

$$|I(\omega)| = I(0)/[1 + (\omega\tau)^2]^{1/2} \tag{28}$$

where $I(0)$ is the intensity (optical power) at zero modulation frequency and τ the total carrier lifetime. The modulation bandwidth Δf (or cutoff frequency f_c) is defined as the frequency at which the detected electrical power $P_{\text{elec}} \sim I^2(\omega)$ is one-half that at zero modulation frequency; that is,

$$\Delta f = \Delta\omega/2\pi = (2\pi\tau)^{-1}, \tag{29}$$

where the carrier lifetime τ, including interfacial recombination and nonradiative recombination in the bulk region, is

$$\tau^{-1} = \tau_r^{-1} + \tau_{\text{nr}}^{-1} + 2s/w. \tag{30}$$

In the absence of bulk nonradiative recombination and interfacial recombination, we have

$$\Delta f = (2\pi\tau_r)^{-1}. \tag{31}$$

The injected-carrier density is related to the carrier lifetime by

$$\Delta n = J\tau_r/ew. \tag{32}$$

Substituting Δn in Eq. (27) by Eq. (32) and solving for τ_r, we obtain

$$\tau_r = (ew/BJ)^{1/2}, \tag{33}$$

$$\Delta f = (2\pi)^{-1}(BJ/ew)^{1/2}. \tag{34}$$

Thus, the bandwidth of a DH LED having a lightly doped active layer and operated in the bimolecular recombination regime increases with $(J/w)^{1/2}$. Figure 7a and b demonstrate this relationship observed in both GaAlAs LEDs (Lee and Dentai, 1978) and InGaAsP LEDs (Wada et al., 1981), respectively.

The measured-bandwidth versus doping concentration is shown in Fig. 8 for GaAlAs LEDs (Lee and Dentai, 1978). The bandwidth increases slowly with doping concentration N_A until $N_A \simeq 10^{18}$ cm^{-3}, above which the bandwidth increases more rapidly with N_A. This occurs because bimolecular recombination is predominant at low N_A. For $N_A > 10^{19}$ cm^{-3}, the bandwidth increases, and the carrier lifetime decreases linearly with increasing N_A in agreement with Eq. (26). Modulation bandwidths greater than 1 GHz have been achieved with $N_A = 1.5 \times 10^{19}$ cm^{-3} at the expense of output power (Heinen et al., 1976).

The active layer in InGaAsP/InP LEDs is usually undoped (with residual n-type concentration varying from $\sim 10^{17}$ cm^{-3} to $\sim 5 \times 10^{17}$ cm^{-3}). The nominal bandwidth of such devices ranges from 50 to 100 MHz (Dentai et al., 1977; Umebu et al., 1978; Goodfellow et al., 1979; Wada et al., 1981; Temkin et al., 1983). With Zn doping at $2-4 \times 10^{18}$ cm^{-3} in the active layer, the bandwidth increases to 170 MHz (Gloge et al., 1980; Goodfellow et al., 1979). Bandwidths as high as 1.2 GHz have been obtained for diodes with a Mg-doped active layer (Grothe et al., 1979).

b. Effect of Parasitics on Risetime

Under pulse-code modulation (PCM), light output in response to the drive pulse is determined by the RC time constants of the LED. The resistance R is the series resistance of the diode. The capacitance C is composed of a space–charge capacitance associated with the $p-n$ junction area over the entire diode chip and a diffusion capacitance, which is related to the carrier lifetime in the small light-emitting area (Fig. 1). As the diode is turned on, the initial injection level is low, and the current spreads in the p layers so that the full area of the diode junction contributes to the device capacitance, which for a typical diode chip amounts to 150–250 pF. As the current is increased, the high spreading resistance of the p layer "crowds" the current so that the portion passing through the emitting region (above the p contact) becomes a larger fraction of the total current. The space–charge capacitance associated with the emitting region C_e is charged rapidly relative to the "perimeter" capacitance of the surrounding $p-n$ junction, C_p. Thus, when a step-current pulse is applied, the resulting light pulse is delayed slightly due to C_e. The delay time decreases from $\sim 3-4$ nsec at 30-mA drive current to 1 nsec at 100-mA drive current (Hino and Iwamoto, 1979). The initial risetime (10–90%) is primarily determined by the carrier-recombination lifetime, as pre-

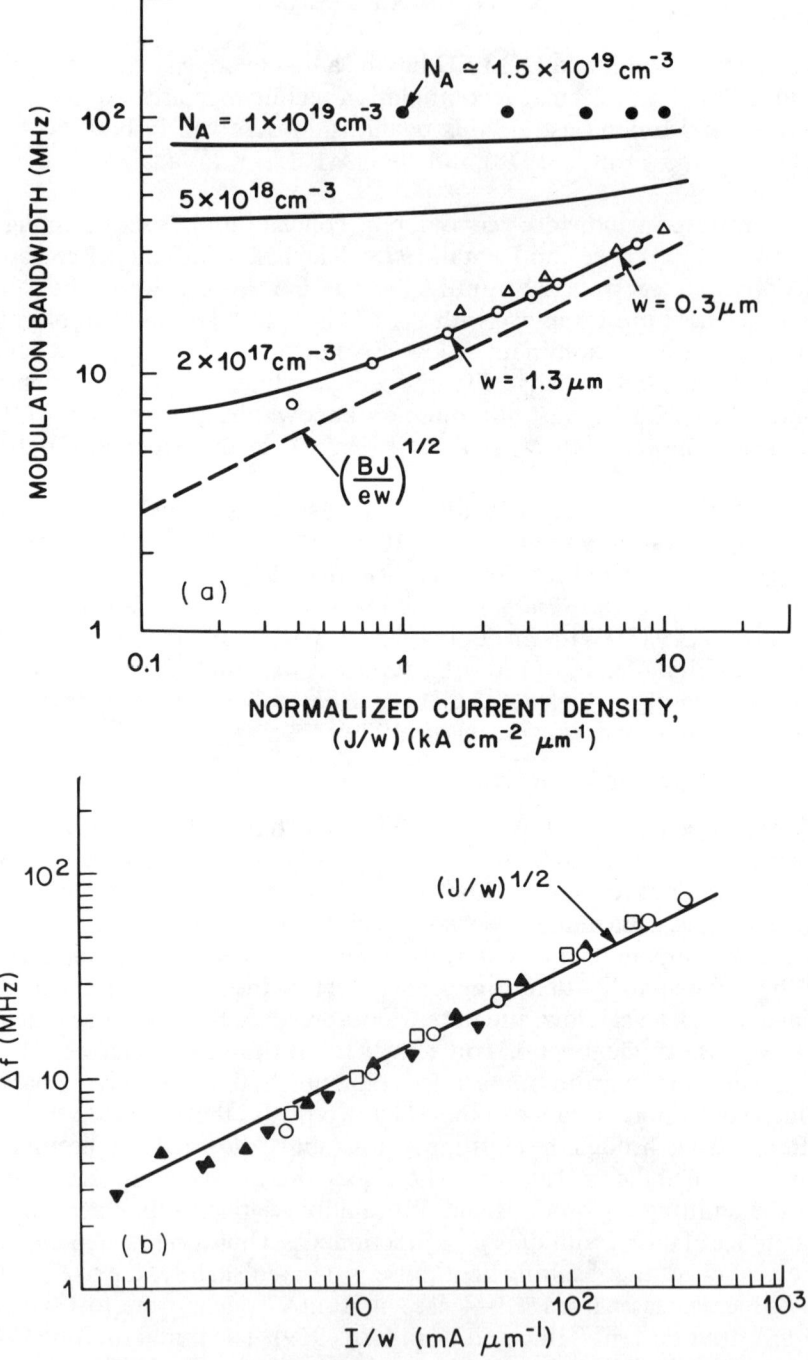

FIG. 7. The modulation bandwidth as a function of (J/w) for (a) GaAlAs [from Lee and Dentai (1978), © 1978 IEEE] and (b) InGaAsP LEDs [$w = 2.7(\blacktriangledown)$, $1.7(\blacktriangle)$, $1.0(\square)$, $0.42(\bigcirc)$] [from Wada et al. (1981), © 1981 IEEE].

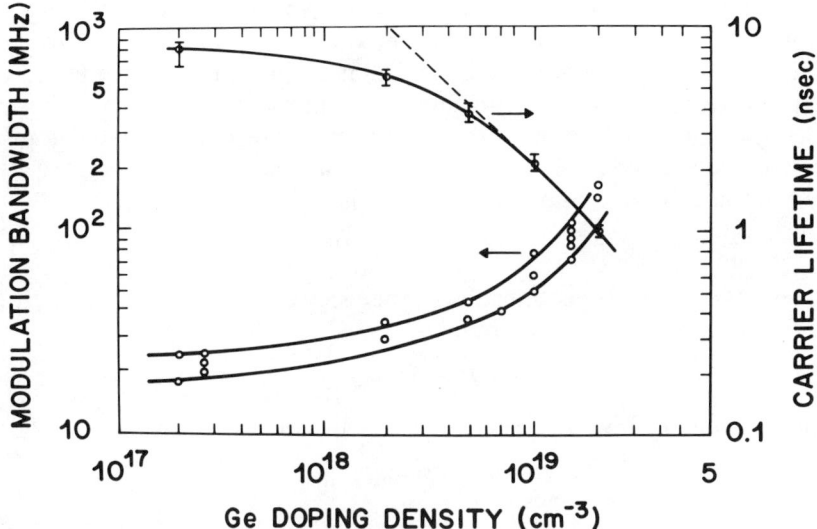

FIG. 8. Modulation bandwidth and carrier lifetime at 100-mA drive current as a function of the doping concentration in the active region.

viously discussed. However, because of the large RC time constant of C_p, the steady state may be reached only slowly. Various improvements in the slow portion of the risetime can be achieved either by applying a small dc prebias (Hino and Iwamoto, 1979) or by rapid injection of a current spike (Lee, 1975; Zucker, 1978). This problem is unique for the oxide-confined and Schottky-barrier-confined structure for which the electrical and optical junction areas are different. Although devices of this type have been particularly easy to fabricate, other structures with equal electrical and optical areas can be made by proton bombardment (Heinen et al., 1976), by etching mesas (Harth et al., 1976), and by use of buried heterostructures (Heinen, 1982a) to eliminate completely the perimeter capacitance. The falltime can be shortened by using a reverse-bias pulse to accelerate the removal of carriers (Dawson, 1980). Thus, by using a variety of drive techniques, the parasitic effects can be made so small that the LED transient response is determined largely by the carrier lifetime discussed previously in Section 3. Detailed transient analyses can be found in previous publications (Lee, 1975; Hino and Iwamoto, 1979).

4. CARRIER LEAKAGE

In the discussions of quantum efficiency in Section 1, the heterojunction barrier height was assumed to be large enough to ensure that injected carriers were well confined in the active region. However, when either the active layer is very thin or the junction temperature rises, the injected carriers may leak through the heterostructure barriers. Lee and Dentai (1978) have considered the height of the heterostructure barriers necessary to confine injected carriers in active regions of various thicknesses at room and elevated temperatures. If barriers are "leaky," an excess current J', in addition to the nominal current J consumed in the active region, must be supplied to produce a given light output. Thus, a confinement factor is defined as

$$C = J/(J + J'). \tag{35}$$

The excess current J' is the sum of the electron and hole currents in the confining layers and is given by

$$J' = J'_n + J'_p, \tag{36}$$

where J'_n and J'_p are the sum of the diffusion and drift currents,

$$J'_n = e\mu_n E'_n + eD_n \, dn'/dx, \tag{37}$$

$$J'_p = e\mu_p E'_p + eD_p \, dp'/dx. \tag{38}$$

In Eqs. (37) and (38), n' and p' are excess electron and hole densities, μ_n and μ_p are electron and hole mobilities, D_n and D_p are diffusion constants, and E is the electrical field, $E = \rho J$.

For a symmetrical DH structure under forward bias, the quasi-Fermi level of the excess electrons in the p confining layer (Fig. 1), denoted by $\delta\phi_N$, and that of the excess holes in the n confining layer, denoted by $\delta\phi_P$, can be written as (Rode, 1974)

$$\delta\phi_N = -E_c + E_a + \phi_P + \phi_n - \phi_p, \quad \delta\phi_P = E_c - E_a + \phi_N - \phi_n + \phi_p, \tag{39}$$

where E_c and E_a are band-gap energies of the confining layers and of the active layer, respectively ($E_c > E_a$). For $Ga_{1-x}Al_xAs$, the band-gap energy at room temperature depends on the aluminum arsenide mode fraction x, according to

$$E(x) = 1.424 + 1.266x + 0.26x^2 \quad \text{eV}. \tag{40}$$

In Eq. (39), ϕ_P corresponds to the Fermi level of holes in the p confining layer and ϕ_N to the Fermi level of electrons in the n confining layer, and ϕ_n and ϕ_p are quasi-Fermi levels of electrons Δn and holes Δp, respectively, in the active region. Under conditions of bimolecular recombination, $\Delta n = \Delta p$ and Δn can be related to the injected current J by the relationship

$$\Delta n = (J/ewB)^{1/2}. \tag{41}$$

In a practical diode of 50-μm-diameter emitting area, the estimated electron density Δn at 100 mA is approximately 1.7×10^{18} cm^{-3} for $w = 2$ μm and 4.5×10^{18} cm^{-3} for $w = 0.3$ μm. At 300 mA, a practical maximum operating current limited by electrical heating, Δn would be 2.8×10^{18} cm^{-3} and 8×10^{18} cm^{-3}, respectively. Using Eqs. (35)–(41), the excess carrier densities can be calculated numerically (Lee and Dentai, 1978). The results of the calculated confinement factor C for two cases at $\Delta n = 2 \times 10^{18}$ cm^{-3} and $\Delta n = 8 \times 10^{18}$ cm^{-3} are shown in Fig. 9a and b, respectively. These figures show (1) that for $w > 1$ μm, good confinement ($C \geq 0.8$) over the temperature range 300–400 K can be achieved with $\Delta x \geq 0.2$ ($\Delta E \geq 264$ meV), Δx being the difference in aluminum mole fraction between the confining layer and the active layer; and (2) that for $0.1 < w < 0.5$ μm, good confinement requires $\Delta x \geq 0.3$ ($\Delta E \geq 400$ meV).

In practice, the carrier leakage in GaAlAs surface-emitting LEDs is usually small as long as $\Delta x \geq 0.3$. The sublinear $L-I$ characteristics are largely due to junction heating (the resistivity of GaAlAs is four times larger than that of GaAs). For InGaAsP LEDs, however, $L-I$ characteristics more sublinear than those of GaAlAs devices often are observed. Carrier leakage (Yamakoshi *et al.*, 1982; Wada *et al.*, 1982c; Chen *et al.*, 1983) is one of many possible mechanisms resulting in this saturation of the light output. More detailed discussions are contained in Part IV.

IV. LED-Device Characteristics

Here we review some salient LED characteristics and, where appropriate, contrast differences between GaAlAs and InGaAs devices. We then compare surface- and edge-emitting LEDs.

5. Current–Voltage Characteristics

Typical forward $I-V$ behavior for a high-radiance LED is shown in Fig. 10. At a given bias level, the total diode current is the sum of a (radiative) diffusion current I_d and a (nonradiative) recombination current I_r. The latter has been attributed (Henry *et al.*, 1977) to surface recombination near the perimeter of the LED chip. The total diode current is the sum of I_d and I_r:

$$I = I'_d \exp[e(V - IR_s)/kT] + I'_r \exp[e(V - IR_s)/2kT], \qquad (42)$$

where R_s is the device series resistance, and I'_d and I'_r are the saturation currents due to diffusion and recombination, respectively. The light output is proportional to I_d. At low bias, the recombination current dominates, and virtually no light is produced. At higher bias, the proportion of the diffusion current becomes significant, and the light output is proportional to I_d. The inflection point A in Fig. 10 corresponds to the previously noted crowding of the current in the junction area above the p contact. The second inflection

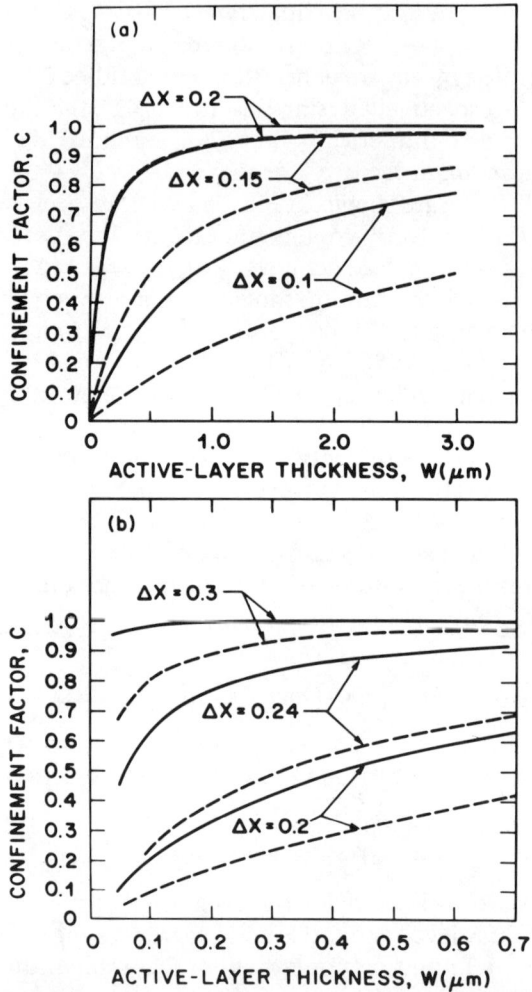

FIG. 9. Confinement factor at 300–400 K as a function of the active-layer thickness W for (a) $\Delta n = \Delta p = 2 \times 10^{18}$ cm^{-3} and for (b) $\Delta n = \Delta p = 8 \times 10^{18}$ cm^{-3}, where Δx is the differential aluminum mole fraction: $P = 5 \times 10^{18}$ cm^{-3}, $N = 1 \times 10^{18}$ cm^{-3}, $T = 400$ K (---) and 300 K (—).

point B results from the device series resistance R_s, which is due primarily to the p-contact resistance. As the p-contact diameter is made smaller to enhance coupling to optical fibers (Section 5), R_s is increased. The associated I^2R power dissipation (device heating) reduces output power, as described next in Section 6.

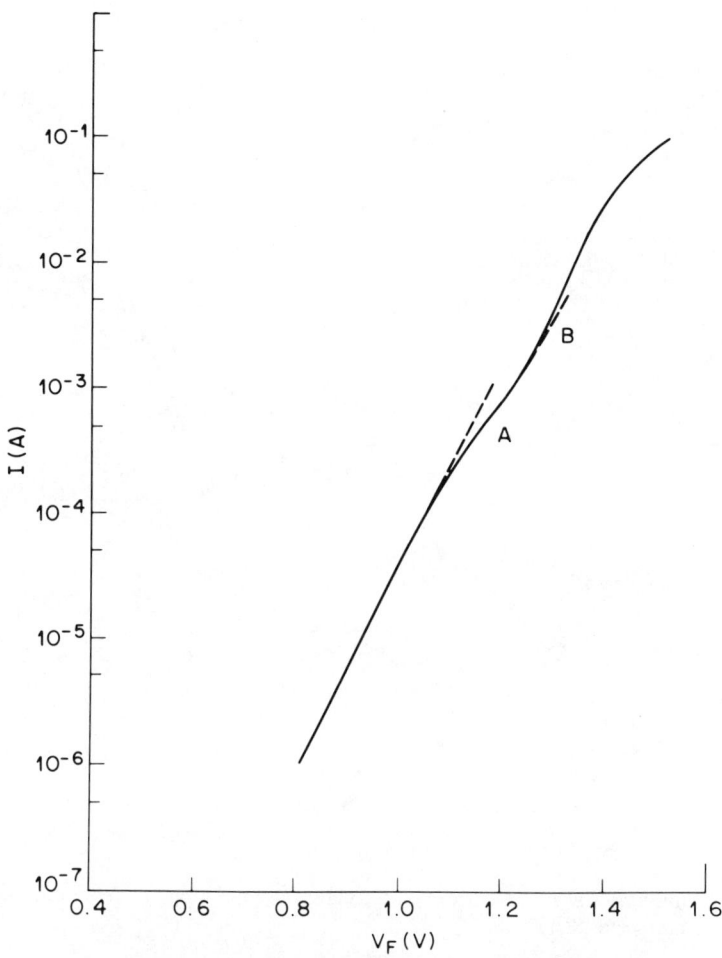

FIG. 10. Forward-current–voltage characteristics for GaAlAs surface-emitting LEDs. The inflection points A and B are discussed in the text. [From Zipfel et al. (1982).]

6. Light Output versus Current and Temperature

Light-emitting-diode luminescence results from direct (band-to-band) recombination of majority carriers with injected minority carriers. Consequently, light output is expected to vary linearly with drive current. However, as shown in Fig. 11 and mentioned previously in Section 5, output power of surface emitters initially increases linearly with current but becomes sublinear at higher currents. The reduction in quantum efficiency with increased temperature (from self-heating at high current) contributes to

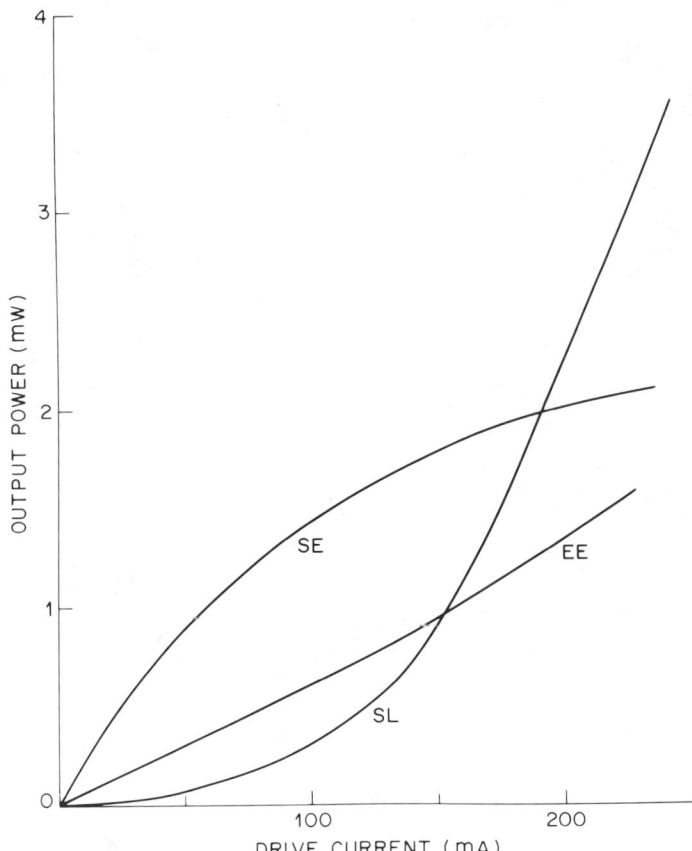

FIG. 11. Output power versus current for typical surface-emitting (SE), edge-emitting (EE), and superluminescent (SL) LEDs.

this sublinear behavior. The sublinearity may be reduced by use of short-duration current pulses and heat sinking to minimize heating effects. The residual sublinearity, termed "radiance saturation," was first reported by Goodfellow *et al.* (1981) who observed various degrees of saturation in both GaAlAs and InGaAsP LEDs. Most work in this field, however, has focused on InGaAsP, where a common mechanism has been sought to explain both the sublinearity of luminescence and a strong temperature sensitivity (above room temperature) of threshold in laser diodes of the same material (Uji *et al.*, 1981; Dutta and Nelson, 1981). Numerous (often conflicting) mechanisms have been proposed to explain these phenomena, some of which are reviewed later.

In-plane superluminescence (Goodfellow *et al.*, 1981), resulting from stimulated emission at high injection levels, has been postulated to account for the observed sublinearity of the surface emission. Observations of the light–current characteristics of faceted (to reflect in-plane luminescence) surface emitters (SpringThorpe et al., 1982) and comparison of the facet (single-pass gain) and side (unpumped) emission of edge-emitting LEDs (Dutta and Nelson, 1982) are consistent with this mechanism. On the basis of this mechanism alone, saturation is expected to be strongly temperature dependent. However, it is found experimentally that the temperature dependence of the surface emission is independent of drive current (Temkin et al., 1981a).

Measurements of diode current (Uji et al., 1981) and carrier lifetimes (Dutta and Nelson, 1982; Sermaze et al., 1983) as a function of injection level indicate that at high carrier density ($\gtrsim 2 \times 10^{18}$ cm^{-3}), the nonradiative component increases with the square of injection level, consistent with an Auger-process contribution to the sublinear $L-I$ characteristic. Nonradiative lifetimes decrease rapidly with increasing free-carrier concentration, also consistent with the Auger effect (Henry et al., 1983b). Further, the enhanced saturation with increased injected-carrier density (observed when the active layer thickness is decreased and the diameter of the emitting region is reduced) has been successfully modeled assuming a nonradiative Auger-type component that depends on injection level (Uji et al., 1982).

For devices with highly doped ($\sim 2 \times 10^{18}$ cm^{-3}) active layers, the current dependence of carrier lifetime and spontaneous emission were also found to be consistent (Su et al., 1982a) with an Auger model. However, for the more common unintentionally doped (mid-10^{17}-cm^{-3}) active layers, the data is consistent (Su et al., 1982b) with a drift-leakage model (Anthony and Schumaker, 1980). Others have pointed to the role of injected carriers that are not confined by the heterobarriers cladding the active layers (Yano et al., 1981). The existence of leakage of injected electrons over the p-cladding interface has been verified (Yamakoshi et al., 1982) in LEDs specially prepared with a thin active layer, and direct measurement of carrier leakage in laser–bipolar-transistor structures has indicated that this leakage is an appreciable fraction (15–30%) of the total current (Chen et al., 1983). Carrier leakage may be enhanced when the temperature of the injected carriers exceeds the lattice temperature (Shah et al., 1981). The rise in carrier temperature at 1.5-μm wavelength was determined to be larger by a factor of ~ 3 than at 1.3-μm wavelength and has been correlated with the enhanced saturation observed at the longer wavelengths (Wada et al., 1982c). Measurements of spontaneous emission and carrier lifetimes in 1.58-μm InGaAsP devices have led some workers (Asada and Suematsu, 1982) to conclude that the major contribution to the nonradiative lifetime is carrier leakage, with a more minor contribution from Auger recombination.

Although there are conflicting data and diverse interpretations, it appears likely to these authors that several interrelated mechanisms are operative, the dominant ones being dependent on the details of the device structure and excitation conditions. For example, hot carriers [if present (Henry et al., 1983a; Manning et al., 1983)] may be the result of Auger recombination, especially in heavily doped active layers and can lead to the observed carrier leakage. At sufficiently high injected-carrier densities ($\gtrsim 3 \times 10^{18}$ cm^{-3}), stimulated emission leads to in-plane superluminescence, which then becomes an important loss mechanism in surface emitters. Additional mechanisms, including interface recombination (Yano et al., 1980) and intervalence-band absorption (Adams et al., 1980), have been proposed to explain the temperature dependence of threshold in lasers, but at present their role in determining the spontaneous emission characteristics of LEDs is unclear.

Representative $L-I$ plots for edge-emitting and superluminescent LEDs are also shown in Fig. 11. Depending on the details of the device design (carrier density, cavity length, facet reflectivities, etc.), substantial gain may be present, giving rise to a strongly superlinear behavior of these devices, and, with increased gain, increased temperature sensitivity is expected. For comparison, Fig. 12 illustrates the light–temperature characteristic for the various LED structures. Surface emitters are characterized by the empirical expression

$$L = L_0 \exp(-T/T_1), \qquad (43)$$

where T_1 is in the ranges 180–220 K and 300–350 K for InGaAsP and GaAlAs LEDs, respectively (Temkin et al., 1983), and is independent of drive current (Temkin et al., 1981a). For edge emitters, however, T_1 is smaller, typically in the ranges 70–80 K and 100– to 120 K, respectively (Kressel et al., 1980). Further, with increased stimulated emission, as in superluminescent LEDs, the effective value of T_1 can be strongly dependent on drive current and can become substantially reduced (increased temperature sensitivity) at high drive current (Fig. 12, inset). Thus, to utilize the high-power potential of these edge-emitting devices at elevated temperatures, thermoelectric coolers may be required. Some care must be exercised, however, since there is always the possibility that laser-geometry devices will lase at low temperatures. In principle, lasing can be avoided by judicious device design, or by use of a thermoelectric heater.

7. Spectral Characteristics

The peak electroluminescence wavelength λ_p is determined primarily by the band gap (composition) of the active layer. Figure 13 shows typical spectra for InGaAsP surface- and edge-emitting LEDs. Modulation on the emission spectra is commonly observed in GaAlAs surface-emitting LEDs (but rarely in InGaAsP devices), apparently due to interference effects re-

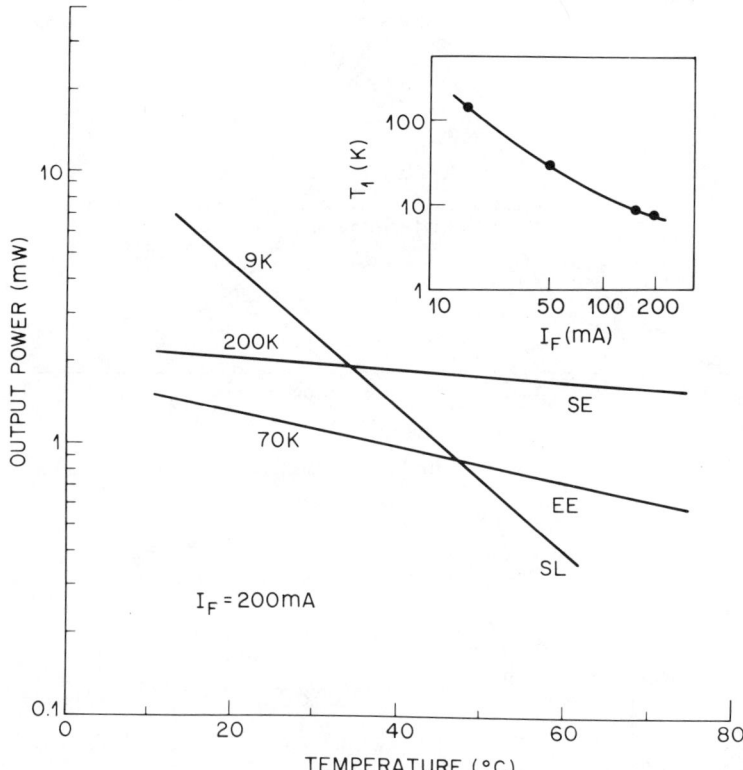

FIG. 12. Temperature dependence of light output for various 1.3-μm LED structures. The inset shows, for an SL LED, the variation of the characteristic temperature T_1 with drive current. [From Dutta *et al.* (1983). © 1983 IEEE.]

sulting from reflections between the emitting surface and the back contact. Under pulsed-current operation, which eliminates Joule heating, increasing the peak current shifts λ_p toward shorter wavelengths due to band filling (Nelson *et al.*, 1963). Under cw drive conditions, where junction heating occurs, however, λ_p shifts toward longer wavelengths with increasing current, since the band gap decreases at higher junction temperatures. For the same reason, λ_p increases with ambient temperature, shifting by 0.35 and 0.6 nm °C^{-1} for GaAlAs (Burrus and Miller, 1971) and InGaAsP (Temkin *et al.*, 1983), respectively. Increased doping of the active layer shifts λ_p to higher values as a result of impurity banding (Casey and Stern, 1976).

The spectral width, taken here to be the full width at half-maximum intensity (FWHM), varies roughly as λ_p^2 for undoped active layers (Fukui and Horikoshi, 1979); this corresponds to a constant energy width of $\sim 2kT$.

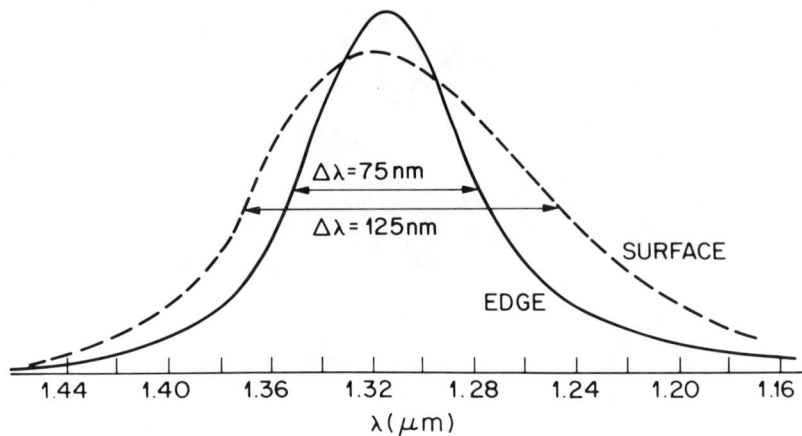

FIG. 13. Spectral output curve for InGaAsP surface- and edge-emitting LEDs.

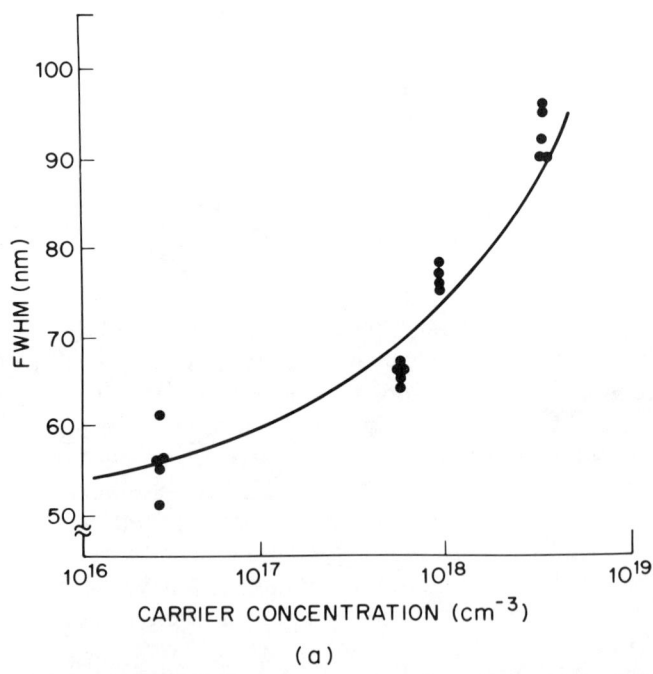

(a)

FIG. 14. Dependence of FWHM spectral width on (a) doping of the active layer ($\lambda_p = 1.08$ μm, $J = 1.2$ kA cm^{-2}) [from Wright et al. (1975), © 1975 IEEE] and (b) current density ($\lambda_p = 1.3$ μm) [from Burton (1983)].

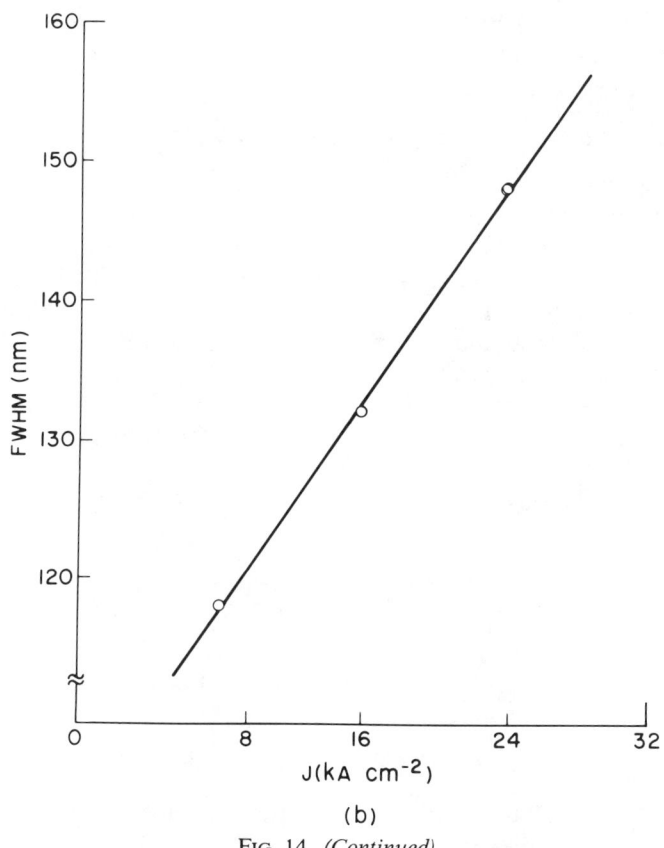

(b)

FIG. 14. *(Continued)*

Although this is borne out in $Ga_{1-x}Al_xAs$ and in $In_{1-x}Ga_xAs_yP_{1-y}$ near the InP and InGaAs ends of the composition range, FWHM is nearly $3kT$ at the composition corresponding to 1.3-μm emission. This has been correlated with compositional grading in the active layer (Temkin et al., 1981b). GaAlAs surface-emitting LEDs operating at 0.85 μm have a spectral width of 40 nm compared with ~110 nm for InGaAsP LEDs emitting at 1.3 μm. As a result of self-absorption along the length of the active layer, FWHM of edge emitters is ~1.6 times smaller than for surface emitters, as indicated in Fig. 13. For superluminescent LEDs, the spectra is further narrowed by the onset of optical gain, and FWHM can be $\gtrsim 4$ times smaller than for surface emitters. The shift toward broader FWHM with increased temperature (~0.3 nm °C^{-1}) and increased (pulsed) drive current (Escher et al., 1982) is attributed to band filling, whereas the shift with increased doping of the active layer (Wright et al., 1975) is attributed to impurity banding. These effects are illustrated in Fig. 14a and b. Thus, a variety of factors contribute to spectral broadening. In general, high-speed LEDs (with doped active layers) driven at high current density (to achieve maximum radiance) will have wide spectra, for example, up to 160 nm for 1.3-μm LEDs.

8. Output Power and Modulation Bandwidth

As described in Part III, a variety of factors affect output and modulation bandwidths of LEDs. State-of-the art results are shown in Fig. 15, where the reciprocal relationship between optical modulation bandwidth and output power is apparent. For GaAlAs surface-emitting LEDs, the highest power reported has been 15 mW with a 3-dB bandwidth of 17 MHz (Lee and Dentai, 1978), whereas the largest modulation bandwidth of 1.1 GHz was achieved at a much lower output power of 0.2 mW (Heinen *et al.*, 1976). Experimental results such as those shown in Fig. 15 indicate that output power P can be related to bandwidth B by $P \sim B^{-\nu}$. For bandwidth $B < 100$ MHz, $\nu \approx \frac{2}{3}$ and for $B > 100$ MHz, $\nu \approx \frac{4}{3}$. For LEDs with small bandwidth (20–30 MHz), the output power can be maximized by using a thick active

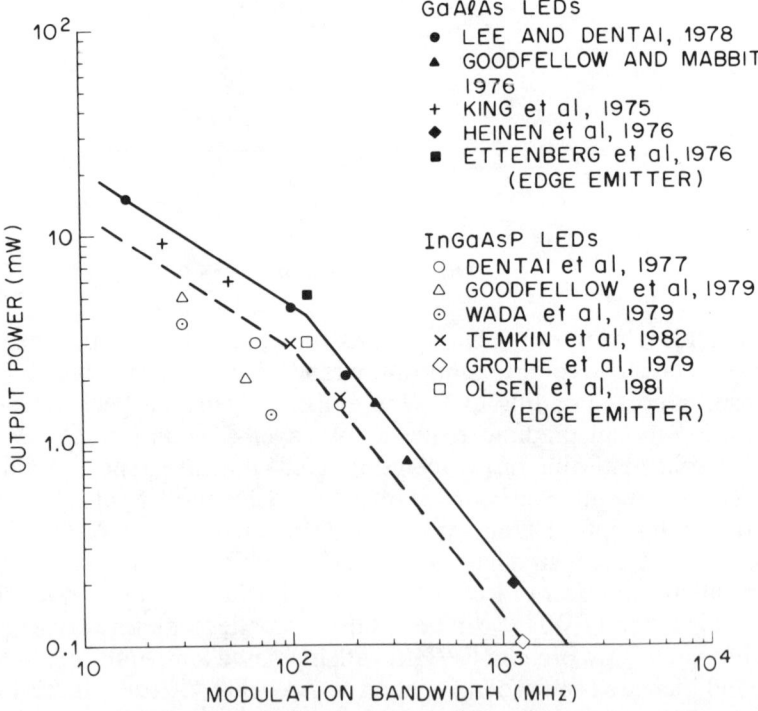

FIG. 15. State-of-the-art output power and bandwidth for 1.3-μm LEDs. [After Lee (1982) and Saul (1983).]

layer (2–2.5 μm) with low doping densities ($<5 \times 10^{17}$ cm^{-3}). Diodes in the intermediate bandwidth region (50–100 MHz) require thinner active layers (1–1.5 μm) and moderate doping densities (0.5–1 × 10^{18} cm^{-3}). An increase in the modulation bandwidth beyond the 100–200-MHz range can be achieved only by doping the active layer in excess of $\sim 5 \times 10^{18}$ cm^{-3}. This heavy doping evidently introduces nonradiative centers, and as a result, the value of v is greater than unity, and the output power is significantly reduced.

Similar power-versus-bandwidth characteristics are observed for InGaAsP 1.3-μm LEDs. The solid and dotted lines in Fig. 15 are contours of best reported results for GaAlAs and InGaAsP LEDs, respectively. The output power of GaAlAs LEDs is a factor of 2 higher than that of InGaAsP LEDs at all bandwidths. Because the photon energy at 1.3-μm wavelength is smaller by a factor of 1.53 than that at 0.85 μm, the output power of InGaAsP LEDs is reduced by the same factor, compared with GaAlAs LEDs of the same quantum efficiency. When this factor is taken into account, the best performance of 1.3-μm LEDs is still slightly below that of 0.85-μm LEDs, probably due both to the fact that the technology is more advanced for GaAlAs than for InGaAsP and to the enhanced radiance saturation in the longer-wavelength material.

The best performance of both GaAlAs and InGaAsP edge emitters is also included in Fig. 15 for comparison. These devices utilize thin ($\lesssim 0.02$ μm), lightly doped active layers, which promotes bimolecular recombination (Section 2). Consequently, edge emitters typically have large bandwidth, generally in the 100–150 MHz range. The thin active layer also narrows the output-beam divergence, which enhances coupling to fibers, as discussed next in Part V.

V. Coupling of LEDs to Optical Fibers

Although light is emitted isotropically within the volume of the active layer of a surface-emitting LED, the geometry is that of a disk with large diameter-to-thickness ratio. It is equivalent to a Lambertian source for which the external radiant intensity (power per unit area per unit solid angle) varies with angle θ from the normal to the emitting plane. Neglecting the small increase in Fresnel reflection at large angles, the external radiant intensity varies as

$$B(r, \theta) = B(r) \cos \theta, \qquad (44)$$

where $B(r)$ accounts for radial variation in radiant intensity within the emitting plane. Equation (44) closely characterizes the far-field pattern for surface-emitting LEDs. This pattern is symmetric and the (full) beam angle to half intensity is $\sim 120°$. On the other hand, in edge emitters, light rays in the active layer are guided parallel to the junction plane by the cladding layers of the double heterostructure, which have a relatively lower index of refraction. With thin ($\lesssim 0.2$ μm) active layers, excitation of the lowest order transverse mode will dominate, which results in a narrowing of the far-field pattern normal to the junction ($\theta_\perp \simeq 30°$). The beam parallel to the junction plane is not guided; i.e., $\theta_\parallel \simeq 120°$.

An optical fiber is a waveguide characterized by a core with index of refraction n_0 and a cladding with index of refraction n_c. Light rays that impinge on the fiber core at a half angle (measured from the fiber axis) that is less than the acceptance angle of the fiber undergo total internal reflection within the fiber core. These rays are guided along the axis of the fiber, suffering minimum attenuation. For a step-index fiber, the critical angle θ_c and numerical aperture (NA) are given by

$$\text{NA} = (n_0^2 - n_c^2)^{1/2}, \qquad \theta_c = \sin^{-1} \text{NA} \simeq \text{NA}. \tag{45}$$

For a typical value of NA = 0.2, $\theta_c \simeq 12°$. Consequently, only a small fraction of the rays emitted from an LED can propagate along the fiber. For a (parabolically) graded-index fiber, NA and θ_c are functions of radial position

$$\text{NA}(r) = \text{NA}[1 - (r/r_f)^2]^{1/2}, \qquad \theta_c(r) = \sin^{-1} \text{NA}(r) \simeq \text{NA}(r), \tag{46}$$

where r_f is the radius of the fiber core. Thus, grading the index, which enhances the bandwidth of the fiber, reduces the effective NA (hence θ_c) of the fiber, as can be seen by comparing Eqs. (45) and (46).

It is apparent that the fraction of emitted light launched into the fiber is a strong function of the angular distribution of the light impinging on the fiber core. This distribution is initially determined by the geometry of the LED, as discussed previously, but it can be modified substantially by the methods used to couple light into a fiber, as discussed later.

9. Direct Coupling to Fibers

The simplest method of coupling an LED to a fiber is direct or butt coupling. When the surface-emitting LED light-spot diameter is greater than or equal to the fiber core diameter, butt coupling gives the highest launched power, independent of the beam divergence of the source. This follows from Liouville's theorem. Of particular interest is the coupling efficiency η_c, which is the fraction of the emitted light launched into the fiber. For the case of an ideal Lambertian light source, i.e., a surface emitter with uniform intensity [$B(r)$ = const], approximate closed-form solutions for coupling efficiency η_c have been obtained by integrating the radiant intensity over the emitting area and the solid angle defined by θ_c given in Eq. (45) or Eq. (46) (Barnoski, 1982; Plihal, 1982; Hudson, 1974).

For step-index (SI) and graded-index (GI) fibers, η_c is given by

$$\eta_c = T(\text{NA})^2 \quad (\text{SI}), \qquad \eta_c = T(\text{NA})^2[1 - (2D^2)^{-1}] \quad (\text{GI}), \qquad (47)$$

where T is the media (air) transmissivity and includes the LED and fiber Fresnel loss. (The former can be minimized by using an antireflection coating.) The parameter D is defined as d_f/d_s, where d_s and d_f are the diameters of the source (generally taken as the p-contact diameter) and fiber core, respectively. Equation (47) is valid for $D \geq 1$, the case of interest for efficient coupling. These analyses consider only meridional rays; skew rays are neglected. The latter undergo only partial reflection at the fiber-core–cladding interface and consequently these "leaky modes" are highly attenuated.

The lower curves in Fig. 16 illustrate the variation of η_c with the ratio d_s/d_f. The maximum coupling efficiency is $(\text{NA})^2$. Note that η_c for a GI fiber

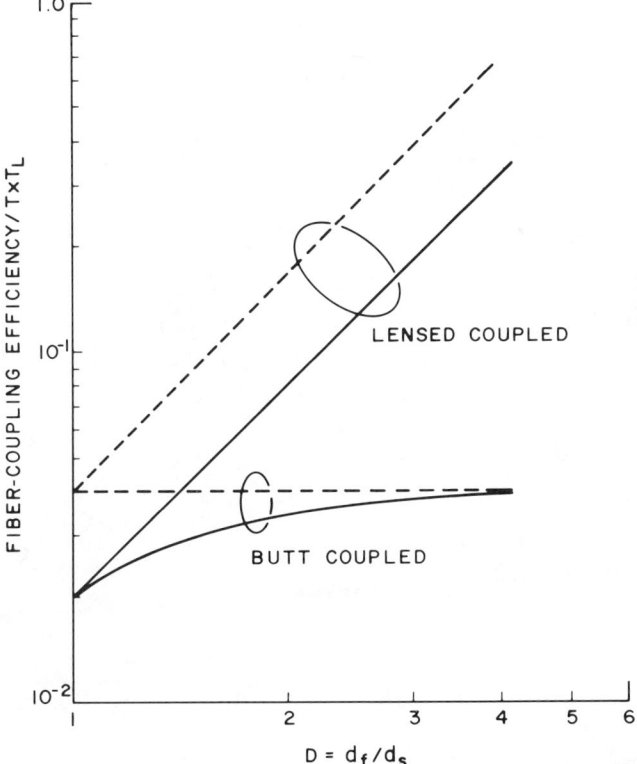

FIG. 16. Maximum coupling efficiency as a function of the ratio of fiber diameter and spot diameter for butt coupling and lensed coupling; T is the media transmissivity and T_L the transmissivity of the lens; NA = 0.2, (- - -) SI, (—) GI.

approaches that of an SI fiber for $d_f \gg d_s$ but is only half that value for $d_f \approx d_s$. It is generally observed that, for small-core fibers with low NA, the experimentally determined value of η_c is appreciably less than that predicted by Eq. (47). Measurements of the near-field pattern reveal that the light intensity from an oxide-defined-contact LED is not constant (Wada *et al.*, 1982a) as is assumed for Eq. (47); rather, the light intensity is uniform only over a small central area and then decreases with a Gaussian-like tail (Wada *et al.*, 1982a). This light-intensity profile reflects the current distribution across the $p-n$ junction: the current is constant directly above the p contact and spreads laterally according to the conductivity of the epitaxial layers. Using numerical integration to account for the measured variation in $B(r)$, good agreement between the experimental and theoretical values of η_c is obtained for small-core fibers (Borsuk, 1983). The effect of the Gaussian tail on η_c is illustrated in Fig. 17, where the abscissa is the standard deviation σ of the Gaussian function. Note that, with typical current spreading η_c is re-

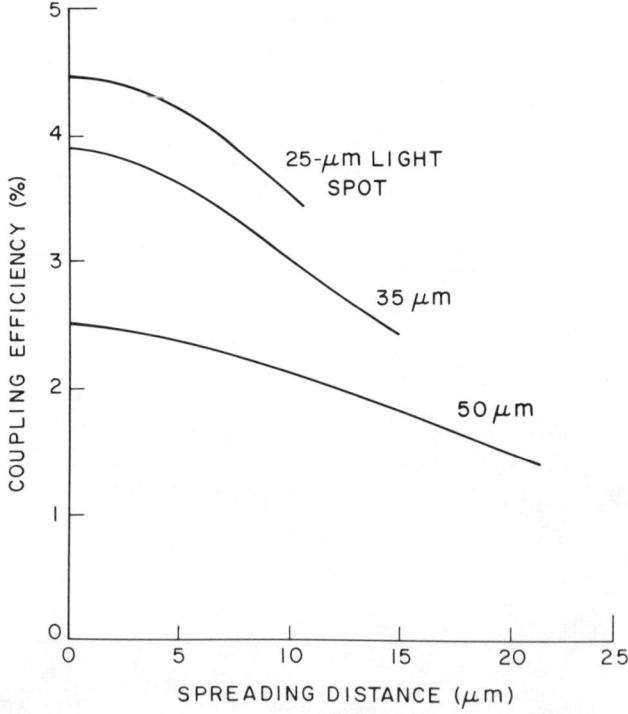

FIG. 17. Butt-coupling efficiency as a function of spreading of the light spot beyond the uniform central portion defined by the contact size: $d_f = 50 \mu m$ (GI); NA = 0.23. [From Borsuk (1983). © 1983 IEEE.]

duced by ~30%, compared to a perfectly confined uniform light spot ($\sigma = 0$). In principle, restricting the junction area by proton bombardment should increase η_c; however, there is some indication that the associated damage reduces total output power so that the launched power is not increased.

Modeling the coupling of edge emitters is more complex because of the asymmetric far-field pattern, but it has been attempted by several authors (Yang and Kingsley, 1975; Marcuse and Kaminow, 1981; Wittke et al., 1976). In general, it is found that, at low NA (small θ_c), the narrower beam divergence of the edge emitter results in appreciably higher coupling efficiencies than is possible with surface emitters. As expected, this advantage becomes less important at high NA (large θ_c). The importance of beam divergence on η_c becomes more pronounced with GI fibers, since the effective θ_c is reduced. The coupling of edge emitters can be further enhanced by using real-index guiding or gain guiding (especially in superradiant devices) to narrow $\theta_\parallel$. Coupling efficiencies into a 50-μm core 0.2-NA fiber as high as 26% have been realized in superradiant 1.3-μm LEDs (Dutta et al., 1983).

10. Coupling with Lenses

It follows from Section 9 that when $d_s \geq d_f$, power launched by surface emitters cannot be improved with lensing. If, however, the source diameter is reduced while maintaining constant drive current, the radiance is increased, and higher launched powers can be achieved using imaging optics. (If the drive current is simultaneously reduced so that the radiance is the same, then lensing will not increase launched power.) For butt coupling, η_c is low because θ_c of the fiber is typically much smaller than the source angular emission. When the source area is smaller than the fiber core, imaging optics can be used to reduce the beam divergence. In the following, we consider only the cases where $d_f > d_s$. The coupling improvement relative to butt coupling has been calculated using conventional optics theory for a variety of situations: lenses placed between the LED and fiber (Barnoski, 1982; Plihal, 1982), fibers with integral lenses (Abe et al., 1977); truncated lenses cemented to the LED surface (Abram et al., 1975); and lenses that are integral with the LED structure (Hasegawa and Namazu, 1980). Ray-tracing techniques also have been used for calculation of coupling gain for a variety of lens shapes (Ackenhusen, 1978).

In general, the largest gain is achieved when the source angular emission θ_s is collimated by the lens so that its magnified image is within θ_c; i.e., $\theta_s = M\theta_c$, where M is the lens magnification. Simultaneously, the magnified image must be no larger than the fiber-core diameter; hence $M = D \equiv d_f/d_s$.

We consider first the case of a lens placed between a surface emitter and a fiber. A critical light-spot diameter d_c can be defined such that all rays leaving the LED surface within the diameter d_c are within θ_c of the fiber; i.e., the rays are launched into the fiber. The parameter d_c is determined by the fiber and lens characteristics, namely,

$$d_c = 2f \text{ NA}, \tag{48}$$

where f is the posterior focal length of the lens (Hasegawa and Namazu, 1980). For a given fiber and source ($d_f > d_s$), coupling gain will increase as d_c is decreased by increased lens magnification. With low magnification $d_c \gg d_s$), coupling gain g is independent of fiber type. For a spherical lens of radius r, coupling gain can be expressed as

$$g_{SI} = g_{GI} = T_L (d_f/d_c)^2, \tag{49}$$

where T_L is the transmissivity of the lens. Coupling gain will continue to increase with increasing M, reaching a maximum value when $d_c = d_s$. The maximum coupling gain is then given by

$$g_{SI} = T_L D^2, \tag{50}$$

$$g_{GI} = T_L \frac{D^2}{2} \left[\frac{1}{1 - (1/2D^2)} \right]. \tag{51}$$

From Eq. (48) and noting $r = f$, this corresponds to spherical lenses for which $r = (d_s/2)$ NA.

Combining Eqs. (47) and (50), the upper curves of Fig. 16 show the maximum coupling efficiency for lensed coupling. Note that η_c of a step-index fiber is twice that of a graded-index fiber and that lensing has more impact for larger D. When $D = 1$ ($d_f = d_s$), butt coupling gives the highest coupling efficiency, and lensing cannot improve on it.

Coupling efficiency also has been enhanced by using monolithic or integral lenses formed on the emitting surface of the LED. In this case, T_L is unity in Eq. (50), yielding a small increase in g. For butt coupling, the enhancement in coupling efficiency of a lensed LED compared to a flat LED has been calculated (Hasegawa and Namazu, 1980). It is found that, for a given fiber and light-spot size, g_{SI} and g_{GI} increase in proportion to β^2, where β is the lateral magnification, up to a limiting value for which $d'_c = d_s$, where d'_c is the critical diameter in the emitting plane. This is entirely analogous to the case of an intervening lens, as described before.

The parameter β is determined by the radius of curvature of the lens and its distance from the light spot. The former is limited by the technology used for fabricating the lens, and the latter is determined by the (minimum) wafer thickness. Batch-fabrication techniques for forming integral lenses have been reported. Epitaxial regrowth onto substrates containing an array of recesses, which define the lens shape, has been reported (King and SpringThorpe, 1975). This particular procedure may be difficult to repro-

FIG. 18. Scanning electron micrograph of batch-fabricated integral lenses on an LED wafer. [From Ostermayer et al. (1983).]

duce over an entire wafer, but fabrication of lenses in conventional epitaxial wafers has been achieved using ion-beam milling (Wada et al., 1981), photoelectrochemical (PEC) etching (Ostermayer et al., 1982), and masked chemical etching (Heinen, 1982b). Figure 18 is a scanning electron micrograph of an LED wafer with batch-fabricated lenses produced by PEC etching. Lens shape (and therefore magnification) is uniform over the entire wafer.

In practice, $1 \lesssim \beta \lesssim 2$, which limits the maximum achieved coupling gain to a value of $\sim 2-3$ (for a 50-μm core fiber). Figure 19 shows the maximum

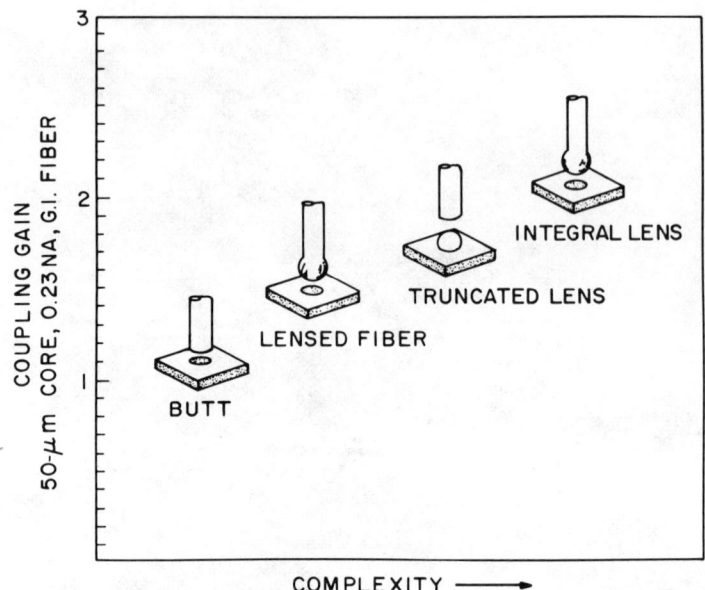

FIG. 19. Achieved coupling gain for various lensing methods (30–40-μm diameter light-spot). [From Saul (1983). © 1983 IEEE.]

reported coupling gain for a variety of lensing schemes (Saul, 1983). In general, the coupling gain achieved is less than that calculated, probably because of spherical aberrations in lenses of small radius, imperfections in optical surfaces, and misalignment of the various components. Additionally, calculations assume a uniform light spot, which tends to overestimate coupling gain. Launched power into a 50-μm-core 0.2-NA fiber as high as 80 μW has been achieved for 1.3-μm LEDs (with integral lens) driven at 150 mA (Burton, 1983).

The previous general principles apply also to edge emitters. However, selection of the optical system must take into account the asymmetry of the far-field pattern of the devices. In particular, although $\theta_\perp$ is relatively narrow (but nevertheless exceeds θ_c of the fiber), $\theta_\parallel$ is not. Using a cylindrical lens whose axis is perpendicular to the junction plane, $\theta_\parallel$ can be narrowed. Unfortunately, the near-field pattern is also nonsymmetric and is widest in the junction plane. This limits the useful magnification. A second cylindrical lens, orthogonal to the first, can be used to narrow $\theta_\perp$. In practice, bulb-ended or tapered fibers are widely used to increase coupling. Coupling gains as high as ∼4 have been realized using lensed 50-μm-core 0.2-NA fibers; launched power as high as 225 μW was achieved (Frahm, 1982). Ure et al. (1982) reported superradiant edge emitters with a truncated spherical lens glued onto the emitting facet. A coupling gain of ~ 5 was obtained; coupled power was 350 μW at 250 mA. Using a ridge-loaded, lateral waveguide (to reduce $\theta_\parallel$) in a 1.3-μm superluminescent LED. Kaminow *et al.* (1983) achieved 0.5 mW of power launched into a lensed 50-μm-core 0.2-NA fiber at 250 mA. In gain-guided GaAlAs superluminescent LEDs, launched power as high as 2.5 mW (~ 60% coupling efficiency) was realized (Davies *et al.*, 1981).

For both surface and edge emitters, achievable coupling efficiency increases as spot size is decreased. In general, however, as spot size is reduced, output power decreases because of enhanced device heating. For surface emitters, radiance saturation compounds the problem. Consequently, the maximum coupled power will not, in general, correspond to the maximum coupling efficiency. For a given fiber and operating current, spot size and magnification must be optimized to give the highest coupled power (Wada *et al.*, 1982b).

11. Sensitivity of Coupling Efficiency to Mechanical Alignment

In the foregoing discussion, it has been tacitly assumed that (1) for direct coupling, the fiber was butted against the LED, and (2) for lensed coupling, the LED and fiber were orthogonally aligned at the focal points of the lens. In practice, mechanical alignment tolerances can lead to substantial deviations from the optimum configuration and therefore lower than optimum coupling efficiencies. For surface emitters, the sensitivity of coupling efficiency to lateral, axial, or angular misalignment has been calculated using numerical techniques (Ackenhusen, 1978; Berg *et al.*, 1981) and has been determined experimentally (Ackenhusen, 1978; Berg *et al.*, 1981; Escher *et al.*, 1982; Johnson *et al.*, 1980). For a given $d_f > d_s$, η_c initially increases as lens diameter decreases (because of increased magnification); simultaneously, however, η_c becomes more sensitive to axial separation and lateral displacement. Figure 20a illustrates the measured coupling gain versus lateral separation for various coupling schemes. Note that a microlens (diameter $\simeq d_f$) not only gives high gain, but it also reduces sensitivity to lateral separation. This results from the reduction in d_c; i.e., $d_c < d_s$. Figure 20b illustrates coupling gain as a function of axial separation. Comparing Fig. 20a and b, it is apparent that η_c is most sensitive to lateral displacement. In Fig. 20b, note that coupling with microlenses significantly increases sensitivity to axial separation relative to direct coupling. This sensitivity can be reduced by using an integral lens that puts the refracting surface as close to the light source as possible. Alternatively, a larger-than-optimum lens can be used to trade coupling efficiency for reduced axial-separation sensitivity. In the latter case, e.g., the macrolens in Fig. 19b, the optimum separation is relatively

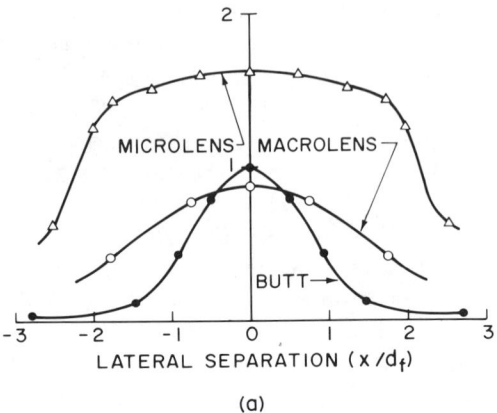

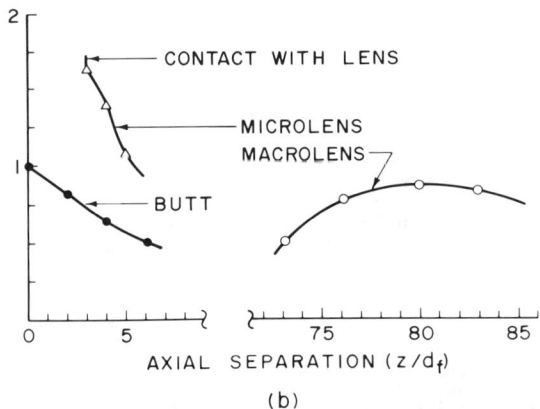

Fig. 20. Coupling gain for various lensing schemes as a function of (a) lateral separation and (b) axial separation of LED and fiber with $d_f/d_s = 1.9$. [From Ackenhusen (1978).]

large, which simplifies the implementation of an inexpensive connectorized package (Johnson *et al.*, 1980).

For edge emitters, the asymmetry of the far-field pattern suggests a greater sensitivity to lateral displacement parallel to the junction plane than orthogonal to it. As in the case of surface emitters, lensing enhances sensitivity to axial separation.

12. Effect of Coupling on Speed

As described in Section 9, near-field intensity profiles are uniform only at the center of the light spot. The radial decrease in intensity near the perimeter of the light spot reflects the fall-off in current density. Since $\tau_r \propto J^{1/2}$ [Eq.

(33), Section 3], it follows that the modulation bandwidth is higher for the central emission than for the peripheral emission (Wada et al., 1982a). Consequently, the effective speed of an LED depends on the portion of the spot from which light is collected, i.e., on the critical diameter d_c, defined by the coupling optics. Figure 21 shows the measured risetime t_r as a function of the light-collection method, where the latter is parameterized in terms of d_c. It is apparent that the effective LED speed is substantially enhanced by a collection system that results in a low value of d_c. For example, comparing coupling to a large-area detector directly without fiber ($d_c \simeq 250$ μm) to a lensed 50-μm-core fiber ($d_c \simeq 35$ μm), t_r is reduced by a factor of ~ 3.8. The common procedure of measuring t_r by direct coupling to a detector (no fiber) can lead to a larger value of t_r than that with coupling to a fiber. The fastest response occurs when d_c is equal to the diameter of the uniform portion of the light spot. (This corresponds to the dashed line for $d_c < 35$ μm in Fig. 21.) As indicated in Section 10, this condition also corresponds to the maximum

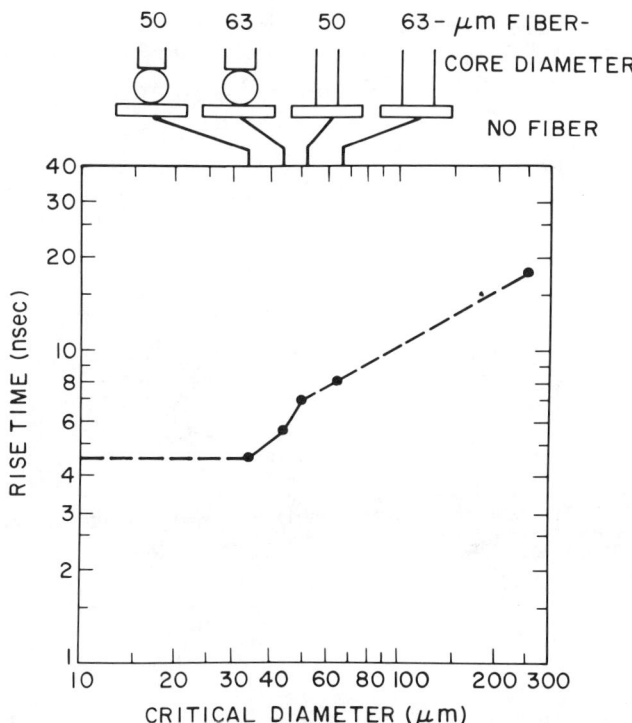

FIG. 21. Measured 10–90% risetime for an LED as a function of critical diameter corresponding to the coupling methods indicated at the top of the figure. [From Borsuk (1982). © 1982 IEEE.]

coupling efficiency. Consequently, optimizing the lens magnification gives the highest power–speed product.

VI. Dual-Wavelength LEDs

The very wide transmission window from 0.8- to 1.6-μm wavelength in present low-loss optical fibers offers opportunities for use of wavelength-division multiplexing (WDM) to increase the transmission capacity of a single optical fiber. The conventional way to accomplish WDM is to combine the signal from each single-wavelength source by means of a passive optical multiplexer, which may be either a grating, filter, or fiber device. A simpler way, however, is to provide a source that emits two (or more), wavelength bands simultaneously from a single chip, where each wavelength band can be modulated independently. Figure 22 shows a schematic cross section of such a dual-wavelength surface-emitting LED (Lee et al., 1980).

Basically, the LED wafer consisted of five epitaxial layers grown by LPE on an InP substrate. A buffer layer of InP was grown first, followed by a 2-μm-thick $In_{0.77}Ga_{0.23}As_{0.5}P_{0.5}$ layer (denoted by Q_1), an InP barrier layer, a 1.5-μm-thick $In_{0.66}Ga_{0.34}As_{0.75}P_{0.25}$ layer (denoted by Q_2), and finally, a top InP (p) layer. The Q_1 layer had a band-gap energy of 1.09 eV, corre-

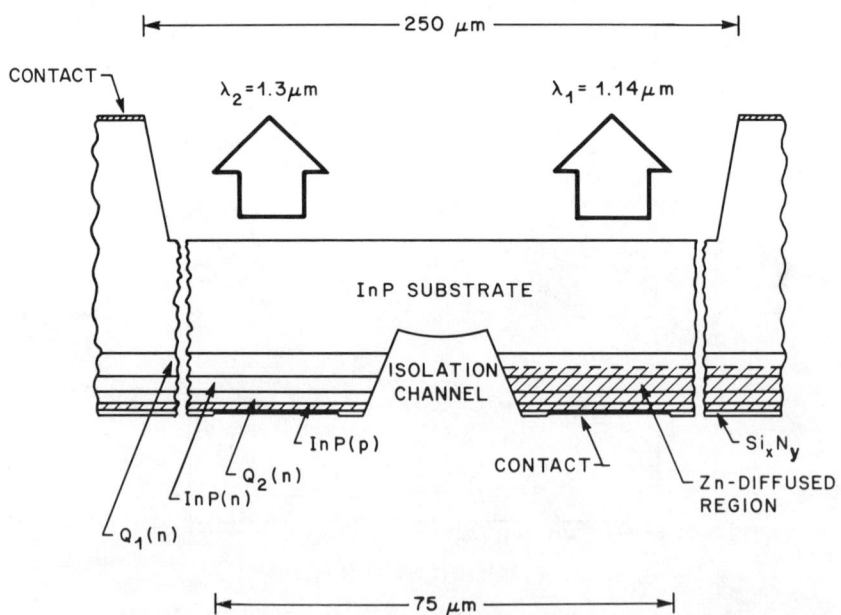

FIG. 22. Schematic illustration of a dual-wavelength LED. [From Lee et al. (1980).]

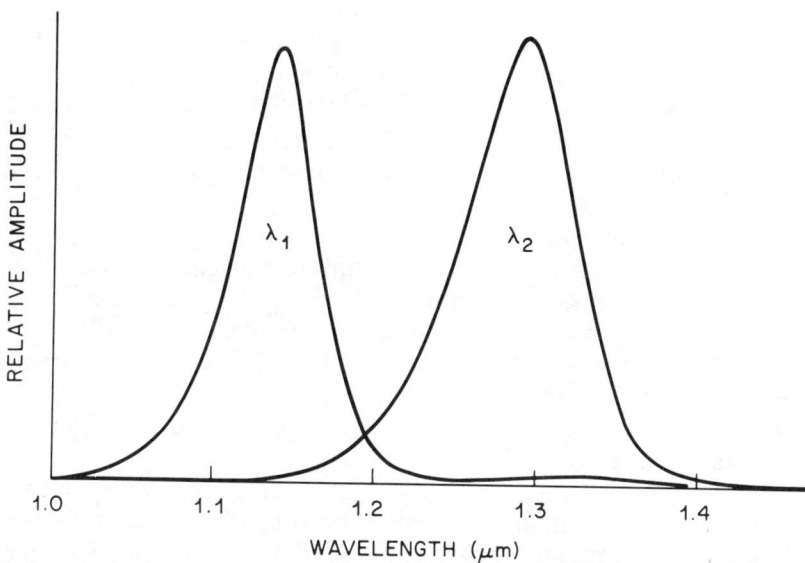

FIG. 23. Emission spectra of dual-wavelength LED. [From Lee et al. (1980).]

sponding to 1.14-μm emission wavelength, whereas the Q_2 layer had a bandgap energy of 0.95 eV, corresponding to 1.3-μm wavelength. The $p-n$ junction in the Q_1 layer was formed by Zn diffusion, and the $p-n$ junction in the Q_2 layer was a grown junction. The contact windows were 25 × 75 μm separated by a 25-μm channel chemically etched through the grown layer to provide electrical isolation; each emitter then was 25 × 26 μm.[3]

The emission spectrum of each junction, superimposed on one graph, is shown in Fig. 23. The measured total cross-talk was −22 to −26 dB. The maximum output of the diode at each wavelength was 1.0 mW when driven at 50 mA dc with a modulation bandwidth of 25 MHz. A dual-wavelength LED of this type was used in conjunction with a dual-wavelength demultiplexing photodetector (Campbell et al., 1980) in a WDM experiment operated at 33 Mbit/sec (Ogawa et al., 1981). A receiver sensitivity of −38.5 and −39.7 dBm (decibels above 1 mW) was achieved for the two wavelength channels, respectively.

VII. LEDs Made by MBE and VPE

The devices discussed in Section 4 were fabricated primarily by liquid-phase epitaxy (LPE). Although the best performance so far has been obtained from devices made of LPE materials, large-volume production is somewhat limited by the LPE growth technique. Both molecular-beam epitaxy (MBE) and vapor-phase epitaxy (VPE) offer alternatives for large-scale wafer growth, and thus they are more suitable for large-volume production. Although MBE technology is more completely developed for GaAs-based materials, present knowledge suggests that VPE growth may be most suitable for the InGaAsP quaternary system. GaAlAs surface-emitting LEDs made from MBE wafers have performed comparably to those made of LPE materials (Lee *et al.*, 1978, 1981).

Molecular-beam epitaxy technology is particularly capable of making sophisticated device structures with multiple thin layers known as quantum-well structures. An edge-emitting LED made from $Ga_{0.47}In_{0.53}As$ and $Al_{0.48}In_{0.52}As$ grown lattice matched to an InP substrate by MBE had emission at 1.60-μm wavelength (Alavi *et al.*, 1983). The active region contained eleven quantum wells of $Ga_{0.47}In_{0.53}As$ 110–120-Å-wide, separated by 55–60-Å-wide barriers of $Al_{0.48}In_{0.52}As$. An external quantum efficiency of 1.5% was obtained with this initial device.

Edge-emitting InGaAsP/InP LEDs fabricated from VPE wafers were comparable in performance to devices from LPE materials (Olsen *et al.*, 1981). The best output power was 2 mW, and the highest modulation bandwidth was 200 MHz.

VIII. Conclusions

Both practical high-radiance surface-emitting and edge-emitting LEDs have demonstrated high-power and large-bandwidth capabilities for use as optical sources in lightwave communications systems. However, there is an inverse relationship between the output power and the modulation bandwidth of LEDs. This relationship dictates the maximum power that can be obtained for a given system data rate. The best power × bandwidth product can be achieved by the optimization of the thickness and doping concentration of the active layer as well as the band-gap energy and doping concentrations of the confining layers in the double-heterostructure wafer. Edge-emitting LEDs and superluminescent diodes have narrower spectral width and higher coupling efficiency to optical fibers, but the output power of these devices decreases more rapidly with increasing temperature than does that of surface-emitting LEDs. A variety of lensing techniques have been developed to increase coupling efficiency to fibers, especially for surface emitters. In general, higher coupled power comes at the expense of greater complexity in device packaging (external lens) or chip fabrication (integral lens). The ability to grow multiple layers with different band-gap energies, all lattice matched, has enabled the fabrication of dual-wavelength emitters in a single chip. The devices may offer a simpler alternative for wavelength-division multiplexing than the conventional WDM system using passive optical multiplexers. Because of the extremely long operating lifetimes of LEDs, even in hostile environments, LED-based systems will prove to be attractive alternatives to laser-based systems for many applications.

REFERENCES

Abe, M., Umebu, I., Hasegawa, O., Yamakoshi, S., Yamaoka, T., Kotami, T., Okada, H., and Takanashi, H. (1977). *IEEE Trans. Electron Devices* **ED-24,** 990.
Abram, R. A., Allan, R. W., and Goodfellow, R. C. (1975). *J. Appl. Phys.* **46,** 3468.
Ackenhusen, J. (1978). *Appl. Opt.* **18,** 3694.
Acket, G. A., Nijam, W., and Lam, H. (1974). *J. Appl. Phys.* **45,** 3033.
Adams, A. R., Asada, M., Suematsu, Y., and Arai, S. (1980). *Jpn. J. Appl. Phys.* **19,** L621.
Alavi, K., Pearsall, T. P., Forrest, S. R., and Cho, A.-Y. (1983). *Electron. Lett.* **19,** 227.
Anthony, P. J., and Schumaker, N. E. (1980). *J. Appl. Phys.* **51,** 5038.
Asada, M., and Suematsu, Y. (1982). *Appl. Phys. Lett.* **41,** 353.
Barnoski, M. K. (1982). *In* "Fundamentals of Optical Fiber Communications" (M. K. Barnoski, ed.), 2nd ed., Chapter 2. Academic Press, New York.
Berg, H. M., Lewis, G. L., and Mitchell, C. W. (1981). *Proc.—Electron. Components Conf.* **31,** 372.
Berkstresser, G. W., Keramidas, V. G., and Zipfel, C. L. (1980). *Bell Syst. Tech. J.* **59,** 1549.
Borsuk, J. A. (1982). *IEEE Spec. Conf. Light Emitting Diodes Photodetectors. Ottawa-Hill, Ontario, Canada* (unpublished).
Borsuk, J. A. (1983). *IEEE Trans. Electron Devices* **ED-30,** 296.
Burrus, C. A., and Dawson, R. W. (1970). *Appl. Phys. Lett.* **17,** 97.
Burrus, C. A., and Miller, B. I. (1971). *Opt. Commun.* **4,** 307.
Burton, R. H. (1983). AT&T Bell Laboratories, unpublished.
Campbell, J. C., Dentai, A. G., Lee, T. P., and Burrus, C. A. (1980). *IEEE J. Quantum Electron.* **QE-16,** 601.
Casey, H. C., Jr., and Stern, F. (1976). *J. Appl. Phys.* **47,** 631.
Chen, T. R., Chin, L. C., Hasson, A., Koren, U., Margabit, S., and Yariv, A. (1983). *Appl. Phys. Lett.* **42,** 1000.
Chin, A. K., Zipfel, C. L., Dutt, B. V., DiGiuseppe, M. A., Bauers, K. B., and Roccasecca, D. D. (1981). *Jpn. J. Appl. Phys.* **20,** 1487.
Davies, I. G. A., Goodwin, A. R., and Plumb, R. G. (1981). *Conf. Lasers Electroopt.* Washington, D.C. (unpublished).
Dawson, R. W. (1980). *IEEE J. Quantum Electron.* **QE-16,** 697.
Dentai, A. G., Lee, T. P., and Burrus, C. A. (1977). *Electron. Lett.* **13,** 484.
Dutta, N. K., and Nelson, R. J. (1981). *Appl. Phys. Lett.* **38,** 407.
Dutta, N. K., and Nelson, R. J. (1982). *IEEE J. Quantum Electron.* **QE-18,** 375.
Dutta, N. K., Nelson, R. J., Wright, P. D., Besomi, P., and Wilson, R. B. (1983). *IEEE Trans. Electron Devices* **ED-30,** 360.
Escher, J. S., Berg, H. M., Lewis, G. L., Moyer, C. D., Robertson, T. V., and Wey, H. A. (1982). *IEEE Trans. Electron Devices* **ED-19,** 1463.
Ettenberg, M., Kressel, H., and Wittke, J. P. (1976). *IEEE J. Quantum Electron.* **QE-12,** 360.
Frahm, R. E. (1982). AT&T Bell Laboratories, unpublished.
Fukui, T., and Horikoshi, Y. (1979). *Jpn. J. Appl. Phys.* **18,** 961.
Gloge, D., Albanese, A., Burrus, C. A., Chinnock, E. L., Copeland, J. A., Dentai, A. G., Lee, T. P., Li, T., and Ogawa, K. (1980). *Bell Syst. Tech. J.* **59,** 1365.

Goodfellow, R. C., and Mabbitt, A. W. (1976). *Electron Lett.* **12**, 50.
Goodfellow, R. C., Carter, A. C., Griffith, I., and Bradley, R. R. (1979). *IEEE Trans. Electron Devices* **ED-26**, 1215.
Goodfellow, R. C., Carter, A. C., Rees, G. J., and Davis, R. (1981). *IEEE Trans. Electron Devices* **ED-28**, 365.
Grothe, H., Proebster, W., and Harth, W. (1979). *Electron. Lett.* **15**, 702.
Harth, W., Huber, W., and Heinen, J. (1976). *IEEE Trans. Electron Devices* **ED-23**, 478.
Hasegawa, O., and Namazu, R. (1980). *J. Appl. Phys.* **51**, 30.
Heinen, J. (1982a). *Electron Lett.* **18**, 23.
Heinen, J. (1982b). *Electron Lett.* **18**, 831.
Heinen, J., Huber, W., and Harth, W. (1976). *Electron. Lett.* **12**, 553.
Henry, C. H., Logan, R. A., and Merritt, F. R. (1977). *Appl. Phys. Lett.* **31**, 454.
Henry, C. H., Logan, R. A., Temkin, H., and Merritt, F. R. (1983a). *IEEE J. Quantum Electron.* **QE-19**, 941.
Henry, C. H., Levin, B. F., Logan, R. A., and Bethea C. G. (1983b). *IEEE J. Quantum Electron.* **QE-19**, 905.
Hino, I., and Iwamoto, K. (1979). *IEEE Trans. Electron Devices* **ED-26**, 1238.
Hudson, M. C. (1974). *Appl. Opt.* **13**, 1029.
Johnson, B. H., Ackenhusen, J. G., and Lorimor, O. G. (1980). *Proc. Electron. Components Conf.* **30**, 279.
Joyce, W. B., Bachrach, R. Z., Dixon, R. W., and Sealer, D. A. (1974). *J. Appl. Phys.* **45**, 2229.
Kaminow, I. P., Eisenstein, G., Stulz, L. W., and Dentai, A. G. (1983). *IEEE J. Quantum Electron.* **QE-19**, 78.
King, F. D., and SpringThorpe, A. J. (1975). *J. Electron. Mater.* **4**, 243.
Kressel, H., Ettenberg, M., Wittke, J. P., and Ladany, I. (1980). *Top. Appl. Phys.* **39**, 9-62.
Kundukhov, R. M., Metreveli, S. G., and Siukaev, N. V. (1967). *Sov. Phys. — Semicond. (Engl. Transl.)* **1**, 765.
Kurbatov, L. N., Shakhidzhanov, S. S., Bystrova, L. V., Krapuhkin, V. V., and Kolonenkova, S. J., (1971). *Sov. Phys. — Semicond. (Engl. Transl.)* **4**, 1739.
Lee, T. P. (1975). *Bell Syst. Tech. J.* **54**, 53.
Lee, T. P. (1980). *Proc. Soc. Photo.-Opt. Instrum. Eng.* **224**, 92.
Lee, T. P. (1982). *Opt. Laser Technol.*, **14**, 15.
Lee, T. P., and Dentai, A. G. (1978). *IEEE J. Quantum Electron.* **QE-14**, 150.
Lee, T. P., Burrus, C. A., Jr., and Miller, B. I. (1973). *IEEE J. Quantum Electron.* **QE-9**, 820.
Lee, T. P., Holden, W. S., and Cho, A.-Y. (1978). *Appl. Phys. Lett.* **32**, 415.
Lee, T. P., Burrus, C. A., and Dentai, A. G. (1980). *Electron. Lett.* **16**, 845.
Lee, T. P., Holden, W. S., and Cho, A. Y. (1981). *IEEE J. Quantum Electron.* **QE-17**, 387.
Liu, Y. S., and Smith, D. A. (1975). *Proc. IEEE (Lett.)* **63**, 542.
Manning, J., Olshansky, R., Su, C. B., and Powazinik, W. (1983). *Appl. Phys. Lett.* **43**, 134.
Marcuse, D., (1977). *IEEE J. Quantum Electron.* **QE-13**, 819.
Marcuse, D., and Kaminow, I. P. (1981). *IEEE J. Quantum Electron.* **QE-17**, 1234.
Namizaki, H., Nagano, M., and Nakahara, S. (1974). *IEEE Trans. Electron Devices* **21**, 688.
Nelson, R. J., and Sobers, R. G. (1978). *Appl. Phys. Lett.* **32**, 761.
Nelson, D. F., Gershenzon, M., Ashkin, A., D'Asaro, L. A., and Saraca, J. C. (1963). *Appl. Phys. Lett.* **2**, 182.
Ogawa, K., Lee, T. P., Burrus, C. A., Campbell, J. C., and Dentai, A. G. (1981). *Electron Lett.* **17**, 857.
Olsen, G., Hawrylo, F., Channin, D. J., Botez, D., and Ettenberg, M. (1981). *IEEE J. Quantum Electron.* **QE-17**, 2130.
Ostermayer, F. W., Jr., Kohl, P. A., and Burton, R. H. (1982). *Appl. Phys. Lett.* **43**, 642.

Plihal, M. (1982). *Siemens Forsch.-Entwicklungsber.* **11**, 221.
Rode, D. L., (1974). *J. Appl. Phys.* **45**, 3887.
Sakai, S., Umeno, M., and Amemiya, Y. (1980). *Jpn. J. Appl. Phys.* **19**, 109.
Saul, S., (1983). *IEEE Trans. Electron Devices* **ED-30**, 285.
Sermaze, B., Eichler, H. J., Heritage, J. P., Nelson, R. J., and Dutta, N. K. (1983). *Appl. Phys. Lett.* **42**, 259.
Shah, J., Laheny, R. F., Nahory, R. E., and Temkin, H. (1981). *Appl. Phys. Lett.* **39**, 618.
SpringThorpe, A. J., Look, C. M., and Emmerstorfer, B. F. (1982). *IEEE Trans. Electron Devices* **ED-29**, 876.
Su, C. B., Schlafer, J., Manning, J., and Olshansky, R. (1982a). *Electron. Lett.* **18**, 595.
Su, C. B., Schlafer, J., Manning, J., and Olshansky, R. (1982b). *Electron. Lett.* **18**, 1108.
Temkin, H., Chin, A. K., and DiGiuseppe, M. A. (1981a). *Appl. Phys. Lett.* **39**, 405.
Temkin, H., Keramidas, V. G., Pollack, M. A., and Wagner, W. R. (1981b). *J. Appl. Phys.* **52**, 1574.
Temkin, H., Joyce, W. B., Chin, A. K., DiGiuseppe, M. A., and Ermanis, F. (1982). *Appl. Phys. Lett.* **41**, 745.
Temkin, H., Zipfel, C. L., DiGiuseppe, M. A., Chin, A. K., Keramidas, V. G., and Saul, R. H. (1983). *Bell Syst. Tech. J.* **62**, 1.
Uji, T., Iwamoto, K., and Lang, R. (1981). *Appl. Phys. Lett.* **38**, 193.
Uji, T., Onabe, K., Hayashi, J., Isoda, Y., Morihisa, Y., Iwamoto, K., and Sakuma, J. (1982). *Electron. Commun. Soc. Natl. Conf., Jpn., 1982,* Pap. 4-34.
Umebu, I., Hasegawa, O., and Akita, K. (1978). *Electron. Lett.* **14**, 499.
Ure, J., Carter, A. C., Goodfellow, R. C., and Harding, M. (1982). *IEEE Spec. Conf. Light Emitting Diodes Photodetectors. Ottawa-Hull, Ontario, Canada* (unpublished).
Wada, O., Yamakoshi, S., Abe, M., Akita, K., and Toyama, Y. (1979). *Proc. Opt. Commun. Conf. 1979,* Pap. 4.6.
Wada, O., Yamakoshi, S., Abe, M., Yishitoni, Y., and Sakwai, T. (1981). *IEEE J. Quantum Electron.* **QE-17**, 174.
Wada, O., Hamaguchi, H., Nishitani, Y., and Sakurai, T. (1982a). *IEEE Electron Device Lett.* **EDL-3**, 129.
Wada, O., Hamaguchi, H., Nishitani, Y., and Sakurai, T. (1982b). *IEEE Trans. Electron Devices* **ED-29**, 1454.
Wada, O., Yamakoshi, S., and Sakurai, T. (1982c). *Appl. Phys. Lett.* **41**, 981.
Wang, C. S., Cheng, W. H., and Hwang, C. J. (1982). *Appl. Phys. Lett.* **41**, 587.
Wittke, J. P., Ettenberg, M., and Kressel, H. (1976). *RCA Rev.* **37**, 159.
Wright, P. D., Chai, Y. G., and Antypas, G. A. (1975). *IEEE Trans. Electron Devices* **ED-26**, 1220.
Yamakoshi, S., Sanada, T., Wada, O., Umebu, I., and Sakurai, T. (1982). *Appl. Phys. Lett.* **40**, 144.
Yang, K. H., and Kingsley, J. D. (1975). *Appl. Opt.* **14**, 288.
Yano, M., Nishi, H., and Takusagawa, M. (1980). *IEEE J. Quantum Electron.* **QE-16**, 661.
Yano, M., Imai, H., and Takusagawa, M. (1981). *J. Appl. Phys.* **52**, 3172.
Zipfel, C. L., Saul, R. H., Chin, A. K., and Keramidas, V. G. (1982). *J. Appl. Phys.* **53**, 1781.
Zucker, J. (1978). *J. Appl. Phys.* **49**, 2543.

CHAPTER 6

Light-Emitting-Diode Reliability

C. L. Zipfel

AT&T BELL LABORATORIES
MURRAY HILL, NEW JERSEY

I.	INTRODUCTION.	239
II.	LED LIFE TESTING	242
III.	GaAs/GaAlAs LEDs: RESULTS OF ACCELERATED AGING	243
	1. *Burrus Structure*	243
	2. *Planar Single Heterostructure*	245
	3. *Planar Double Heterostructure*	246
	4. *Edge-Emitting Structure*	249
IV.	RAPID DEGRADATION MECHANISMS IN GaAs/GaAlAs LEDs	249
	5. *DLD Formation in LEDs and Lasers*	249
	6. *Phenomenology of DLD Growth*	250
	7. *Effect of Dislocation Density on DLD Growth*	255
	8. *Effect of Stress and Damage on DLD Growth*	255
	9. *Burn-In to Eliminate Early Failures*	258
	10. *Growth-Related Effects*	259
V.	GRADUAL DEGRADATION IN GaAs/GaAlAs LEDs	260
	11. *Reliability Projections*	260
	12. *Mechanisms in Long-Term Aging*	261
VI.	InP/InGaAsP LEDs: RESULTS OF ACCELERATED AGING	264
	13. *Burrus Structure*	264
	14. *Planar Double Heterostructure*	264
	15. *Edge-Emitting Structure*	268
VII.	DEGRADATION MECHANISMS IN InP/InGaAsP LEDs	268
	16. *Indium Inclusions*	268
	17. *Dark-Spot-Defect Formation*	269
	18. *Dark-Line-Defect Formation*	272
	19. *Misfit Dislocations*	273
	20. *Gradual Degradation and Reliability Projections*	274
VIII.	CONCLUSIONS	275
	REFERENCES	276

I. Introduction

The wide temperature range of operation, well-defined degradation characteristics, and high reliability even at high temperatures as well as high fabrication yields make light-emitting diodes (LEDs) the source of choice in many transmission systems (Saul, 1983). Two types of LED structures are used for lightwave transmission: surface emitters and edge emitters. The commonly used materials systems are GaAs/GaAlAs emitting in the 0.85–0.9-μm range and InP/InGaAsP emitting in the 1.3–1.5-μm range. A major concern in the design of lightwave LEDs is to maximize the power that can be launched in a fiber. This is done in surface emitters by using thin p layers and a small-diameter p contact to promote current crowding so that the resulting light spot is less than or equal to the size of the fiber core ($\sim 50\ \mu$m). The fiber is then placed as close as possible to the light spot, either by etching a well in the substrate (the Burrus LED in Fig. 1a of Chapter 5), by etching off the entire substrate (the planar GaAs/GaAlAs LED in Fig. 1 of this chapter), or by thinning the substrate (the InP/InGaAsP LED in Fig. 2b of Chapter 5). In edge-emitting LEDs (shown in Fig. 2 in Chapter 5), light is launched into the fiber parallel to the junction plane as in a laser. To launch the maximum power and therefore to keep repeater spacings as wide as possible, LEDs of both types are operated at the highest current densities that are consistent with reliability. In addition, LEDs are often operated at high ambient temperatures in systems for which thermoelectric coolers are not feasible. For these reasons, LED failure rate is a crucial factor in transmission-system design, and published reports of new structures usually include lifetime projections. For LEDs to be chosen for a system they must be an economical alternative to lasers, and, therefore, fabrication and burn-in yields must be high. Unfortunately, yield figures are seldom reported.

Light-emitting-diode failures can be classified into three populations: infant failures due to random fabrication defects, freak failures, and the main population of long-lived devices. Infant failures result from localized crystal defects that can never be completely eliminated. Typically they constitute a small percentage of devices from each wafer, and can be screened out in a burn-in of about 100 hr. Freak failures in semiconductor devices are usually defined as those units which pass burn-in but which fail earlier than would be expected for the main population, due to an extreme combination of the same effects that cause failure in the main population (Peck and Zierdt, 1974). Here the definition of freak is broadened to include any devices that fail by preventable or controllable failure mechanisms not significant in the main population. These can result from wafer-wide fabrication defects such as lattice misfit. Once infant failures are eliminated by a burn-in, LEDs in the main population degrade only gradually in light output by processes with well-defined time dependence. Since repeater spacings in transmission systems are limited by LED light output, a trade-off can be made between the power margin allotted for LED end of life and reliability (Saul, 1983). Although end of life is usually given as a 50% (-3 dB) drop in light output, it has also been defined as a 20% (-1 dB) drop (Zipfel *et al.*, 1983). In the latter case MTTF is reduced, but repeater spacing can be increased.

It has been the custom in reporting LED reliability to give the mean time to a 50% drop in light output but not the spread in the failure distribution or a consideration of how effectively infant failures can be eliminated in a burn-in and freaks prevented in fabrication. Since most projections of MTTF for LEDs are longer than typical service lives (<20 year or 2×10^5 hr), the important quantities to systems designers are failure rates or cumulative failures. A long MTTF means nothing unless the spread in the distribution is narrow enough to make failure rates small. For example, for a log-normal failure distribution with MTTF $= 10^6$ hr and standard deviation $\sigma = 0.6$, only 0.5% will fail during a 20-year service life. However, for MTTF $= 10^6$ hr and $\sigma = 1.4$, 30% of the devices will fail. The application of log-normal statistics to semiconductor-device reliability is discussed by Peck and Zierdt (1974) and by Jordan (1978). Newman and Ritchie (1981) discuss its application to lasers and LEDs.†

Although the device structures used for GaAs- and InP-based LEDs are similar, the short-term failure mechanisms in the two materials are very different. Dislocationlike defects are the dominant cause for early failure in GaAs-based devices, whereas precipitatelike defects are dominant in InP-based devices. In the early development of GaAs/GaAlAs lasers and LEDs, large percentages of devices developed nonradiative regions, observed in electroluminescence as dark-line defects (DLDs) (DeLoach *et al.*, 1973). Since DLDs are lossy regions absorbing light and acting as carrier sinks, their appearance in a laser cavity can sharply reduce gain in very short times. DLD formation can also cause rapid degradation in light output in high-radiance LEDs, which typically operate at current densities even higher than those used in lasers. However, since LED light output is reasonably linear in current, DLDs and other dark defects have a less drastic effect in them than in lasers.

Most DLDs propagate by dislocation climb, a mechanism requiring only preexisting crystal defects and electron–hole recombination (Petroff and Kimerling, 1976). Failures due to DLD formation have been largely eliminated in GaAs/GaAlAs LEDs by using low-dislocation-density substrates and eliminating other sources of dislocations and strain. The small percentage of devices that still fail can be readily eliminated by a burn-in (Yamakoshi *et al.*, 1977). State-of-the-art planar double-heterostructure LEDs, which remain clear in burn-in, remain clear and degrade only gradually in light output under accelerated aging by processes whose time dependences are well defined mathematically (Zipfel *et al.*, 1982). No catastrophic failure mechanisms relevant at operating temperatures have been reported in these devices. Because degradation in light output is the only long-term failure

† Some reports give median life ML. For the log-normal distribution MTTF = ML $\times \exp(\frac{1}{2}\sigma^2)$.

mode in LEDs, their reliability is easier to predict than that of lasers. LEDs are not subject to the additional failure modes that can plague lasers, such as changes in threshold current, asymmetry between front and back facets, kinks in the light–current characteristic, and facet damage.

Lattice-matched InP/InGaAsP double-heterostructure (DH) LEDs were developed to exploit the low transmission loss and low material dispersion in silica fibers at 1.3 µm. Happily, several properties of this materials system have tended to promote high reliability. Unlike GaAs-based LEDs, DLD growth by climb has not been observed in InP/InGaAsP LEDs (Yamakoshi et al., 1979). This is fortunate because low-dislocation-density InP substrates are not yet readily available. Wright et al., (1979) cite the exact lattice match possible between InP and InGaAsP and good heat sinking through the binary InP confining layer rather than through the ternary GaAlAs as additional reasons for the enhanced reliability. Starting with the earliest reports, the predicted mean times to failure (MTTF) have been extraordinarily long. For example, Yamakoshi et al. (1979) estimate MTTF at 25°C to be 5×10^9 hr, about three orders of magnitude longer than for GaAs/GaAlAs LEDs.

Ettenberg and Neuse (1975) have suggested that the degradation rate for LEDs and lasers is strongly dependent on the band gap. Degradation mechanisms such as DLD growth involve recombination–enhanced defect reactions (Kimerling, 1978). In lower band-gap materials such as InGaAsP, there is less energy available from electron–hole recombination. In contrast to GaAs devices, however, which remain clear in long-term aging, InP/InGaAsP LEDs develop precipitatelike dark-spot defects (DSDs) during gradual degradation (Yamakoshi et al., 1979). The origin of DSDs and their ultimate effect on reliability is a major concern in these devices.

Part II deals with light-power and junction-temperature measurements in life testing. In Part III the results of accelerated aging of GaAs-based LEDs, including observed failure modes and estimates of MTTF, are presented. In Part IV, rapid degradation mechanisms in these devices are discussed with emphasis on the elimination of infant and freak failures. In Part V, reliability projections for GaAs-based LEDs are considered, and possible long-term degradation mechanisms are discussed. In Part VI the results of accelerated aging of InP-based LEDs are given, and in Part VII the failure mechanisms and reliability projections are discussed. Conclusions about the reliability of lightwave LEDs are drawn in Part VIII.

II. LED Life Testing

Most reports of LED life testing have been at constant dc current and at constant, usually elevated, temperature. Light-emitting diodes are brought to room temperature for light-power measurements. Typically, the total emitted light (i.e., light emitted into a broad area detector) is monitored during aging. Only about 2% of the total light from a surface-emitting LED can be butt-coupled into standard optical fibers, and this power is very sensitive to small changes in the light-intensity profile (Borsuk, 1983). Thus, it cannot be assumed that the degradation in total power and butt-coupled power will be proportional over the course of life testing. Therefore butt-coupled light-power measurements during aging are preferable to broad-area measurements, because they resemble the use condition.

Since high-radiance LEDs are operated at high currents, the power P dissipated in the device can cause a large rise in junction temperature T_J over the ambient. The power dissipated can be written as $I(IR_S + V)$, where I is the applied current, R_S is electrical resistance, and V the junction voltage. Thermal impedance Z_{th} is defined as the increase in T_J per unit power dissipated ($Z_{th} = \Delta T_J/\Delta P$). Surface-emitting LEDs have $10-20$-μm plated-gold heat sinks to decrease the thermal impedance between chip and header (King et al., 1975). However, T_J can rise because of high thermal impedance between the header and the outside world. There are two temperature-dependent quantities that can be used as "thermometers" to determine T_J: the peak emission wavelength λ_p and the forward voltage V_F at some low current such as 1 μA, where self-heating is negligible. In the first case, $\Delta\lambda_p/\Delta T_J$ measured at low duty cycle, where heating is negligible and $\Delta\lambda_p/\Delta P$ measured at full power can be used to find Z_{th}. In the second case, $V_F(1\ \mu A)$ is first measured as a function of temperature. Then, the LED is operated at full current until it reaches thermal equilibrium; the current is turned off abruptly; $V_F(1\ \mu A)$ is measured a short time (< 100 nsec) after turn off; and T_J is calculated from the temperature dependence of $V_F(1\ \mu A)$.

In operation, LEDs are put in packages where they can be well heat sunk to a large thermal mass, with the result that Z_{th} can be kept as low as 15°C W^{-1} (Johnson et al., 1980). In life testng, where heat sinking is often impractical, Z_{th} can be of the order of 300°C W^{-1} (Zipfel et al., 1983). For an LED operating at 200 mA and 1.5 V, this results in $\Delta T_J = 90$°C. The temperature rise ΔT_J is simply treated as an increase in the effective aging temperature, and LED aging results are usually given in terms of T_J.

III. GaAs/GaAlAs LEDs: Results of Accelerated Aging

1. BURRUS STRUCTURE

The first high-radiance LED developed for fiber optics was a homostructure with a thin (3–5 μm) p-diffused active layer on an n-GaAs wafer (Burrus and Dawson, 1970). A well was etched in the thick, highly absorbing substrate, and the optical fiber was held in place in the well with epoxy resin. An interesting effect was observed in the homostructure, which has received little attention since then. Degraded devices could be restored to their initial light output by an applied dc reverse bias at elevated temperature. Similarly, diodes could be kept from degrading by superimposing a periodic reverse pulse during normal dc forward-bias operation. It was concluded that degradation was caused by rapidly diffusing metallic impurity ions, which drifted into the junction region as a result of the diminished fields in forward bias and decreased the radiative recombination efficiency of the device. The next generation structure, shown in Fig. 1a of Chapter 5, was an $Al_xGa_{1-x}As$ DH with the etched well extending all the way to the n-confining layer (Burrus and Miller, 1971). These devices could last for several thousand hours at 25°C and $J = 7500$ A cm^{-2}. These early reports proved the feasibility of LEDs for fiber optics, not only in terms of light power P_F that could be launched in a fiber but also in terms of reliability.

A major cause of failure in etched-well LEDs was identified as the strain in the thin epilayers due to the differential thermal expansion between the semiconductor and the hard UV-cured epoxy used to hold the fiber in place (King *et al.*, 1975). The devices were given a 24-hr burn-in at 130°C and 3 kA cm^{-2}. Devices passing the burn-in were then encapsulated in the UV-cured epoxy and given an additional burn-in at 80°C. About 75% of the encapsulated devices failed the second burn-in due to DLD formation, and all devices degraded rapidly in thermal cycling. It was found that most of these failures could be eliminated by applying a soft, compliant conformal coat to the LED before encapsulation in epoxy to reduce strain. With this coat, second burn-in yields increased to 75%, and the devices behaved similarly to unencapsulated devices during accelerated aging with projected lifetimes > 10^5 hr at 25°C.

The reliability of Burrus LEDs using proton bombardment at the p surface to achieve current confinement has been compared with devices using conventional oxide-masking techniques (Dyment *et al.*, 1977). Both types of devices were operated at 100°C and 3 kA cm^{-2}. Fifty percent of the devices degraded 20% in the first 1000 hr and were continuing to degrade. The other 50% of the devices degraded catastrophically. Since there were no significant differences observed in the reliability of the two types of devices, it was concluded that proton bombardment *per se* introduced no new degradation modes. However, the large drop in power in only 1000 hr and the large freak population suggest that the reliability of these devices is marginal for transmission-system requirements. Zucker and Lauer (1978) similarly report a significant drop in light outut in Burrus LEDs operated at 3.7 kA cm^{-2} and 50% duty cycle at room temperature.

Other reports, however, have shown that Burrus LEDs can be exceptionally reliable (Hersee and Goodfellow, 1976; Hersee and Stirland, 1977). Several device structures were studied, including those with $p-n$ junctions formed by Zn diffusion or grown by VPE and those with proton isolation or with oxide isolation. Aging experiments were done for $T_J = 130-210\,°C$ and $J = 2.5-20$ kA cm^{-2}. All devices showed an initial drop of ~15% followed by a much lower decay rate characterized by an activation energy $E_A = 0.8 \pm 0.1$ eV. The estimated time to a 50% drop in light output for devices that remained free of DLDs was 4×10^6 hr at room temperature. Devices fabricated with a 6-μm contact and operated at 140 kA cm^{-2} were shown to be as reliable as those with larger contact areas operated at low current densities. During aging, the slope of the I-V characteristic decreased, indicating an increase in the space-charge recombination component of current. The falltime of the optical pulse emitted by the LED decreased during degradation. These results show that degradation is due to the creation of nonradiative recombination centers at the junction that reduce the minority-carrier lifetime, thereby decreasing the internal quantum efficiency. Dark-spot defects (DSDs) in these devices were shown by O'Hara *et al.* (1977) to result from dislocation generation around Zn precipitates. Devices developing DLDs and DSDs had lifetimes of between 10^4 and 10^5 hr. The possibility that there is an ultimate limit to MTTF because of the breakdown of the gold-based p contact (Hersee, 1977) is discussed in Section 12.

SpringThorpe *et al.* (1982) fabricated a DH Burrus LED in which a ring etched into the p surface and concentric with the p contact provides 45° mirror surfaces to reflect back into the fiber light that would otherwise be lost. The high percentage of failures (40%) due to DLD formation during a 100-hr, 100°C burn-in suggests that the added stress in complicated chip geometries such as this may promote DLD growth and limit the usefulness of such devices.

The inconsistencies in the reports of Burrus LED reliability by these various groups are most likely due to problems inherent in fabrication of the well. The epitaxial material left when the well is etched away is very thin, with nothing to support or protect it. The active area can be easily damaged during assembly by being poked with the fiber. In addition, the lattice mismatch between the GaAs substrate and the high-aluminum GaAlAs n-confining layer is a source of stress, particularly at the edges of the etched well. For these reasons, it is not surprising that large numbers of devices can fail catastrophically by dark-line defect formation.

2. PLANAR SINGLE HETEROSTRUCTURE

The variable reliability and difficulty in fabricating etched wells with high yield make Burrus LEDs somewhat impractical for transmission systems. The next generation of devices were planar single heterostructures (SH) with the entire substrate etched off. These devices have the advantage that the LED consists only of material grown by LPE. Yamakoshi *et al.* (1978)

speculate that these devices will have enhanced reliability because the bulk substrate with its higher defect and impurity density cannot act as a source of defects during aging. Stress due to lattice-constant mismatch between the substrate and the n-confining layer (window layer) is eliminated. Typically, planar devices are not encapsulated in epoxy as are Burrus LEDs, eliminating a major source of strain in the active layer.

Abe *et al.* (1977) report a planar device in which an n-$Ga_{1-x}Al_xAs$ graded confining layer and an n-GaAlAs cap layer are grown on GaAs substrate. Zinc is diffused through the cap layer and part of the way into the window layer to form p confining and p active layers. The devices were operated at 100 mA through a 35-μm contact ($J \cong 10$ kA cm^{-2}) at 20°C for 100 hr (Yamakoshi *et al.*, 1977). All devices which degraded rapidly (104 of 834) showed DLDs, and all failures occurred in the first 20 hr. Samples that were clear at 100 hr were subjected to high-temperature aging. None of these devices developed DLDs, and all degraded gradually in light output with $E_A = 0.57$ eV and an estimated half-life at 40°C of 5×10^6 hr. The results were independent of current density in the range 3.5–10 kA cm^{-2} and were the same for a mesa structure. The authors conclude therefore that gradual degradation in these devices is a bulk effect that does not depend on device geometry or current density.

In a particularly simple version of the planar SH LED (Keramidas *et al.*, 1980), an n-$Ga_{1-x}Al_xAs$:Te graded band-gap layer is grown first. After this, a 5-μm-thick p-GaAs:Ge layer is grown, and the substrate is etched off. In most reports that have appeared, current density is estimated simply by dividing device current by contact area and ignoring current spreading. Here, current spreading due to the finite conductivity of the 5-μm p layer is taken into account, and J is estimated to be 700 A cm^{-2} (versus 3×10^3 A cm^{-2} from simple geometrical considerations). These devices proved exceptionally reliable with no catastrophic failures, and $E_A = 0.75$ eV and a MTTF at $T_J = 74°C$ of 2.7×10^7 hr. The main limitation of the planar SH LED is the power that can be launched in a fiber.

3. Planar Double Heterostructure

The next generation of devices was planar DHs grown by LPE with the substrate etched off as in the SH to expose the window layer (Abe *et al.*, 1978). The DH is used to confine carriers and reduce nonradiative surface recombination thereby increasing the power that can be launched in a fiber. However, it should also have increased reliability over the SH because the active area is entirely isolated from any exposed surface. It has been shown that dispersed low-radiative-efficiency regions can move in from laser facets in LED-mode operation (Ettenberg and Kressel, 1980). Increasing the edge-to-volume ratio of the diode (and thus the proportion of recombination at the edges) increased the degradation rate in LEDs (Kressel et al., 1977). It was concluded from these studies that exposed surfaces act as sources of point defects that can move by recombination-enhanced defect reactions to internal sinks within the diode.

The planar DH LED is shown in Fig. 1 (Yamakoshi et al., 1978; Chin et al., 1980). The 50-μm-thick n-$Ga_{1-x}Al_xAs$:Te graded band-gap confining layer, 0.7-μm-thick p-GaAlAs:Ge active layer, and 2-μm-thick p-GaAlAs:Ge confining layer result in a very thin chip once the substrate is etched off. The n layer must be grown as thick as possible since it is the mechanical support for the chip. The thickness of this layer can cause crystal-growth problems that can have a major impact on reliability, as is discussed later in Section 10.

As expected these devices exhibit a certain percentage of early failures due to DLD formation, which is directly related to the dislocation density in the substrate (Zipfel et al., 1981). Early failures are effectively screened out in a 100-hr, 100-mA room-temperature burn-in followed by visual inspection of the light-emitting spot for dark defects. DLD growth has been shown to be strongly current-density dependent but only weakly thermally activated (Yamakoshi et al., 1978). These effects, as well as the influence of crystal defects and stress on DLD formation are discussed in Part IV.

Devices that remained clear in burn-in have proved in long-term accelerated aging to be exceptionally reliable. Degradation in these devices can be described by simple mathematical expressions. For example, Yamakoshi et al. (1978) find that the relative light output L/L_0 decreases with time as $\exp(-t/\beta)$, with activation energy $E_A = 0.56$ eV and MTTF at 25°C = 4×10^7 hr, whereas Zipfel et al. (1981) find that it varies as $\exp[-(t/\tau)^{1/2}]$ with $E_A = 0.65$ eV and MTTF at 25°C = 9×10^7 hr. The failure distributions are very tight, with σ typically about 0.5. Aging results are shown in the Arrhenius plot in Fig. 2. (Also shown in Fig. 2 is the rate constant α for an annealing process that can cause the light output to increase. This is covered

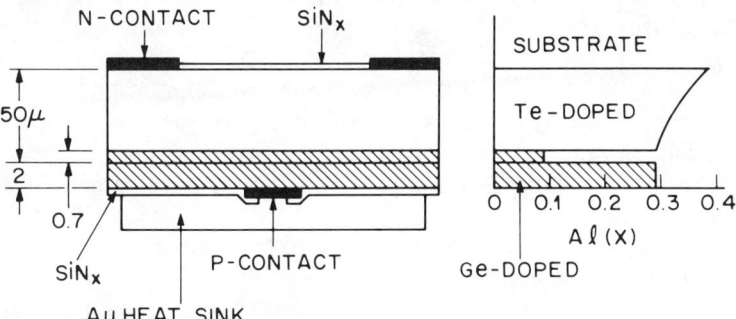

FIG. 1. Schematic diagram of planar DH $Ga_{1-x}Al_xAs$ LED. [From Chin et al. (1980).]

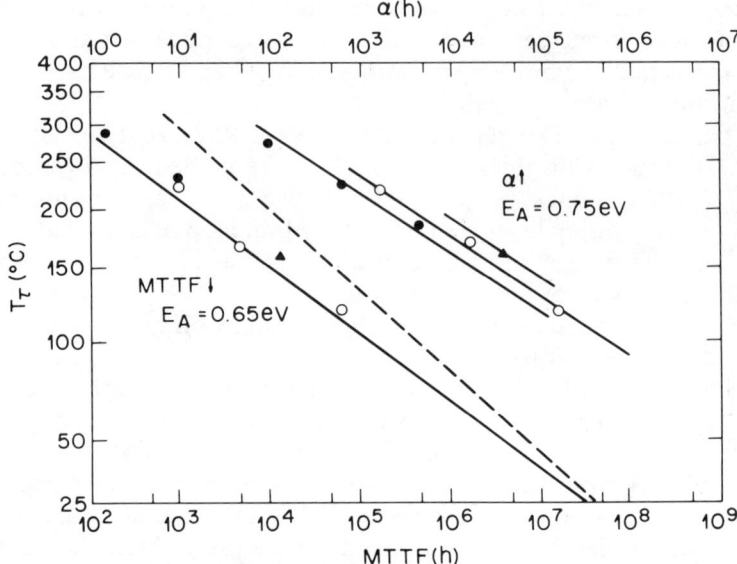

FIG. 2. Arrhenius plot for planar DH $Ga_{1-x}Al_xAs$ LED: (---) results of Yamakoshi et al. (1978); (—) results of Zipfel et al. (1982); α, is the rate constant for recombination-enhanced annealing; stress current: 30 (▲), 60 (○), 100 (●) mA.

in Section 12.) The closeness of these estimates indicates that gradual degradation is a well-controlled, reproducible process in the planar DH LED. The lifetime estimates are one to two orders of magnitude longer than for the Zn-diffused SH (Yamakoshi et al., 1978) and for the Burrus DH (King et al., 1975).

Although there has been no reported comparison of reliability for Burrus and planar LEDs for which identical substrates and processing techniques are used, the data presented in these last three sections suggest that there can be many more infant and freak failures in etched-well devices.

Light-emitting diodes with junction-current confinement achieved by a regrown buried heterostructure have been reported (Escher et al., 1982; Speer and Hawkins, 1980). Hawkins (1984) has reported aging results for two variations of this structure: a lightly doped (1×10^{18} cm^{-3}) high-power, low-speed LED and a heavily doped (5×10^{18} cm^{-3}) lower-power, high-speed LED. Aging data over the junction temperature range from 48 to 160°C give an activation energy of 0.43 eV and a median life at 25°C of 3×10^5 hr for the low-speed device. Data for the high-speed devices suggest similar values of median life. The times to failure for the low-speed device are log-normally distributed with $\sigma = 1.5$. For this large value of σ, at 25°C about 20% of the LEDs will fail by 3×10^4 hr. This large number of early failures and low value of median life indicate that the regrown buried heterostructure is significantly less reliable than the planar DH LED. The activation energy of 0.43 eV is far enough from the findings of Yamakoshi and Zipfel to suggest that a different failure mechanism other than gradual degradation is in play. Defects and dislocations at the regrown surface may limit the life of this structure. The advantage of the buried heterostructure LED is that current confinement is grown-in. Thus, the complicated processing required to produce the small p contact in the planar DH LED is not necessary.

4. Edge-Emitting Structure

There have been few reports on the reliability of edge-emitting LEDs. Kressel *et al.* (1975) reported four LEDs with oxide-stripe geometry that did not degrade for 10^4 hr at 1000 A cm^{-2}. Davies *et al.* (1980) reported edge-emitting LEDs that degrade only gradually at current densities from 3 to 11 kA cm^{-2} but which fail catastrophically in 1000 hr at 16 kA cm^{-2}. They found no systematic dependence of degradation on stripe width, length, or facet coating.

IV. Rapid Degradation Mechanisms in GaAs/GaAlAs LEDs

5. DLD Formation in LEDs and Lasers

Since DLD growth was a severe problem in laser reliability in the early days of development, a large number of studies have been made on this phenomenon, and several excellent reviews exist (Kressel and Butler, 1977; Casey and Panish, 1978; Hayashi, 1980; Newman and Ritchie, 1981; Ettenberg and Kressel, 1980). Depending on threshold current, lasers operate in the current density range $2-5 \times 10^3$ A cm^{-2} (Ettenberg and Kressel, 1980). Laser structures can degrade by DLD formation whether being operated in the LED or laser modes (Kressel and Butler, 1977). Dark-line defects can also form by optical excitation in bulk material (Johnston and Miller, 1973) and by electron-beam excitation (Petroff and Kimerling, 1976). Dark-line defects are observed to form readily at defects such as dislocations, stacking faults, and damage and in regions of high strain (Johnston *et al.*, 1974; Ito *et al.*, 1974). It was concluded from these observations that DLD formation requires only the energy from nonradiative electron–hole recombination and some type of crystal defect.

Many elegant observations have been made by transmission electron microscope (TEM) of DLDs in laser structures (Petroff and Hartman, 1973; Hutchinson *et al.*, 1975). Although preparation of samples for TEM studies on the nature of DLDs in LEDs is more difficult than in lasers because of the thick active layers, equivalent observations have been made in LEDs (O'Hara *et al.*, 1977; Ueda *et al.*, 1977, 1979). Dark-line defects were shown to consist of networks of dislocations that originated from threading dislocations. The networks were made up of ragged helices elongated primarily in the ⟨100⟩ and ⟨210⟩ directions, and occasionally in the ⟨110⟩ direction (Petroff and Hartman, 1973).

Two different mechanisms have been proposed for ⟨100⟩ and ⟨210⟩ DLD growth: one in which dislocations extend themselves by climb and another in which they grow by glide. In the climb model, dislocations grow from an existing dislocation by the absorption of interstitial point defects and the extension of the dislocation out of the plane in a reaction that is enhanced by the energy from electron–hole recombination at the defect (Petroff and Kimerling, 1976). This model requires a very high density of native point defects. Hutchinson et al. (1975) proposed that climb proceeds not by the absorption of existing defects but by the emission of vacancy pairs at the dislocation, again by a recombination-enhanced defect reaction. The source of point defects required for climb is still a major controversy. It was clear, however, from observations of DLDs in regions of high stress that the climb model is not sufficient and that stress plays an important role in DLD growth. An alternative model was proposed in which ⟨100⟩ dislocations extend themselves by glide (Matsui et al., 1975).

6. Phenomenology of DLD Growth

Reviewed here are some of the studies that have been done on DLDs in surface-emitting LEDs where the active area can be readily observed through the window layer with an IR microscope. Essentially the same phenomena have been seen in lasers, although IR observations are much harder because special window lasers must be made (DeLoach et al., 1973).

a. Current Dependence of DLD Growth

Figure 3 shows the typical history of a DH diode developing DLDs (Zipfel et al., 1981). After 15 min at 100 mA and room-temperature ambient, a DLD is first visible growing in the ⟨100⟩ direction, although there is no measurable effect yet on light output. At 23 hr, a ⟨210⟩ line appears, and by 46 hr, a large fraction of the emitting area has DLDs, and the light output has fallen by 46%. Beyond this time very little additional DLD growth is observed.

Yamakoshi et al. (1978) describe many features of DLD growth. In both SH and DH Ge-doped LEDs, ⟨100⟩ DLDs were observed to grow at constant velocity and then to stop growing for intervals of time, sometimes for thousands of hours. The growth velocity was found to vary as J^2 in the range 2×10^3 to 5×10^4 A cm^{-2}. No DLDs are observed when LEDs are aged without bias for temperatures up to 250°C, and DLD growth under bias is only weakly dependent on temperature, with $E_A \leq 0.2$ eV. Dark-line defect

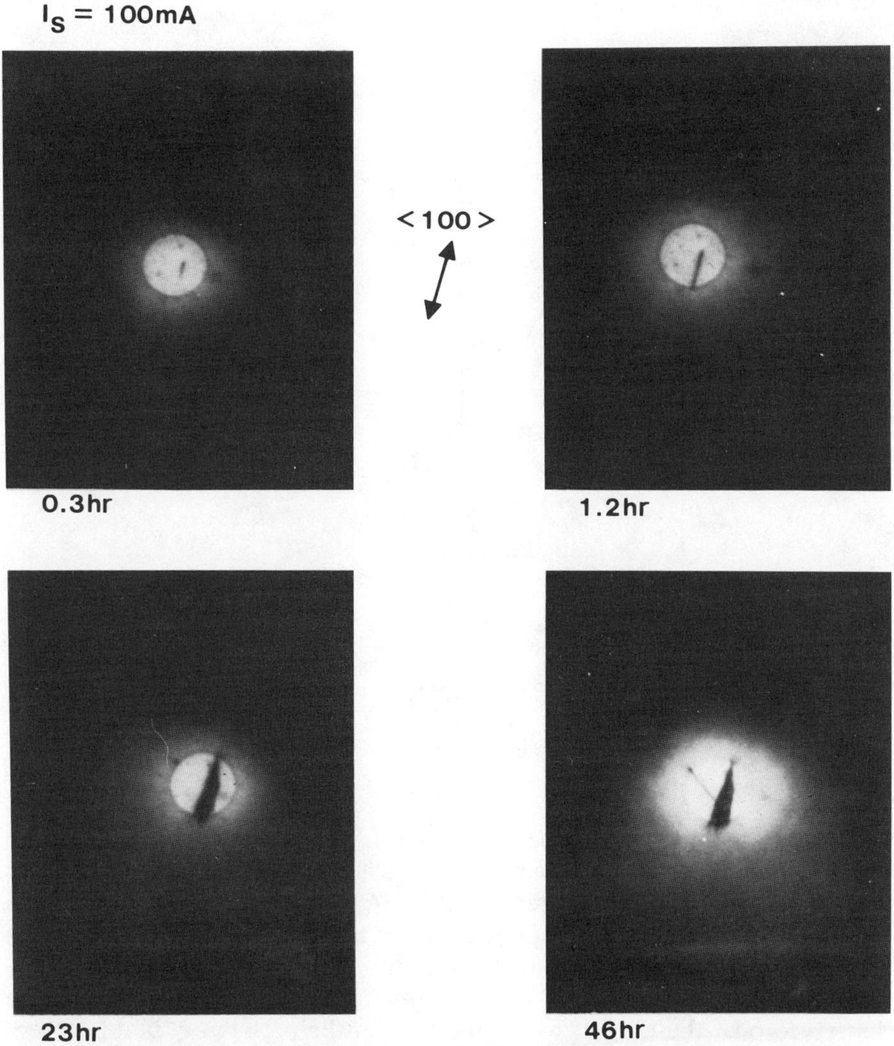

$I_S = 100$ mA

<100>

0.3 hr

1.2 hr

23 hr

46 hr

FIG. 3. The EL image of a planar DH $Ga_{1-x}Al_xAs$ LED developing DLDs at 100 mA and 25°C. [From Zipfel et al. (1981).]

velocity is weakly dependent on Ge-doping density up to $p = 5 \times 10^{18}$ cm^{-3}, whereas above this value it drops off very rapidly as p^{-3}. At high doping density, Ge tends to enter the crystal in precipitates or other defect states (Roszytoczy and Wolfstirn, 1971), and the drop-off in DLD velocity is attributed to dislocation pinning at these defects.

Figures 4 and 5 are cumulative probability plots showing, respectively, the

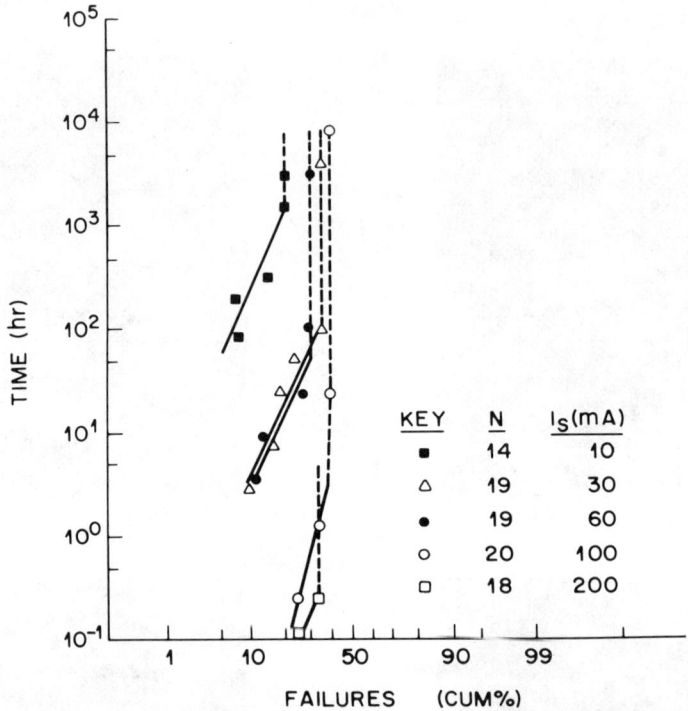

FIG. 4. Log-normal probability plot showing current dependence of DLD initiation time at room temperature. The number of diodes in each sample is given by N. [From Zipfel et al. (1981).]

current and temperature dependence of DLD initiation time t_{DLD}; i.e., the time at which the first DLD in an LED is just visible. The samples used were all taken from the same wafer. At 200 mA, all of the devices that would ultimately fail showed at least one DLD by 1 hr, whereas, at 10 mA, DLDs continue to appear for 10^3 hr. Beyond these times, no further devices develop DLDs for times $> 10^4$ hr. The time at which all failures have been observed (i.e., the time at which the line on the probability plot becomes vertical) is a strong inverse function of J ($t \propto J^{-3}$). However, the same percentage of devices (30 ± 10% on this wafer) ultimately develop DLDs at each current.

b. Temperature Dependence of DLD Growth

Figure 5 shows that when t_{DLD} is studied as a function of temperature, two distinct populations are observed. There is the same early failure population of about 30% as in Fig. 4. The DLD initiation times are seen to be independent of temperature in the range 60–185°C. Unlike Fig. 4, however, at the

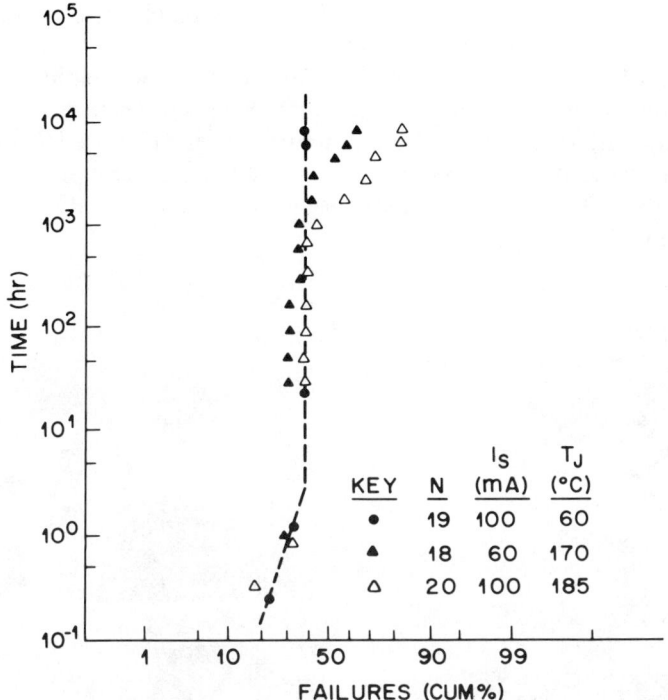

FIG. 5. Log-normal probability plot showing temperature dependence of DLD initiation time. [From Zipfel et al. (1981).]

higher temperatures a freak DLD population appears after $\sim 10^3$ hr. No freak failures were observed below 150°C (Zipfel et al., 1981). The size of the freak population is highly variable from wafer to wafer. Since MTTF for gradual degradation in the planar DH LED is typically very long, the most important factor in long-term reliability is the elimination of unexpected freak failures. Therefore, the causes for these freak failures observed at high temperatures should be well understood so that they may be eliminated in fabrication. (A burn-in to eliminate freaks would probably have to be so severe in current and temperature that the LED's useful life would be shortened.)

Several causes for these late-developing DLDs have been identified:

1. The $\langle 100 \rangle$ and $\langle 210 \rangle$ DLDs can begin to grow slowly at dislocations far from the contact area where the current density is very low. When they reach the contact area where current density is high, they grow fast and the device fails. The effect of dislocation density is discussed in Section 7.

2. The $\langle 110 \rangle$ DLDs can begin to grow at mechanical damage far from the

contact area and cause the device to fail if they reach the contact area. An example of this is given in Section 8.

3. The ⟨100⟩ DLDs can form at regions of high mechanical or thermal stress, such as the dielectric–metal interface. This is discussed in Section 8.

4. Dark patches formed at long times have been attributed to gold migration from the contact (Hersee, 1977). The gold-rich precipitates formed may contribute to DLD formation in the long term. Gold migration is discussed in Section 12.

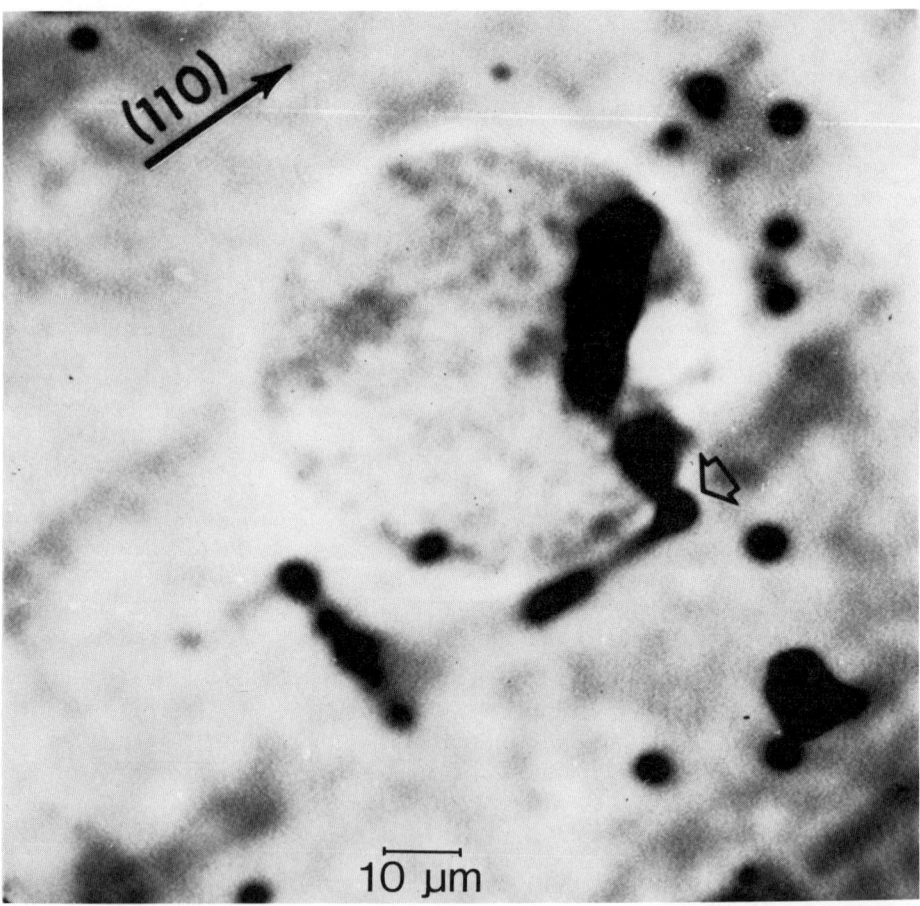

FIG. 6. Transmission cathodoluminescence image of ⟨100⟩ and ⟨110⟩ DLDs growing from threading dislocations in a planar DH $Ga_{1-x}Al_xAs$ LED after operation at 100 mA and 25°C. Heat sink, dielectric layer, and contact have been stripped off. The circular region defines the contact position. [From Chin et al. (1980).]

7. Effect of Dislocation Density on DLD Growth

Dislocations cannot be seen in electroluminescence (EL), and neither dislocations nor DLDs can be imaged in either the electron-beam-induced current (EBIC) or cathodoluminescence (CL) modes of the scanning electron microscope (SEM) in planar LEDs with the electron beam striking the thick n layer. However, both defects can be readily observed in planar LEDs, with the electron beam striking the p confining and active layers. Figure 6 is a transmission cathodoluminescence (TCL) (Chin et al., 1979) image of the p-surface of an LED that degraded during the first 100 hr of operation at room temperature by the formation of a $\langle 100 \rangle$ DLD and two $\langle 110 \rangle$ DLDs (Chin et al., 1981a). The dark spots are the points at which threading dislocations that originate in the substrate intersect the p-surface. Dislocations such as these are the sources of the DLDs. Typically, LEDs that degrade early exhibit a high density of dislocations, such as seen in Fig. 6. In devices that do not fail early by DLD formation, the contact area as seen in TCL is invariably free of dislocations. Chin et al. (1980) showed that DLDs can be formed readily at these dark spots by using the electron beam of an SEM to inject carriers.

For a substrate dislocation density ρ of 10^4 cm^{-2}, about 20% of LEDs with a 50-μm-diameter contact will have a dislocation threading the active area. (This is a rough estimate, since the dislocations tend to be clustered as in Fig. 6.) The devices represented in Fig. 4 were grown on a substrate with $\rho \sim 10^4$ cm^{-2}, and 20–40% of them developed DLDs. The percentage with DLDs tends to be somewhat higher than ρ would predict, because processing damage also causes DLD formation, as is discussed in Section 8. For $\rho = 2 \times 10^3$ cm^{-2} (typical of high quality GaAs substrates currently available) only 4% of the LEDs are expected to have a dislocation. With careful handling during processing, the percentage forming DLDs can be consistently kept at 4%, although the DLD initiation times are the same as for higher ρ (Zipfel et al., 1981).

8. Effect of Stress and Damage on DLD Growth

It is well established that stress from a variety of sources can accelerate DLD growth and result in enhanced degradation in both LEDs and lasers. An early experiment by Ikeda et al. (1975) showed that $\langle 100 \rangle$ DLDs primarily formed in the high-current regime ($J > 1000$ A cm^{-2}) but that only $\langle 110 \rangle$ DLDs formed in the low-current regime ($J < 1000$ A cm^{-2}). The $\langle 110 \rangle$ DLDs started at the p surface and extended into the substrate. Using the photoelastic effect, they observed much more strain in LEDs bonded with Si/Au solder than in those bonded with In. Degradation due to DLD formation was correlated with strain. Strain from the wire bond could be minimized by forming the bond as far from the window as possible. Whereas it is often considered imperative to bond lasers with In (Hartman and Hartman, 1973), today most LEDs are bonded with high-melting-point Si/Au solder (Yamakoshi et al., 1977) or epoxy (Johnson et al., 1980), with no apparent impact on reliability. This is because the thick gold-plated heat sinks commonly used in LEDs, but not in lasers, take up the bonding stress.

It was shown that ⟨110⟩ DLDs in a DH wafer could be induced by mechanical bending to a radius of curvature of ∼8 cm (Kishino et al., 1976). These DLDs formed at a critical stress of ∼2×10^9 dyne cm^{-2}, long before wafer cracks appeared. Kamejima et al. (1977) determined that both uniaxial stress and carrier injection are necessary for DLD generation. Kishino et al. (1976) showed that the ⟨110⟩ DLDs formed by stress alone were the same as those induced optically or under bias. It was therefore conjectured that ⟨110⟩ DLDs can grow by glide, at a rate that can be enhanced by energy from electron–hole recombination (Kamejima et al., 1977).

Another significant source of degradation in GaAs LEDs is ⟨110⟩ DLD growth by recombination-enhanced glide at mechanical damage (Monemar and Woolhouse, 1976). This has been shown to account for some of the freak failures observed in long-term accelerated aging (Section 6). Figure 7 shows the secondary-electron and TCL images of an LED that failed catastrophically after 5000 hr of operation at 60°C and 6 kA cm^{-2} (Chin et al., 1981a). The TCL image shows that two scratches were initially present (the two lines indicated by arrows). During operation, a ⟨110⟩ DLD grew from one of the scratches toward the contact. Once it reached the contact, it grew fast primarily in the ⟨100⟩ direction, and the device failed rapidly. Small DLDs are seen growing from the other scratch. Because of current crowding, the current density away from the active area is much less than that on it. It is not clear whether the ⟨110⟩ DLD far from the active area requires electron–hole recombination to grow or whether dislocation glide proceeds in the strain field around the crack alone. The growth of ⟨100⟩ DLDs from glided ⟨110⟩ DLDs was observed by Ueda et al. (1980c). Similarly, DLDs grow readily from cracks or pits on the bottom p surface. Because of this, great care must be taken not to damage the LED surface during growth or processing.

Detailed observations of the effect of high mechanical and thermal stress were made by Ueda et al. (1980c). They used TEM to study LEDs that had degraded catastrophically due to DLD formation under very severe conditions. Two phenomena were observed. At 30 kA cm^{-2} and 120°C ambient temperature, a dark ring of defects appeared in the emitting region corresponding to the circular opening in the SiO$_2$ mask. This dark ring was attributed to the generation of a high density of dislocations and dislocation loops (in some cases, stacking faults) by dislocation-glide motion due to the relaxation of the stress in this region. In other diodes aged under conditions

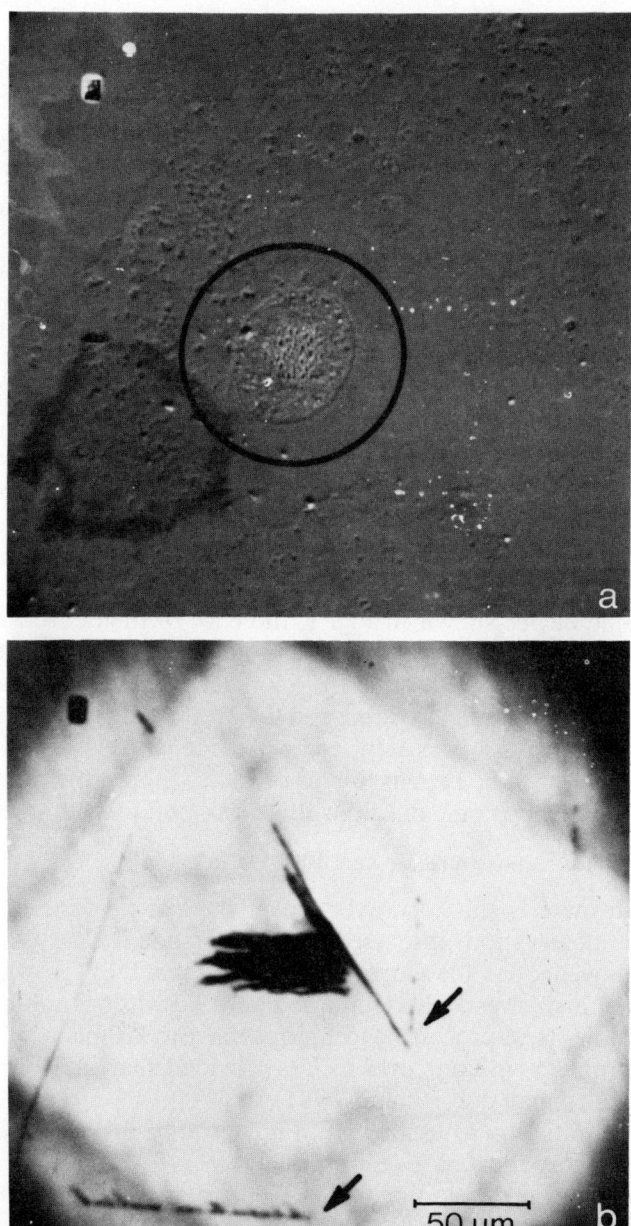

FIG. 7. (a) Secondary-electron image of the p surface of a planar DH $Ga_{1-x}Al_xAs$ LED that failed during long-term aging. Pits can be seen within the small circle on the contact defined by the opening in the dielectric. (b) TCL image of same region showing $\langle 110 \rangle$ and $\langle 100 \rangle$ DLDs. The sources of the DLDs are the two scratches indicated by arrows. [From Chin *et al* (1981a).]

such as 4×10^3 A cm^{-2} and 300°C, a cross-hatched dark region with many ⟨100⟩ and ⟨110⟩ DLDs appeared. This was attributed to dislocation glide followed by climb of dislocation dipoles and loops. A threshold for this second process in terms of some parameter, such as current density or defect density, was postulated. Precipitates were observed but not identified.

Other processing steps can result in DLD growth from highly stressed regions. LEDs with $p-n$ junctions formed by Zn diffusion are apt to form DLDs and DSDs (Hersee and Goodfellow, 1976). Strain at Zn precipitates resulting from heavy Zn diffusion can cause networks of misfit dislocations that appear as DSDs from which ⟨100⟩ DLDs form (O'Hara et al., 1977).

The dielectric layer used in most LEDs to isolate the semiconductor electrically from the heat sink can be a significant source of stress. The dielectric (typically SiO_2 or SiN) is deposited at high temperature, and because of differential thermal expansion between it and the semiconductor, the epilayers are under compression at room temperature. At the edges of the circular opening in the dielectric, there are very high local strain fields. Strain fields in the oxide-stripe laser have been calculated in detail (Goodwin et al., 1979). In the device shown in Fig. 1, the edge of the dielectric is a circle concentric with the 35-μm contact and of half the diameter. In this geometry, the dielectric is deposited after the p contact so that the contact metal can relieve the stress at the dielectric edges. However, DLDs frequently appear at about this position as well as at the edges of oxide-stripe lasers. Therefore the dielectric layer must be kept as thin as possible without the introduction of pinholes, and deposition conditions must be carefully controlled.

9. Burn-In to Eliminate Early Failures

Although there is still controversy over the exact mechanism of DLD growth, the observations discussed here have made it clear that threading dislocations were often the source of DLDs and that DLD growth was faster in regions of high stress. Therefore the use of low-dislocation-density substrates and low-stress-processing techniques have today made DLD growth a minimal problem in lasers and LEDs, provided that they are effectively burned-in (Newman and Ritchie, 1981).

The DLD initiation time shown in Fig. 4 is to be distinguished from the actual TTF. From the data given in Zipfel et al. (1981) it can be shown that the time for an LED developing DLDs to drop 50% in light output is proportional to J^{-4}, whereas t_{DLD} was proportional to J^{-2}. At 100 mA, the TTF for devices with DLDs is 100 hr, whereas at 60 mA it is 10^4 hr (O'Hara et al., 1977; King et al., 1975). Operated without screening at 60 mA the early failures would come at about 1 year, which would be intolerable in most systems. Therefore, a 100-hr burn-in at 100 mA is often reported (Yamakoshi et al., 1977; Zipfel et al., 1981). This is a practical length of time for a burn-in, since at 100 hr the DLD is well developed and easily detected by an operator using an IR microscope. A burn-in at 200 mA would be faster, but there is evidence that at large overcurrents additional failure modes are accelerated unnecessarily. For example, Ueda et al. (1982b) has reported failures due to gold migration after very large current pulses.

Since early DLD formation is not appreciably temperature dependent, nothing is gained by performing the burn-in above room temperature. For highest reliability, the visual inspection must use vertical illumination to screen out devices with cracks or scratches on the bottom surface, as well as DLDs.

10. GROWTH-RELATED EFFECTS

As discussed in Sections 1, 2, and 3, the planar DH LED has a great advantage in reliability over the Burrus LED because the substrate is etched off, eliminating it as a source of stress. The disadvantage to the planar structure is that the crystal growth is much more difficult because a thick (~ 50 μm) graded band-gap n-GaAlAs window layer must be grown to act as mechanical support for the chip. This may explain why the planar device is not widely used.

Chin et al. (1980) observed that even with low-dislocation-density substrates ($\rho \cong 10^3$ cm^{-2}), the burn-in yield for the planar DH LED could be almost zero on some wafers. A TCL observation of these low-yield wafers showed a high density of defects ($> 10^4$ cm^{-2}) that could not be seen in Nomarski contrast microscopy. These defects were localized in the active and last-to-grow layers. A TEM study by the same authors revealed that the dark defects were dislocation tangles ~ 2 μm in size not associated with substrate dislocations. The observation of these dislocation tangles led to this explanation: at the temperature at which the growth of the thick window layer is completed, the active-layer melt is supercooled more than 5°C. Faceted growth on $\{111\}$ planes is initiated locally with different compositions on different growth fronts, and dislocation networks form to accommodate these growth fronts. The networks can be eliminated by adjusting the melt compositions so that the p layers will grow under equilibrium conditions. Transmission cathodoluminescence (TCL) can be used to inspect wafers for these growth defects before processing.

In addition to the reduction of dislocation density and strain, cleanliness of materials and gasses during epitaxy has served to decrease the incidence of DLD growth and to improve reliability (Casey and Panish, 1978). It has been shown that particulate matter can generate stacking faults and dislocations in epilayers (Woolhouse et al., 1976). Stacking faults in GaAs epilayers can be the sources of DLDs (Johnston et al., 1974). Stacking faults have been observed in GaAlAs layers grown by LPE on Syton-polished GaAs substrates

(Dutt et al., 1981). The faults were found to result from calcium contamination on the substrate surface. Epitaxial layers grown on bromine–methanol-polished substrates were largely free of faults.

Ishii et al. (1977) have shown that oxygen contamination in the ambient gas during LPE results in crystal defects such as dark spot defects and saucer pits. Lasers with these defects degraded rapidly due to DLD formation, whereas those grown with extremely low oxygen concentrations were long-lived. The addition of aluminum to the active layer of lasers (Ishii et al., 1977) and LEDs (Ettenberg et al., 1974) has been shown to increase reliability significantly. There have been several explanations: that stress from the mismatch at the heterojunctions has been reduced (Olsen and Ettenberg, 1977); that the addition of aluminum produces better stoichiometry (Ettenberg and Kressel, 1980); or that the aluminum getters oxygen (Ishii et al., 1977). Since LEDs with no aluminum in the active layer have been shown to be exceptionally reliable (Keramidas et al., 1980), a plausible explanation is that these devices were grown in an oxygen-free environment and therefore require no aluminum.

V. Gradual Degradation in GaAs/GaAlAs LEDs

11. RELIABILITY PROJECTIONS

It can be concluded from the evidence presented in Parts III and IV that once failures due to DLD growth are eliminated by proper device design, fabrication, and burn-in, GaAs/GaAlAs LEDs remain clear in accelerated aging and degrade only gradually in light output by processes that can be well defined mathematically. Estimates of E_A ranged from 0.56 and 0.57 for the SH and DH planar LEDs of Yamakoshi et al. (1978) to 0.65 eV for the DH planar LED of Zipfel et al. (1982), to 0.6 ± 0.16 eV for early decay (Hersee, 1977), and to 0.8 ± 0.1 eV for long-term decay (Hersee and Goodfellow, 1976) in Burrus LEDs. Both Yamakoshi et al. (1977) and Zipfel et al. (1982) report no J dependence for gradual degradation. It is difficult to compare these results with those for lasers where the degrading parameter often chosen is the threshold current I_{th}. Imai et al. (1982) find $E'_a = 0.5$ eV and little evidence of J dependence, in reasonable agreement with LED results. However, Mizuishi et al. (1979) find that I_{th} degrades exponentially with the initial value of I_{th}. This suggests that the laser degrades faster when more defects are initially present, which is not inconsistent with LED results.

Straightforward reliability estimates can be made for planar DH LEDs, since long-term degradation is well defined with MTTF $> 10^7$ hr at 25°C and since there is no significant freak population at this temperature. Zipfel et al. (1981) report that the failures in this structure are log-normally distributed with $\sigma = 0.5$. It can therefore be concluded that failure rates for this LED will be very small in systems use, i.e., less than 1 FIT, where a FIT is defined as 1 failure per 10^9 device hours.

12. Mechanisms in Long-Term Aging

Little is known about gradual degradation mechanisms in GaAlAs/GaAs LEDs. One reason is that they are small effects, probably involving point–defect interactions, which are hard to detect. There have been no reports clearly identifying different crystal-growth or processing techniques as reasons for variations in MTTF once catastrophic failures due to DLDs were eliminated. Three effects that have been seen in long-term aging are now discussed.

a. Electromigration of Gold from the Contact

The experiments of Hersee (1977) may have implications for the ultimate reliability of GaAs devices with gold-based contacts. The LEDs studied were GaAs Burrus structures operated at 15 kA cm^{-2} at ambient temperatures from 20 to 140°C. Four distinct stages were observed in the degradation. During the first two stages the light output degraded exponentially by 15% in 10^3 hr at 25°C and 50% at 135°C with an activation energy of 0.6 eV and a decay rate proportional to J. During this process, the slope of the I–V characteristic decreased, indicating an increase in the space-charge recombination current due to the electromigration of charged defects to the junction.

During the third stage, the light output stops degrading for $\sim 10^3$ hr, but faint dark patches develop in the EL image. During the fourth stage, the patches become very dark and dense, and the light output begins to degrade again. Gold-bearing precipitates were found in the contact area near the junction in degraded devices. It was concluded that fourth-stage degradation was due to breakdown of the Ti/Au p contact and subsequent electromigration of gold to the junction. Using the diffusion constant for gold in GaAs, the authors calculate that at 25°C it will take $\sim 10^6$ hr for a significant amount of gold to reach the junction in the best devices, with many devices failing earlier. Other reports (Yamakoshi *et al.*, 1977; Zipfel *et al.*, 1981) show only one stage of degradation, possibly because the first stage occurs over a much longer time scale.

Further evidence that gold can cause degradation comes from Chin *et al.* (1982a) who studied LEDs that failed catastrophically by the growth of dark regions after remaining clear for 5×10^3 hr at 6×10^3 A cm^{-2} and $T_J = 150°C$. They found a high density of small pits in the p surface elongated in the $\langle 110 \rangle$ direction. These pits can be seen in the secondary-electron image of the p surface in Fig. 7a. During high-temperature aging under bias, gold from the massive heat sink penetrates the Ti layer and interacts strongly with the GaAlAs. Pits do not form in the region protected by SiO_2, where only the small amount of gold in the contact is available. Diodes fabricated using a Ti/Au Schottky barrier for isolation rather than dielectric isolation failed rapidly due to Au penetration of the Ti. Gold migration could be reduced by the use of a barrier metallization such as Pt or Pd.

b. Creation of Point Defects

A persistent idea is that gradual degradation in LEDs is due to the creation of or migration of nonradiative recombination centers to the junction with a consequent decrease in internal quantum efficiency. Deep-level transient spectroscopy (DLTS) (Lang, 1974) has proved to be a useful technique for detecting traps in semiconductors. Newman and Ritchie (1981), however, make a distinction between traps and nonradiative recombination centers and suggest that this technique might have limited usefulness in correlating defects with degradation in light output in LEDs and lasers. In addition, the surface-emitting LED structure is not well suited to investigation by DLTS, which probes only the depletion layer in reverse bias. Since the p active layer is highly doped in LEDs, the DLTS signal comes primarily from the n layer and not from the p layer, where recombination-enhanced defect reactions take place during aging. Additionally, the active area (light-emitting region) is only a small fraction of the junction area. Kondo et al. (1983) surmounted these difficulties with the fabrication of a special structure with low Ge-doping density and a broad-area contact. A two-stage increase in deep levels during aging was observed by DLTS. Small dislocation loops were observed in the active region of aged devices by TEM. Because the concentration of deep levels observed by DLTS is greater than that of defects seen in TEM by a factor of 10, the authors conclude that the loops are not the direct image of the deep levels. They propose a model for gradual degradation in which the loops are sinks for point defects created continuously at the deep levels by energy from nonradiative recombination. This suggests that LED lifetime is limited by the number of preexisting deep levels.

c. Recombination-Enhanced Annealing

Germanium-doped LEDs can show an annealing process under accelerated aging (Zipfel et al., 1982). Figure 8 shows that at the lowest aging temperatures, light output initially decreases and then begins to increase. This increase appears earlier in time and is of greater magnitude as the aging temperature increases, until at the highest temperatures only an increase is seen. The data are interpreted as two concurrent processes: (1) the expected gradual degradation in light output with $E_A = 0.65$ eV and a rate indepen-

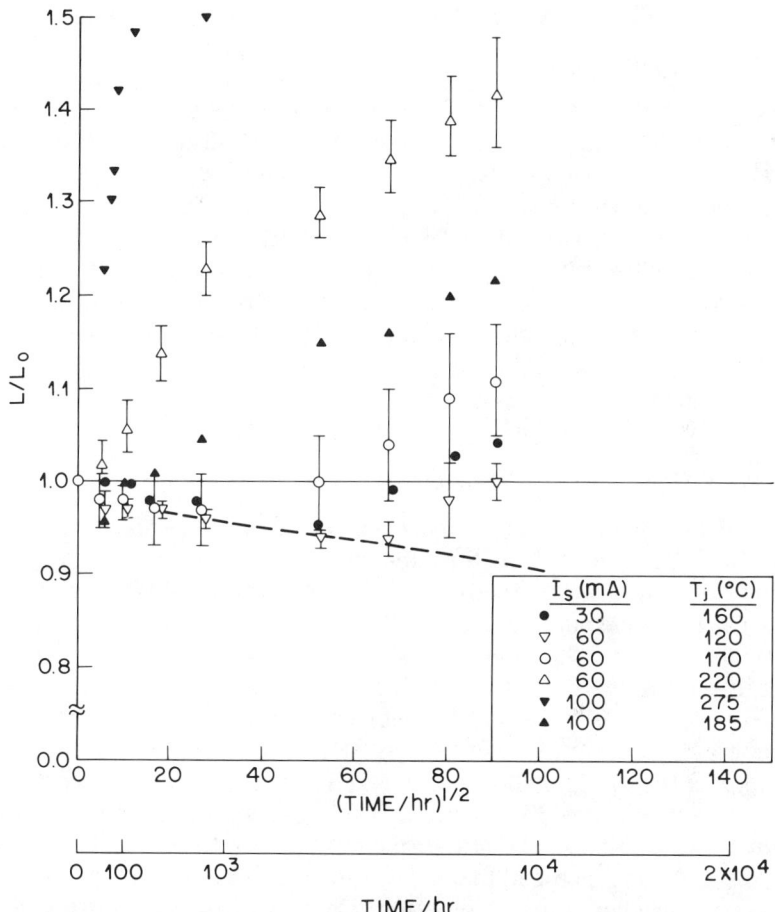

FIG. 8. Relative change in light output L/L_0 for DH $Ga_{1-x}Al_xAs$ LED plotted versus $t^{1/2}$ for various aging conditions. Recombination-enhanced annealing masks the degradation process except at the lowest temperatures. [From Zipfel et al. (1982).]

dent of J; and (2) an increase in injection efficiency due to the annealing out of nonradiative centers by a recombination-enhanced defect reaction with $E_A = 0.75$ eV and a rate constant α linear in J. The temperature and current dependence of α are shown in Fig. 2. The nonradiative centers are probably related to Ge defects. Since an annealing effect such as this can mask the degradation process, it is important to pay careful attention to the time dependence of light output during aging. Taking both processes into account, the MTTF is estimated to be 9×10^7 hr at 25°C.

VI. InP/InGaAsP LEDs: Results of Accelerated Aging

13. BURRUS STRUCTURE

Even though the InP substrate is transparent to 1.3 μm light, the earliest reports were for LEDs with a well etched in the substrate. This allowed placement of the fiber close to the light spot to increase coupling efficiency. Dentai *et al.* (1977) reported that Zn-diffused devices of this type operated at 5 kA cm^{-2} and 35°C for > 10^3 hr with no degradation. Wright *et al.* (1979) published similar results for devices with a well etched part of the way into the substrate. The only physical change associated with reduced LED output was an increase in leakage current. Some devices changed in forward voltage with no change in output power.

14. PLANAR DOUBLE HETEROSTRUCTURE

a. Gradual Degradation Characteristics

Most of the reliability studies have been done on devices of the type shown in Fig. 1b of Chapter 5. Here, the substrate is simply thinned uniformly by polishing with a bromine-methanol solution to allow close placement of the fiber to the light spot (Yeats *et al.*, 1981; Wada *et al.*, 1981; Temkin *et al.*, 1983).

The important features of the reliability of this device were described by Yamakoshi *et al.* (1979, 1981). Devices were bonded with Au/Ge solder and were operated at 100 mA dc (8 kA cm^{-2}). A large number (378) of diodes were operated at 20°C for 100 hr. In marked contrast to GaAs-based devices, no DLDs were observed even though high-dislocation-density substrates (10^5 cm^{-2}) were used, and dislocations were shown to penetrate the active area. The authors concluded that dislocations in this material are not strong regions of nonradiative recombination. Devices with lattice mismatch appeared clear initially, but under bias aging at elevated temperature ⟨110⟩ lines appeared. The ⟨110⟩ lines were shown by TEM to be misfit dislocations. The most surprising observation was that whereas small DSDs or ⟨110⟩ DLDs appeared during aging, little decrease in light output was seen. The DSDs did not correspond to dislocation etch pits. The time to the appearance of DSDs varied as J^{-2} with an activation energy of 1.2 eV. The output power of LEDs operating at < 170°C remained unchanged for 10^4 hr, as shown in Fig. 9. For T_J, from 170° to 230°C the light output decayed gradually in an exponential fashion. From the Arrhenius plot in Fig. 10, Yamakoshi estimated E_A and MTTF at 60°C to be 0.9 to 1.0 eV and 10^9 hr, respectively.

Yeats *et al.* (1981) used a simple screening to choose devices for long-term aging. Light-emitting diodes with high series resistance or high leakage current at low voltage were eliminated, and all devices had to withstand 400 mA (20 kA cm^{-2}) for several minutes. Aging was done from $J = 5$ to 20 kA cm^{-2} for oven temperatures from 75° to 175°C. The number of DSDs observed increased with aging time. Dark-spot defects were seen not to affect light output consistent with Yamakoshi's findings. However, just before failure they became large and light output dropped catastrophically. Significant numbers of devices failed at the highest current density by short circuiting. This is different from the results of Yamakoshi who saw only gradual degradation at 8 kA cm^{-2} and no freak population. At both 125 and 175°C, there was a well-defined failure distribution with $\sigma = 1.6$. The activation energy was estimated to be 1.45 eV and the median life at 25°C to be 2.7×10^6 hr at 20 kA cm^{-2}. The authors use the result of Yamakoshi that DSD appearance time goes as J^{-2} to assume that MTTF goes as J^{-2} and estimate MTTF to be 8×10^{10} hr. This use of the J^{-2} dependence, which results in this extraordinarily long MTTF, is probably erroneous, since Yamakoshi made no connection between DSD appearance and decrease in light output.

b. Freak Behavior

Whereas the reliability studies published so far have led to predictions of extremely long lifetimes, it has not always been clear whether there is a freak

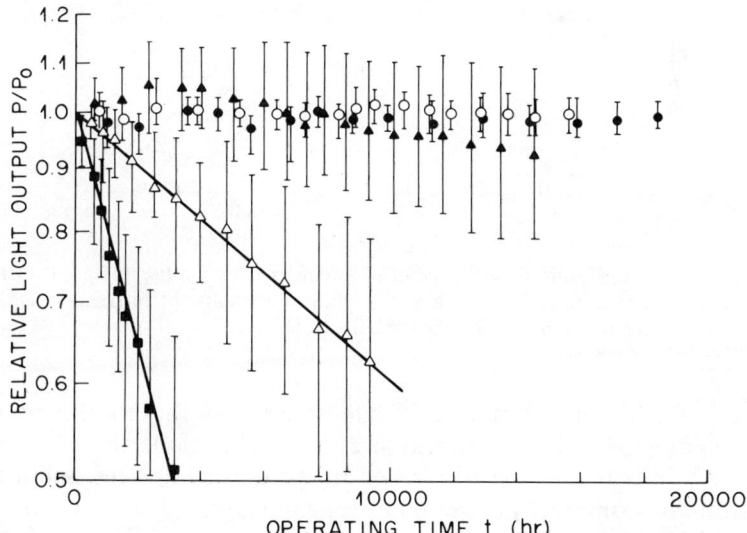

FIG. 9. Degradation in light output as a function of operation time for InP/InGaAsP LEDs. The symbol and error bar indicate arithmetic mean and spread in the data. Temperatures: ●, 20°C; ○, 120°C; ▲, 170°C; △, 200°C; ■, 230°C. [After Yamakoshi et al. (1981). © 1981 IEEE).]

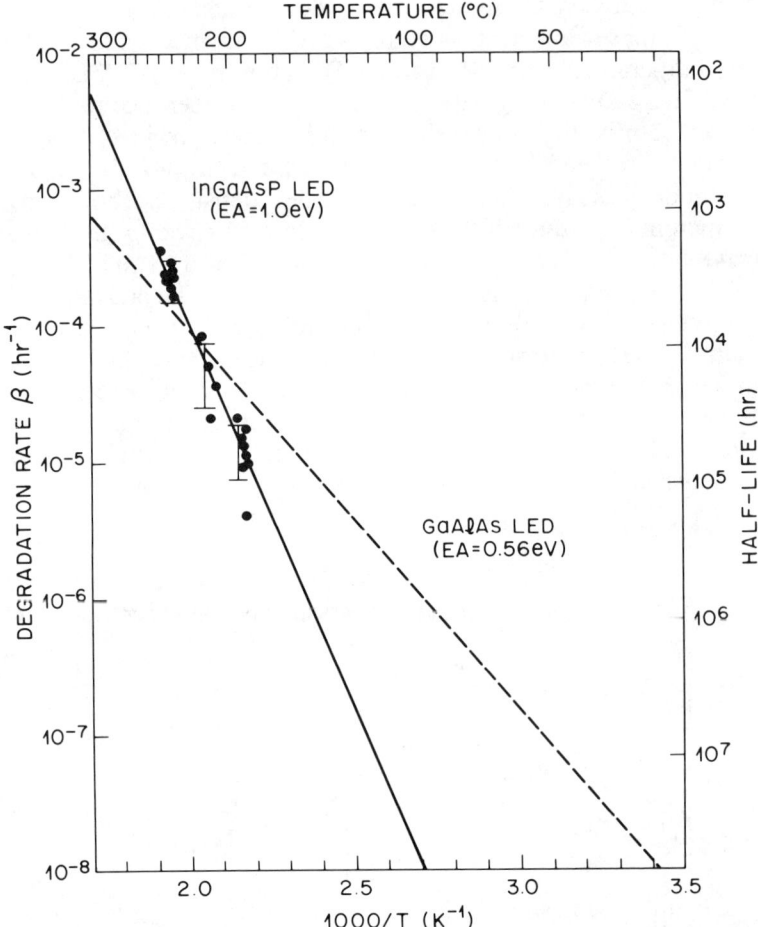

FIG. 10. The temperature dependence of degradation rates for InP/InGaAsP LED; (—) calculated using the method of least squares; (---) shows the results obtained for GaAs/GaAlAs LEDs. [After Yamakoshi et al. (1981). © 1981 IEEE).]

population. Although all studies to date have shown that there is no infant mortality due to DLD formation at threading dislocations, there is a form of infant failure that must be eliminated in quaternary devices. Indium-rich inclusions are commonly observed in epitaxial layers of InP and InGaAsP. Depending on the quality of the wafer, a certain percentage of devices exhibit instabilities in the I–V characteristic. These instabilities result from shunt paths across the junction caused by In inclusions. These shunt paths cause initially high leakage currents at fixed voltage. Under local heating at high current density, shunt paths can form at inclusions. If the shunt resistance is small, the light output at operating current will be affected. A burn-in of 100 hr at 150 mA is effective for screening out leaky or unstable devices (Zipfel et al., 1983). In wafers with good morphology, the yield can be close to 100%.

Temkin et al. (1983) have shown that crystal growth or processing-induced defects can result in freak behavior. Two groups of diodes were studied: The first was made from material (grown early in a development program) that contained many inclusions up to 10 μm in size, whereas the second group was made from a wafer with few inclusions. The first group of LEDs was operated at $J = 5-10$ kA cm^{-2}. Two different effects were identified: $\langle 110 \rangle$ DLDs and rapid decrease in power in ~50% of the LEDs above 140°C even without bias, and DSD formation at all temperatures under bias. It was shown by TCL that the DLDs originated at microscopic inclusion-like defects and were confined to the InP buffer layer (Temkin et al., 1981). It was concluded that these DLDs grew by thermal stress and bonding stress by dislocation glide on $\{111\}$ slip planes. Because 140°C is a threshold temperature for this effect, it is not a relevant failure mode at operating temperatures. However, it can complicate high-temperature aging results and make estimates of E_A too high. The other 50% of LEDs that did not form $\langle 110 \rangle$ dark lines developed DSDs and dropped gradually in light output with $E_A = 0.85$ eV and the median life (ML) $\cong 10^6$ hr at 70°C.

The second group of LEDs studied by Temkin et al. had good morphology and few inclusions. They were aged at J up to 40 kA cm^{-2}. The anomalous DLD failure mode discussed earlier was absent, and few devices developed DSDs. The activation energy for both groups was 0.85 eV, in reasonable agreement with Yamakoshi et al. The MTTF was a factor of 4 higher in the group with good morphology even though current density was also a factor of 4 higher. Elimination of inclusions in crystal growth is discussed in Section 16.

c. High Current Density

Another study showed the effects of processing differences on the reliability of these LEDs (Zipfel et al., 1983). The standard LED structure where the p surface is isolated with a dielectric layer was compared with a structure that used a Schottky barrier for isolation (Chin et al., 1981b). The LEDs were aged at J up to 40 kA cm^{-2}. DSDs formed in a dense ring in the dielectric isolated devices, and some of the devices failed catastrophically with the formation of $\langle 110 \rangle$ DLDs and crystal damage at the DSDs. This limited the MTTF at 70°C to 2×10^5 hr.

In contrast, devices with Schottky-barrier isolation formed only a few small DSDs and had a projected MTTF at 70°C of $> 10^7$ hr, independent of J. These results suggest that stress from the dielectric layer, especially the high stress around the opening in the dielectric, promotes DSD growth.

These results show that the devices can be reliable at the high current densities needed in transmission applications. These results are similar to laser-aging studies of Imai *et al.* (1982); V-grooved substrate buried-heterostructure lasers were aged in the LED mode for 5000 hr at 160 mA and 70°C and found to be stable. These lasers have a stripe width of 2.5 μm and a length of 250 μm and therefore a comparable current density to 25-μm-diameter LEDs operated at 150 mA.

15. Edge-Emitting Structure

Olsen *et al.* (1981) and Ettenberg and Olsen (1981) have compared the reliability of edge emitters grown by LPE and by vapor-phase epitaxy (VPE) operating at 10 kA cm^{-2}. One set of devices proved the feasibility of VPE by operating undegraded for 1.7×10^4 hr at 70°C. However, Ettenberg and Olsen report a significant difference between the two growth techniques. In most cases, the VPE wafers degraded at a relatively fast rate independent of lattice parameter mismatch with a small activation energy (0.45 eV) and projected MTTF at 25°C equal to $\sim 10^4$ hr. The LPE-grown LEDs behaved much better with an activation energy of 0.8 eV and MTTF at 25°C of $\sim 10^6$ hr. They report an apparent order of magnitude improvement in MTTF for LPE-grown LEDs mounted with Au/Sn solder rather than In. It is possible that the metallization was flawed and that In went through and dissolved the InP, resulting in failure (Nash, 1984).

VII. Degradation Mechanisms in InP/InGaAsP LEDs

16. Indium Inclusions

The most common defects present in epitaxial layers of InP and InGaAsP are In-rich inclusions (Temkin *et al.*, 1983). In contrast, inclusions are rare in GaAs epilayers. An InGa-rich phase can be expected to have a lower melting point and under thermal or strain gradients to move through the crystal lattice as a microdroplet (Johnston, 1983). The inclusions can thus cause instabilities in diode $I-V$ characteristics and can therefore affect reliability. Occasionally, inclusions are visible as dark defects in EL. They can readily be seen in optical microscopy using Nomarski contrast (Mahajan *et al.*, 1983).

The inclusions originate at either the meltback or wipe-off stages of the growth cycle or by thermal decomposition of the substrate (Temkin *et al.*, 1983). When a buffer layer is not completely wiped off, In inclusions left at the buffer-layer surface locally prevent growth from the InGaAsP melt and thus produce holes in the active layer (Temkin et al., 1983). This melt carryover can occur at any stage of epitaxial growth and can cause holes to propagate through subsequent layers (Mahajan et al., 1982, 1983). Holes transcending all four layers can form if wipe-off after In meltback is incomplete. Another source of In inclusions is contamination with particulate matter such as carbon particles generated during prolonged equilibration baking (Temkin et al., 1983).

In LPE growth the substrate is usually melted back with In to remove the thermally degraded surface. Chin et al. (1982c) have developed a double meltback procedure which maintains flat surface morphology, thereby reducing defects which can cause freak failure. DiGiuseppe et al. (1984) discuss the role of melt carryover in the LPE growth of planar buried double heterostructures. The etching of In charges prior to growth was found to decrease significantly In-rich inclusions that propagate through the layers.

17. DARK-SPOT-DEFECT FORMATION

The greatest puzzle in InGaAsP LED degradation is the source of, nature of, and growth mechanism of DSDs. In virtually all aging experiments on InGaAsP/InP LEDs and lasers where observations of the light-emitting spot could be made, DSDs were reported. Unlike GaAlAs LEDs where the appearance of a DLD was followed by rapid degradation, quaternary devices can have many DSDs and show little degradation.

Ueda et al. (1980b) used TEM to study DSDs that appeared in EL to be $2-3$ μm in diameter. They found bar-shaped precipitates $0.5-1.0$ μm in length lying in the $\langle 100 \rangle$ and $\langle 110 \rangle$ directions. The defect regions were In rich compared with the crystal matrix, and therefore they concluded that the DSDs were formed by precipitation of the host atoms at nucleation sites (yet unidentified) during operation (Ueda et al., 1981).

Some interesting results have been reported for lasers that are relevant to DSD growth in LEDs. When lasers were aged at $J = 10$ kA cm^{-2} and $T_J = 250\,°$C (LED mode aging), $\langle 100 \rangle$ DLDs and DSDs appeared, and the lasing threshold current increased abruptly (Fukuda et al., 1981). The local temperature at the DSD was above that of the surrounding area, suggesting that current concentrates at DSDs. DSD generation time was found to be strongly J dependent but only weakly T dependent (Fukuda et al., 1983). Fukuda found that degradation could be divided into two stages. In the initial stage DSDs form but do not absorb light. During this stage, the $2kT$ current increases, suggesting that DSDs act as nonradiative recombination centers. In the second stage DSDs increase in size but not in number and become strong absorbers of light. It was shown that the DSDs penetrate from the p cladding layer through the active layer to the n cladding layer. The host atoms As and Ga were observed by an electron microprobe analyzer to segregate at the DSDs (Seki et al., 1982).

It was shown by TEM that dislocation loops enlongated in the [1$\bar{1}$0] direction and platelike precipitates lying in the {111} planes and associated with the dislocation loops were observed to correspond to the DSDs (Wakita et al., 1982). Some wafers form no DSDs, and lasers forming DSDs could be screened out in a short high-current burn-in (Fukuda et al., 1983). The authors suggest that devices degrade because they contain "the origins of dark defects," and that DSDs are not an inevitable result of aging.

The results of these studies differ in several significant ways from those of Ueda et al. (1980b, 1981), suggesting that more than one phenomenon may manifest itself as DSDs. Ueda saw concentrations of In at DSDs, whereas Fukuda saw Ga and As; Ueda saw bar-shaped precipitates and no dislocations, whereas Fukuda saw dislocation loops and platelike precipitates. Ueda found a strong T dependence for DSD generation time and saw little decrease in light output in LEDs when DSDs form. This could be explained by the fact that Fukuda was measuring lasing threshold, which is very sensitive to the presence of light absorbers.

Although a lot is now known about the phenomenology of DSD growth, and their appearance in quaternary devices is taken for granted, few models have been proposed for their growth. A model has been proposed in which DSDs form due to electrothermomigration of gold from the p contact (Chin et al., 1982b, 1983c). This model is based on a series of experiments in which DSDs in LEDs at various stages of degradation were imaged in successive epilayers using EL, CL, and EBIC, as shown in Fig. 11. The findings were these: during short-term aging ($\sim 10^2$ hr), defects corresponding to DSDs in EL were observed by CL in either the InGaAsP contact layer or the p confining layer. In contrast with the results of Seki et al. (1982), defects causing DSDs are found in the active layer only after long times ($\sim 10^3$ hr). Chin suggests that even in the early stages of degradation when DSDs are located only in the contact or confining layers, local heating can lower the quantum efficiency at the DSDs enough so that the contrast with surrounding areas is visible in EL. Only the host atoms were found by energy dispersive x-ray spectroscopy (EDS) at the defects, in agreement with Ueda et al. (1981). However, Chin argues that EDS would not be sensitive enough to detect gold at the 0.5-μm bar-shaped defects seen by Ueda.

The gold contact was found to interact strongly and nonuniformly with the semiconductor (Camlibel et al., 1982). In localized regions the alloyed material extends into the p-InP confining layer, and under extended heat treatment gold-rich particles were found in the active layer, thus simulating a process at high temperature that can take place at lower temperatures under bias. Chin et al. (1983b) found direct evidence of gold in the active layer in

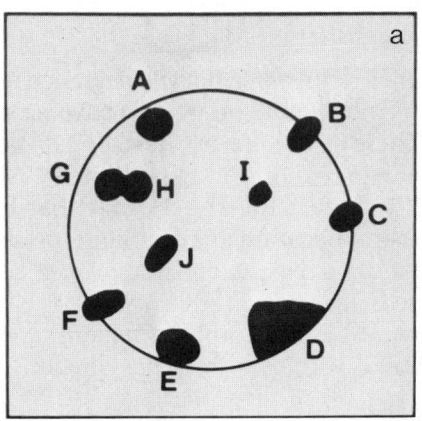
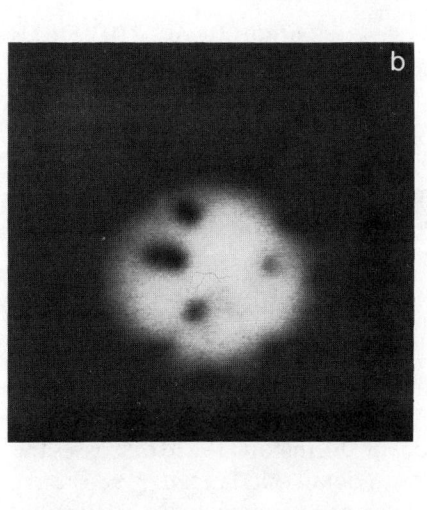
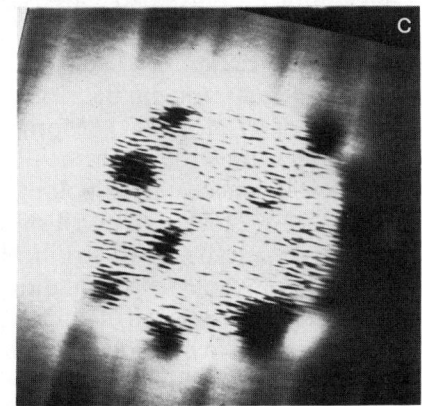
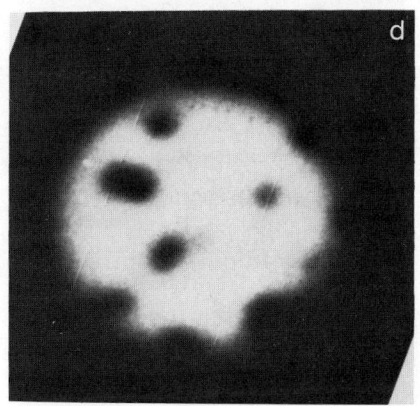
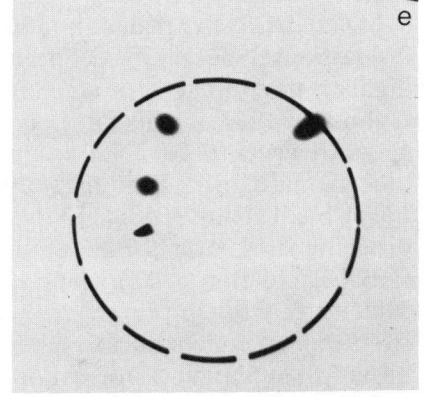

FIG. 11. (a) Schematic of DSDs in degraded InP/InGaAsP LED. (b) EL image of light-emitting spot. (c) EBIC image of p contact region with all four epitaxial layers intact. (d) EBIC image of p contact region after removal of contact layer. (e) EBIC image of p-contact region after removal of the p-InP confining layer. [After Chin et al. (1983c).]

the form of gold-rich inclusions when heavily degraded devices were cleaved through the contact region. Chin et al. (1984) have produced LEDs with a platinum p contact. These devices do not develop DSDs during accelerated aging, thus confirming the role of gold migration in DSD formation.

Ueda et al. (1982b) studied InGaAsP LEDs that had degraded catastrophically during the application of very large pulsed current (~2 A). Dark-spot defects and dark regions extending from the dark spots were observed in EL. These defects were shown by TEM to consist of damaged matrix crystal (micrograins and amorphous regions) and regions where the matrix crystal was alloyed with gold from the contact.

18. Dark-Line-Defect Formation

In complete contrast to GaAs/GaAlAs devices, no $\langle 100 \rangle$ or $\langle 210 \rangle$ DLDs attributable to climb have been observed in InP/InGaAsP LEDs. These DLDs have occasionally been observed in InP/InGaAsP lasers, but they are an insignificant problem because the growth rates are very slow. For example, Fukuda et al. (1981) report a $\langle 100 \rangle$ DLD in an InP/InGaAsP laser growing at a rate 100 times slower than in GaAs, and Ishida et al. (1982) report $\langle 100 \rangle$ dislocation dipoles growing by climb from a $\langle 110 \rangle$ DLD at a rate 10^3 times slower than in GaAs. There have been several arguments to explain why climb is not a significant process in these materials. The energy from electron–hole recombination is only 2/3 that in GaAlAs. This may be too small for the creation of point defects at a dislocation or for the migration of point defects to a dislocation, the two processes that have been proposed for climb in GaAlAs. Another possibility is that the recombination rate at a dislocation may be too low for climb to occur. Johnston (1983) discusses the differences between the materials systems in detail. Ueda et al. (1984) have shown, however, that dislocation climb can occur when InGaAsP/InGaP material is grown on a GaAs substrate.

In GaAs/GaAlAs material, $\langle 110 \rangle$-oriented DLDs resulting from glide of dislocations under plastic deformation are seen infrequently because of the high stress required to produce plastic deformation at room temperature (Johnston, 1983). Deliberate bending was required to promote dislocation glide. However, $\langle 110 \rangle$ DLDs were observed to grow rapidly at scratches or other damage. In contrast, the stress to produce plastic deformation in InP is much less. Temkin et al. (1983) report that $\langle 110 \rangle$ DLDs can be produced by dielectric stress over the entire area of the p-InP confining layer, causing rapid degradation in LED light output. The SiO_2 film must therefore be sufficiently thick to prevent pinholes, yet as thin as possible to minimize stress on the semiconductor. Glide of $\langle 110 \rangle$ DLDs has also been produced by optical excitation at dislocation tangles around inclusions in wafer material (Mahajan et al., 1979; Johnston et al., 1978). Unlike GaAs, however, dislocation glide does not occur even under forward bias in InGaAsP LEDs at mechanical damage such as at scratches made in the epitaxial material when the melt is wiped off (Ueda et al., 1980a).

Ueda et al. (1982a) studied several types of dark defects generated during crystal growth or processing (misfit dislocations, stacking faults, precipitate-like spherical defects, mechanical damage, and alloyed regions produced by penetration of the electrode metals). It was found that both the recombination-enhanced dislocation glide and climb processes, which readily occur in GaAlAs devices with such defects, occur only with difficulty in InGaAsP devices. They explain the differences between these materials in terms of the interaction between impurity atoms and dislocations and/or the electronic states of defects and nonradiative recombination rates at defects.

19. MISFIT DISLOCATIONS

Attention must be paid in LPE to the reproducible growth of quaternary active layers with a composition corresponding to the 1.3-μm emission wavelength (e.g., $In_{0.7}Ga_{0.3}As_{0.64}P_{0.36}$) while also being lattice matched to the InP substrate (Temkin et al., 1983). The lattice match $\Delta a/a$ must be in the range 0 to -0.08% at room temperature to make the area without misfit dislocations as wide as possible (Yamazaki et al., 1982). This criterion depends upon confining-layer thickness. Grown-in misfit dislocations act as nonradiative recombination centers and are easily seen in TCL or EL as a crisscross pattern in the $\langle 110 \rangle$ directions.

Yamakoshi et al. (1979) reported that devices with misfit dislocations do not degrade at 25°C and 8 kA cm^{-2} for times up to 10^3 hr. At higher current densities, however, the growth of misfit dislocations becomes an important failure mode (Chin et al., 1983a). As LEDs with an intentional misfit of $\Delta a/a = 0.10\%$ are aged, the EL contrast of the $\langle 110 \rangle$ misfit dislocation lines increases, and the light output degrades as $\exp(-t/\tau)$. The half-life τ is strongly dependent on current density J with $\tau \sim J^{-3}$, but only weakly dependent on temperature with $E_A \leq 0.3$ eV. The devices do not degrade when operated unbiased at high temperature. In unaged devices examined layer by layer by defect etching and by CL and EBIC imaging, misfit dislocations were found only in the p-InP confining layer close to the active-layer interface. This agrees with the x-ray topograph results of Yamazaki et al. (1982). Initially, the dislocations are in weak contrast in EL because, being in the confining layer, they can reduce EL efficiency only by a local reduction in carrier confinement or injection efficiency. In heavily degraded LEDs, misfit dislocations are found in both the confining and active layers. The increased contrast in EL is directly due to nonradiative recombination. The strong J dependence of degradation suggests that the growth of dislocations into the active layer results from recombination-enhanced glide. In the device structure studied (0.7-μm-thick active layer and 2-μm-thick p-InP confining layer), a lattice misfit of $\Delta a/a \lesssim 0.05\%$ was found to ensure misfit-dislocation-free diodes.

20. Gradual Degradation and Reliability Projections

It is not clear from the reports published so far whether gradual degradation of the type observed in GaAs devices, in which the light-emitting spot remains clear, is occurring in InGaAsP LEDs. Virtually all investigators have reported that DSDs form during accelerated aging, and it may be that the slow degradation observed is due to the increasing influence of the DSDs. All reports have estimated the activation energy to be in the range of 1 eV. Because of this high activation energy, extrapolations of high-temperature data to room temperature have led to projections of MTTF that are extremely long; e.g. Yeats *et al.* (1981) predicted 2.7×10^6 hr at 20 kA cm^{-2}, and Yamakoshi *et al.* (1979) predicted $> 5 \times 10^9$ hr at 8 kA cm^{-2} both at 25°C for a 3-dB drop in power, and Zipfel *et al.* (1983) predicted $> 10^7$ hr at 70°C and 40 kA cm^{-2} for a 1-dB drop. The relevant question then for systems reliability is how well the freak population, if any, is eliminated, and what is the spread in the failure distribution. The failure distribution of Yeats *et al.* included many freaks (catastrophic failures). However, the fail-

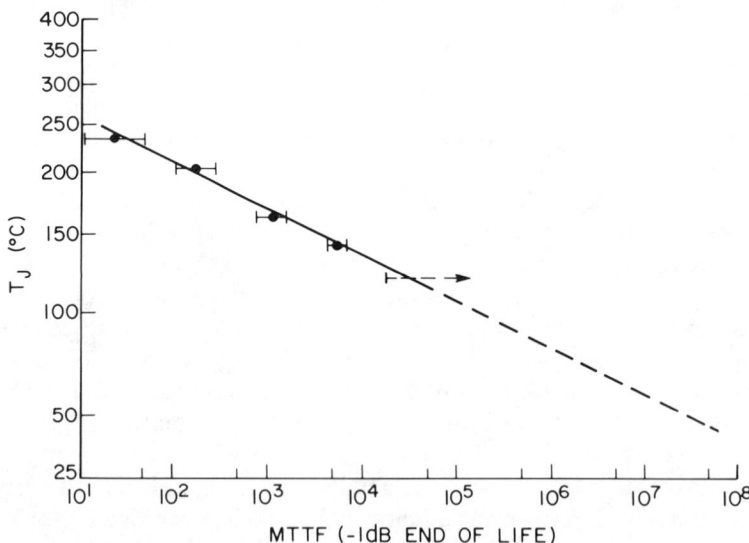

FIG. 12. Arrhenius plot for InP/InGaAsP LED aged at 40 kA cm^{-2}. MTTF (-1 dB) and spread in data for diodes at each aging condition are shown. The dashed line indicates an estimated lower limit on time to failure at the lowest aging temperature. [C. L. Zipfel, unpublished work (1983).]

ure distribution for the whole population was well defined with $\sigma = 1.6$ and ML $= 2.7 \times 10^6$ hr, which predicts only 2% cumulative failures at 20 years. Zipfel et al. (1983) find $\sigma = 0.6$ for $J = 10$ kA cm^{-2} with freak failures removed. Similarly, the times to a 1-dB drop in light power for the data of Yamakoshi at 8 kA cm^{-2} (Fig. 9) can be shown to fit a narrow log-normal distribution $\sigma \cong 0.6$. The exceptional reliability of InP/InGaAsP LEDs even at 40 kA cm^{-2} and at high-junction temperature and allowing only 1 dB for end of life is shown in Fig. 12. There is a single well-defined activation energy of ~ 1 eV and extrapolated MTTF in excess of 10^6 hr at 70°C, which is a typical maximum operating temperature in an uncooled environment (C. L. Zipfel, unpublished work, 1983).

VIII. Conclusions

Degradation in light output is the only important failure mode in lightwave LEDs. Most reports of reliability studies give only MTTF, and in most cases this is much longer than any conceivable service life of a system. However, the suitability of an LED for a lightwave system must be assessed in terms of the failure rate or the cumulative number of expected failures that have to be replaced. To do this it is necessary to know not only the MTTF but the spread in the failure distribution and how effectively infant failures are removed by burn-in. The fabrication techniques necessary to prevent freak behavior must be well understood.

Extensive research has been done on early failure modes in GaAs/GaAlAs lasers and LEDs with the result that DLD formation has been largely eliminated by use of low-dislocation-density substrates and by eliminating other sources of dislocations and stress. These require constant attention to crystal growth and processing. Certain device structures, such as those with Zn-diffused junctions or with etched wells, were shown to be prone to early failure. The most successful LED in terms of reliability has been the planar DH structure. Infant failures in these structures can be effectively eliminated in a 100-hr burn-in. Long-term gradual degradation is a well-controlled process that can be described mathematically, and there are no freak failures at operating temperatures. The long MTTF (4×10^6 hr) and narrow failure distribution ($\sigma = 0.5$) lead to very small failure rates. The disadvantages of this device are that near-equilibrium growth of the thick-graded Al n layer is more difficult than for a thin layer and with the substrate etched off, the chip is susceptible to cracking unless carefully handled.

The development time for the long-wavelength LED has been very short. However, from the earliest reports, estimates of MTTF for these devices have been exceptionally long, and they have already been introduced into systems. Because the InP/InGaAsP material system is not susceptible to dislocation climb, early failures due to DLD growth have not been observed. However, there are several causes for rapid degradation that have been observed. These include instabilities in the $I-V$ and $L-V$ due to shunt paths at In inclusions in the material, glide of misfit dislocations, and formation of $\langle 110 \rangle$ DLDs due to dielectric stress or other stresses. These effects increase in importance as J increases. The elimination of these effects requires extreme care in crystal growth and processing. Indium inclusions are eliminated by strict wipe-off tolerances. Misfit dislocations result if $\Delta a/a$ exceeds 0.05%. Low dielectric stress requires controlled deposition conditions. Once these failures are eliminated, devices typically degrade slowly and form DSDs. The role played by DSDs in degradation is not understood, since LEDs can exhibit many DSDs but little degradation in power. There is some evidence, however, that at high current densities DSDs can grow large and result in catastrophic degradation. Evidence suggests that gold migration from the p contact causes DSDs. Nonetheless, all reports published so far suggest that these devices are exceptionally reliable over a wide range of current densities and temperatures. For example, MTTF is projected to be $> 10^6$ hr at 70°C and 40 kA cm^{-2} for a 1-dB end of life and $\sim 10^9$ hr at 60°C and 8 kA cm^{-2} for a 3-dB end of life.

Acknowledgments

The author thanks A. K. Chin and R. H. Saul for their help and F. R. Nash for a critical reading of this chapter.

References

Abe, M., Umebu, I., Hasegawa, O., Yamakoshi, S., Yamaoka, T., Kotani, T., Okada, H., and Takanashi, H. (1977). *IEEE Trans. Electron Devices* **ED-24**, 990.
Abe, M., Hasegawa, O., Komatsu, Y., Toyama, Y., and Yamaoka, T. (1978). *IEEE Trans. Electron Devices* **ED-25**, 1344.
Borsuk, J. A. (1983). *IEEE Trans. Electron Devices* **ED-30**, 296.
Burrus, C. A., and Dawson, R. W. (1970). *Appl. Phys. Lett.* **17**, 97.
Burrus, C. A., and Miller, B. I. (1971). *Opt. Commun.* **4**, 307.
Camlibel, I., Chin, A. K., Ermanis, F., DiGiuseppe, M. A., Lourenco, J. A., and Bonner, W. A. (1982). *J. Electrochem. Soc.* **129**, 2585.
Casey, H. C., and Panish, M. B. (1978). "Heterostructure Lasers," Part B, Chapter 8. Academic Press, New York.
Chin, A. K., Temkin, H., and Roedel, R. J. (1979). *Appl. Phys. Lett.* **34**, 476.
Chin, A. K., Keramidas, V. G., Johnston, W. D., Jr., Mahajan, S., and Roccasecca, D. D. (1980). *J. Appl. Phys.* **51**, 978.
Chin, A. K., Temkin, H., and Mahajan, S. (1981a). *Bell Syst. Tech. J.* **60**, 2187.
Chin, A. K., Zipfel, C. L., Dutt, B. V., DiGiuseppe, M. A., Bauers, K. B., and Roccasecca, D. D. (1981b). *Jpn. J. Appl. Phys.* **20**, 1487.
Chin, A. K., Zipfel, C. L., and Dutt, B. V. (1982a). *Jpn. J. Appl. Phys.* **21**, 1308.

Chin, A. K., Zipfel, C. L., Mahajan, S., Ermanis, F., and DiGiuseppe, M. A. (1982b). *Appl. Phys. Lett.* **41**, 555.
Chin, B. H., DiGiuseppe, M. A., Temkin, H. and Bonner, W. A. (1982c). *Mater. Lett.* **1**, 81.
Chin, A. K., Zipfel, C. L., Chin, B. H., and DiGiuseppe, M. A. (1983a). *Appl. Phys. Lett.* **42**, 1031.
Chin, A. K., Zipfel, C. L., Camlibel, I., Singh, S., and Ermanis, F. (1983b). *Proc. NATO Conf. Indium Phosphide, 2nd. 1983*, p. 60.1.
Chin, A. K., Zipfel, C. L., Ermanis, F., Marchut, L., Camlibel, I., DiGiuseppe, M. A., and Chin, B. H. (1983c). *IEEE Trans. Electron Devices* **ED-30**, 304.
Chin, A. K., Zipfel, C. L., Geva, M., Camlibel, I., Skeath, P., and Chin, B. H. (1984). *Appl. Phys. Lett.* **45**, 37.
Davies, I. G. A., Bricheno, T., Goodwin, A. R., and Plumb, R. G., (1980). *IEE Conf. Publ.* **190**, 199.
DeLoach, B. C., Jr., Hakki, B. W., Hartman, R. L., and D'Asaro, L. A. (1973). *Proc. IEEE* **61**, 1042.
Dentai, A. G., Lee, T. P., and Burrus, C. A. (1977). *Electron. Lett.* **13**, 484.
DiGiuseppe, M. A., Chin, A. K., Chin, B. H., Lourenco, J. A., and Camlibel, I. (1984). *J. Cryst. Growth* **67**, 1.
Dutt, B. V., Mahajan, S., Roedel, R. J., Schwartz, G. P., Miller, D. C., and Derick, L. (1981). *J. Electrochem. Soc.* **128**, 1573.
Dyment, J. C., SpringThorpe, A. J., King, F. D., and Straus, J. (1977). *J. Electron. Mater.* **6**, 173.
Escher, J. S., Berg, H. M., Lewis, G. L., Moyer, C. D., Robertson, T. U., and Wey, H. A. (1982). *IEEE Trans. Electron Devices* **ED-19**, 1463.
Ettenberg, M., and Kressel, H. (1980). *IEEE. Quantum Electron.* **QE-16**, 186.
Ettenberg, M., and Neuse, C. J. (1975). *J. Appl. Phys.* **46**, 2137.
Ettenberg, M., and Olsen, G. H. (1981). *Tech. Dig.—Int. Electron Devices Meet.*, p. 280.
Ettenberg, M., Kressel, H., and Lockwood, H. F. (1974). *Appl. Phys. Lett.* **25**, 82.
Fukuda, M., Wakita, K., and Iwane, G. (1981). *Jpn. J. Appl. Phys.* **20**, L87.
Fukuda, M., Wakita, K., and Iwane, G. (1983). *J. Appl. Phys.* **54**, 1246.
Goodwin, A. R., Kirkby, P. A., and Davies, I. G. A. (1979). *Appl. Phys. Lett.* **34**, 647.
Hartman, R. L., and Hartman, A. R. (1973). *Appl. Phys. lett.* **23**, 147.
Hawkins, B. M. (1984). *Proc. Electron. Component Conf., 34th, New Orleans, 1984*, p. 239.
Hayashi, I. (1980). *J. Phys. Soc. Jpn.* **49**, Suppl. A, 57.
Hersee, S. D. (1977). *Tech. Dig.—Int. Electron Devices Meet.*, p. 567.
Hersee, S. D., and Goodfellow, R. C. (1976). *Proc. Eur. Conf. Opt. Fiber Commun., 2nd, 1976*, p. 213.
Hersee, S. D., and Stirland, D. J. (1977). *Conf. Ser.—Inst. Phys.* **33a**, 370.
Hutchinson, P. W., Dobson, P. S., O'Hara, S., and Newman, D. H. (1975). *Appl. Phys. Lett.* **26**, 250.
Ikeda, K., Tanaka, T., and Ito, A. (1975). *Conf. Ser.—Inst. Phys.* **24**, 174.
Imai, H., Ishikawa, H., and Hori, K. (1982). *Appl. Phys. Lett.* **41**, 583.
Ishida, K., Kamejima, T., and Matsumoto, Y. (1982). *Appl. Phys. Lett.* **40**, 16.
Ishii, M., Kan, H., Susaki, W., Nishiura, H., and Ogata, Y. (1977). *IEEE J. Quantum Electron.* **QE-13**, 600.
Ito, R., Nakashima, H., and Nakada, O. (1974). *Jpn. J. Appl. Phys.* **13**, 1321.
Johnson, B. H., Ackenhusen, J. G., and Lorimor, O. G. (1980). *IEEE Trans. Components, Hybrids Manuf. Technol.* **CHMT-3**, 488.
Johnston, W. D., Jr. (1983). *Mater. Res. Soc. Symp. Proc.* **14**, 453.
Johnston, W. D., Jr., and Miller, B. I. (1973). *Appl. Phys. Lett.* **23**, 192.
Johnston, W. D., Jr., Callahan, W. M., and Miller, B. I. (1974). *J. Appl. Phys.* **45**, 505.

Johnston, W. D., Jr., Epps, G. Y., Nahory, R. E., and Pollack, M. A. (1978). *Appl. Phys. Lett.* **33**, 992.
Jordan, A. S. (1978). *Microelectron. Reliab.* **18**, 267.
Kamejima, T., Ishida, K., and Matsui, J. (1977). *Jpn. J. Appl. Phys.* **16**, 233.
Keramidas, V. G., Berkstresser, G. W., and Zipfel, C. L. (1980). *Bell Syst. Tech. J.* **59**, 1549.
Kimerling, L. C. (1978). *Solid-State Electron.* **21**, 1391.
King, F. D., SpringThorpe, A. J., and Szentesi, O. I. (1975). *Tech. Dig.—Int. Electron Devices Meet.,* p. 480.
Kishino, S., Chinone, N., Nakashima, H., and Ito, R. (1976). *Appl. Phys. Lett.* **29**, 488.
Kondo, K., Ueda, O., Isozumi, S., Yamakoshi, S., Kita, K. A., and Kotani, T. (1983). *IEEE Trans. Electron Devices* **ED-30**, 321.
Kressel, H., and Butler, J. K. (1977). "Semiconductor Lasers and Heterojunction LEDs," Chapter 16. Academic Press, New York.
Kressel, H., Ladany, I., Ettenberg, M., and Lockwood, H. F. (1975). *Tech. Dig.—Int. Electron Devices Meet.,* p. 477.
Kressel, H., Ettenberg, M., and Lockwood, H. F. (1977). *J. Electron. Mater.* **6**, 467.
Lang, D. V. (1974). *J. Appl. Phys.* **45**, 3023.
Mahajan, S., Johnston, W. D., Jr., Pollack, M. A., and Nahory, R. E. (1979). *Appl. Phys. Lett.* **34**, 717.
Mahajan, S., Brasen, D., DiGiuseppe, M. A., Keramidas, V. G., Temkin, H., Zipfel, C. L., Bonner, W. A., and Schwartz, G. P. (1982). *Appl. Phys. Lett.* **41**, 266.
Mahajan, S., Temkin, H., Zipfel, C. L., DiGiuseppe, M. A., Brasen, D., Bonner, W. A., and Schwartz, G. P. (1983). *Proc. NATO Conf. Indium Phosphide, 2nd, 1983,* p. 60.8.
Matsui, J., Ishida, K., and Nannichi, Y. (1975). *Jpn. J. Appl. Phys.* **14**, 1555.
Mizuishi, K., Chinone, N., Sato, H., and Ito, R. (1979). *J. Appl. Phys.* **50**, 6668.
Monemar, B., and Woolhouse, G. R. (1976). *Appl. Phys. Lett.* **29**, 605.
Nash F. R. (1984). Personal communication.
Newman, D. H., and Ritchie, S. (1981). *In* "Reliability and Degradation" (M. J. Howes and D. V. Morgan, eds.), Chapter 6. Wiley, Chichester.
O'Hara, S., Hutchinson, P. W., Davis, R., and Dobson, P. S. (1977). *Conf. Ser.—Inst. Phys. Ser. No.* **33a**, 379.
Olsen, G. H., and Ettenberg, M. (1977). *J. Appl. Phys.* **48**, 2543.
Olsen, G. H., Hawrylo, F. Z., Channin, D. G., Botez, D., and Ettenberg, M. (1981). *IEEE J. Quantum Electron.* **QE-17**, 2130.
Peck, D. S., and Zierdt, C. H. (1974). *Proc. IEEE* **62**, 185.
Petroff, P. M., and Hartman, R. L. (1973). *Appl. Phys. Lett.* **23**, 469.
Petroff, P. M., and Kimerling, L. C. (1976). *Appl. Phys. Lett.* **29**, 461.
Roszytoczy, F. E., and Wolfstirn, K. B. (1971). *J. Appl. Phys.* **42**, 426.
Saul, R. H. (1983). *IEEE Trans. Electron Devices* **ED-30**, 285.
Seki, M., Fukuda, M., and Wakita, K. (1982). *Appl. Phys. Lett.* **40**, 115.
Speer, R. S., and Hawkins, B. M. (1980). *IEEE Trans. Components, Hybrids Manuf. Technol.* **CHMT-3**, 480.
SpringThorpe, A. J., Look, C. M., and Emmerstorfer, B. F. (1982). *IEEE Trans. Electron Devices* **ED-29**, 876.
Temkin, H., Zipfel, C. L., and Keramidas, V. G. (1981). *J. Appl. Phys.* **52**, 5377.
Temkin, H., Zipfel, C. L., DiGiuseppe, M. A., Chin, A. K., Keramidas, V. G., and Saul, R. H. (1983). *Bell Syst. Tech. J.* **62**, 1.
Ueda, O., Isozumi, S., Kotani, T., and Yamaoka, T. (1977). *J. Appl. Phys.* **48**, 3950.
Ueda, O., Isozumi, S., Yamakoshi, S., and Kotani, T. (1979). *J. Appl. Phys.* **50**, 765.
Ueda, O., Yamakoshi, S., and Yamaoka, T. (1980a). *Jpn. J. Appl. Phys.* **19**, L251.

Ueda, O., Yamakoshi, S., Komiya, S., Akita, K., and Yamaoka, T. (1980b). *Appl. Phys. Lett.* **36**, 300.
Ueda, O., Imai, H., Fujiwara, T., Yamakoshi, S., Sugawara, T., and Yamaoka, T. (1980c). *J. Appl. Phys.* **51**, 5316.
Ueda, O., Komiya, S., Yamakoshi, S., and Kotani, T. (1981). *Jpn. J. Appl. Phys.* **20**, 1201.
Ueda, O., Umebu, I., Yamakoshi, S., and Kotani, T. (1982a). *J. Appl. Phys.* **53**, 2991.
Ueda, O., Yamakoshi, S., Sanada, T., Umebu, I., and Kotani, T. (1982b). *J. Appl. Phys.* **53**, 9170.
Ueda, O., Wakoro, K., Yamaguchi, A., Komiya, S., Isozumi, S., Nishi, H., Umebu, I. (1984). *Appl. Phys. Lett.* **44**, 861.
Wada, O., Yamakoshi, S., Abe, M., Nishitani, Y., and Sakurai, T. (1981). *IEEE J. Quantum Electron.* **QE-17**, 174.
Wakita, K., Takaoka, H., Seki, M., and Fukuda, M. (1982). *Appl. Phys. Lett.* **40**, 525.
Woolhouse, G. R., Blakeslee, A. E., and Shih, K. K. (1976). *J. Appl. Phys.* **47**, 4349.
Wright, P. D., Chai, Y. G., and Antypas, G. A. (1979). *IEEE Trans. Electron Devices* **ED-26**, 1220.
Yamakoshi, S., Hasegawa, O., Hamaguchi, H., Abe, M., and Yamaoka, T. (1977). *Appl. Phys. Lett.* **31**, 627.
Yamakoshi, S., Sugahara, T., Hasegawa, O., Toyama, Y., and Takanashi, H. (1978). *Tech. Dig. — Int. Electron Devices Meet.,* p. 642.
Yamakoshi, S., Abe, M., Komiya, S., and Toyama, Y. (1979). *Tech. Dig. — Int. Electron Devices Meet.,* p. 122.
Yamakoshi, S., Abe, M., Wada, O., Komiya, S., and Sakurai, T. (1981). *IEEE J. Quantum Electron.* **QE-17**, 167.
Yamazaki, S., Kishi, Y., Nakajima, K., Yamaguchi, A., and Akita, K. (1982). *J. Appl. Phys.* **53**, 4761.
Yeats, R., Chai, Y. G., Gibbs, T. D., and Antypas, G. (1981). *IEEE Electron Device Lett.* **EDL-2**, 234.
Zipfel, C. L., Chin, A. K., Keramidas, V. G., and Saul, R. H. (1981). *Proc. 19th Ann. IEEE Int. Reliab. Phys. Symp.,* p. 124.
Zipfel, C. L., Saul, R. H., Chin, A. K., and Keramidas, V. G. (1982). *J. Appl. Phys.* **53**, 1781.
Zipfel, C. L., Chin, A. K., and DiGiuseppe, M. A. (1983). *IEEE Trans. Electron Devices* **ED-30**, 310.
Zucker, J., and Lauer, R. B. (1978). *IEEE J. Solid-State Circuits* **SC-13**, 119.

CHAPTER 7

Light-Emitting-Diode-Based Multimode Lightwave Systems

Tien Pei Lee* and Tingye Li

AT&T BELL LABORATORIES
CRAWFORD HILL LABORATORY
HOLMDEL, NEW JERSEY

I.	INTRODUCTION	281
II.	MULTIMODE-FIBER PROPERTIES	282
	1. Transmission Loss	282
	2. Transmission Bandwidth.	283
III.	SYSTEM CONSIDERATIONS.	286
	3. Receiver Sensitivity	286
	4. Degradation in Receiver Sensitivity Due to LED Modulation Bandwidth	288
	5. Degradation in Receiver Sensitivity Due to Fiber Chromatic Dispersion	288
	6. Transmission Margin	290
IV.	SYSTEM EXPERIMENTS AND FIELD TRIALS.	292
V.	CONCLUSIONS.	296
	REFERENCES	297

I. Introduction

Optical fiber communication systems employing light-emitting diodes (LEDs) as transmitting sources and multimode fibers as transmission media have the appeal of simplicity, reliability, and economy. They are among the first commercial lightwave systems to be deployed in the field. These first systems operated with AlGaAs surface-emitting (Burrus-type) LEDs at wavelengths near 0.85 μm, where material (chromatic) dispersion of the fiber limited the repeater spacing of a moderate-speed system to a few kilometers (bit-rate–distance product of about 140 Mbit $\times$ km sec^{-1} (Miller *et al.*, 1973). However, this limitation does not impose hardship on many short-distance applications such as data links in a building or within a room. Indeed, data links employing 0.85-μm LEDs are still being sold and installed for various "on-premises" applications today.

---

* Present address: Bell Communications Research, Murray Hill, New Jersey.

The situation is quite different at wavelengths near 1.3 μm where fiber transmission losses are much lower and chromatic dispersion approaches zero. Here, a 100-Mbit/sec LED-based multimode-fiber system can have a repeater span exceeding 20 km (Gloge *et al.*, 1980a). Such systems would have applications that include interoffice trunks, subscriber loops, local-area networks (LAN), as well as on-premises data links. In addition to simplicity, reliability, and economy, the temperature insensitivity of the LED makes LED-based multimode systems especially attractive for subscriber-loop and LAN applications. At present, 1.3-μm multimode systems that operate with InGaAsP LEDs and InGaAs $p-i-n$ photodiodes are being developed, manufactured, and installed for commercial use (Davis *et al.*, 1982; Greco and Stone, 1983; Nemchic and Buckler, 1983; Olsen and Schepis, 1984). Many such systems are expected to be depoyled in the near future.

This chapter is a review of LED-based multimode-fiber transmission systems. Properties of multimode fibers pertaining to systems operation, and systems considerations that include receiver sensitivity and various performance impairments are discussed; systems experiments and field trials are described. Special emphasis is placed on digital transmission technology for operation around 1.3-μm wavelength.

II. Multimode-Fiber Properties

Transmission loss and pulse dispersion are the two key properties of the fiber that determine the maximum distance between repeaters in a lightwave transmission system operating at a given bandwidth or bit rate. Other systems parameters of importance include source spectral width, power launched into the fiber, and receiver sensitivity. The interplay among fiber properties and source and receiver parameters in optimizing systems performance will be treated here in Part II.

1. Transmission Loss

The transmission loss of multimode fibers has been progressively reduced ever since the first "low-loss" 20-dB km^{-1} fiber was achieved in 1970 (Kapron *et al.*, 1970). Figure 1 summarizes the progress made in the reduction of transmission loss in doped-silica multimode fibers in the past decade (Li, 1983). The transmission window of interest in the 1970s was between 0.8 and 0.9 μm, the wavelength of emission of AlGaAs lasers and LEDs. However, as is discussed later, the large chromatic dispersion in this wavelength region prevented the use of LEDs in medium-to-long-haul systems that operated at speeds beyond a few tens of megabits per second. When the loss was reduced drastically in the wavelength region beyond 1 μm, it was recognized immediately that the ideal wavelength of operation for an LED-based

7. LED-BASED MULTIMODE LIGHTWAVE SYSTEMS

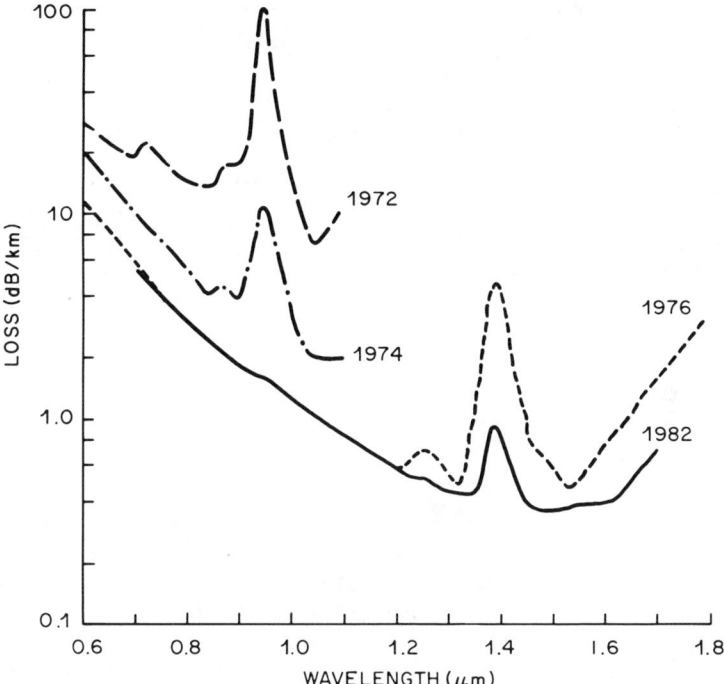

FIG. 1. Loss spectra of representative multimode fibers (doped silica core), illustrating progress made in the past decade. [Compiled by Li (1983) from Keck *et al.* (1972), French *et al.* (1974), Osanai *et al.* (1976), and Nagel *et al.* (1982). © 1983 IEEE.]

multimode system was 1.3 μm, where chromatic dispersion is minimum (Payne and Gambling, 1975; Muska *et al.*, 1977). Typical multimode fibers commercially available today exhibit losses in the vicinity of 0.6–0.8 dB/km at $\lambda = 1.3$ μm (Jablonowski *et al.*, 1982).

2. Transmission Bandwidth

The transmission bandwidth of a multimode fiber is determined by pulse spreading caused by propagation-delay differences among (1) the various modes and (2) the various spectral components of the signal. The former, known as intermodal dispersion, is governed by the refractive index profile of the fiber, whereas the latter, known as chromatic dispersion, is determined principally by material effects. Since, in the absence of mode-coupling effects, pulse spreading increases linearly with distance, it is appropriate to characterize the transmission capacity of a fiber in terms of its bandwidth–distance product, in megahertz-kilometer or megabits-kilometer per second, for example.

Intermodal dispersion in a multimode fiber with a step-index profile is very large (e.g., the bandwidth–distance product is 13 MHz km for a fiber with a relative refractive index difference of $\Delta = 0.01$), thus rendering it useless in all but the shortest distance applications. By suitably grading the refractive index profile, it is possible to minimize intermodal dispersion. For the class of graded-index profiles described by the power law

$$n^2(r) = n_0^2[1 - 2\Delta(r/a)^g], \tag{1}$$

where $n(r)$ is the refractive index at radius r ($r \leq a$); a is the radius of the core, n_0 the maximum refractive index at the center of the core, Δ the relative refractive index difference between core and cladding, and g is the power-law profile parameter. Intermodal dispersion is minimized when g is approximately two (Gloge *et al.,* 1979). At this optimum value of $g = g_0$, the rms intermodal pulse spreading of a fiber of length L is given by

$$\sigma_i(g_0) = (n_0 L \Delta^2 / 20 \sqrt{3} c) \tag{2}$$

(where c is the speed of light in vacuum), which is $(10/\Delta)$ times smaller than that of a comparable step-index fiber (with the same Δ). Thus for $\Delta = 0.01$, a three-order-of-magnitude enhancement in bandwidth is expected. However, in order to achieve the enormous improvement, very precise control of the index profile of the fiber is necessary during manufacture because the bandwidth, which is inversely proportional to $\sigma_i(g)$, peaks very sharply at $g = g_0$. Index-profile perturbations resulting from uncontrolled variables in the fabrication process can result in dramatic bandwidth degradations. To determine the optimal value of g, g_0, the wavelength dependence of Δ must be taken into account, giving a g_0 that varies with wavelength (Olshansky and Keck, 1976; Marcatili, 1977). This effect, known as profile dispersion, causes the bandwidth of an optimized fiber to vary with wavelength (in a highly peaked manner just as the bandwidth varies with g). Hence, a graded-index fiber optimized for broadband at one wavelength may not be suitable at another wavelength (Yokota *et al.,* 1979). Bandwidths of some of the best fibers produced to date fall in the range from several GHz km to almost 10 GHz km. (Keck and Bouillie, 1978; Lin *et al.,* 1981; Horiguchi *et al.,* 1982). Multimode fibers with bandwidths around 1 GHz km are presently commercially available (Jablonowski *et al.,* 1982).

An effect which ameliorates intermodal dispersion is that associated with mode mixing (Personick, 1971). Index variations or perturbations of fiber geometry with longitudinal periodicities equal to the beat wavelengths of the propagating modes will cause power to transfer among modes, resulting in reduced pulse spreading at the output end of the fiber. When mode-mixing effect dominates, pulse spreading does not increase with length L but increases with $\sqrt{LL_c}$, where L_c is a coupling length over which mode mixing has resulted in a steady-state distribution of power among the modes. Unfortunately, mode mixing introduces additional loss because power also transfers to radiation modes (Marcuse, 1972). High-performance fibers produced today exhibit very little mode mixing (L_c is greater than or equal to a few kilometers). Mode mixing may also occur at fiber splices, which are likely to appear in intervals considerably less than a few kilometers.

Because the spectral widths of the emission of LEDs are broad, the chromatic dispersion of the fiber, which is proportional to the bandwidth of the signal, tends to dominate over other factors in setting the maximum repeater spacing. Pulse spreading due to chromatic dispersion is given by (Gloge et al., 1979);

$$\sigma_m = \sigma_s L |M(\lambda)|, \tag{3}$$

where σ_m is the rms value of pulse spreading due to chromatic dispersion, σ_s the rms spectral width of the source, L the fiber length, and $M(\lambda)$, the specific material dispersion of the fiber, is defined as

$$M(\lambda) = (\lambda/c)/(d^2n/d\lambda^2), \tag{4}$$

where λ, c, and n are wavelength, speed of light, and refractive index, respectively, as defined previously. Specific material dispersion M is measured conveniently in units of picoseconds per kilometer (of fiber length) per nanometer (of source spectral width). At $\lambda = 0.85$ μm, the wavelength of operation of first-generation lightwave systems, M is about 100 psec km^{-1} nm^{-1} for germanium-doped silica fibers, and σ_m/L is about 1.8 nsec km^{-1} for operation with surface-emitting AlGaAs LEDs, corresponding to a transmission bandwidth of 140 Mbit km/sec. As wavelength increases, M decreases and crosses zero near $\lambda = 1.3$ μm. It is exactly at this wavelength (λ_0) of vanishing M that high-speed LED-based systems should operate. Pulse spreading at λ_0 is governed by second-order effects and is given by (Kapron, 1977)

$$\sigma_m(\lambda_0) = L(\sigma_s^2/\sqrt{2})|dM/d\lambda|_{\lambda_0}. \tag{5}$$

for a Gaussian source spectrum. The slope of M at λ_0, $|dM/d\lambda|_{\lambda_0}$, is approximately 0.1 psec km^{-1} nm^{-2} (Lin, et al., 1978).

Figure 2 (Muska et al., 1977) summarizes the previous discussion by showing pulse spreading or transmission bandwidth for typical step and graded-index fibers as functions of source spectral width. The left ordinate is pulse spread per unit length, $2\sigma_f = 2\sigma_m/L$ in nanoseconds per kilometer and the right ordinate is transmission-bandwidth expressed as (data rate) × (distance) in megabit-kilometers per second. The upper (horizontal) curves are for a step-index fiber of $\Delta = 0.01$. The middle set of curves is for a graded-index germanium-doped silica fiber with a maximum transmission bandwidth of 1 Gbit km/sec, as determined by intermodal dispersion; the set

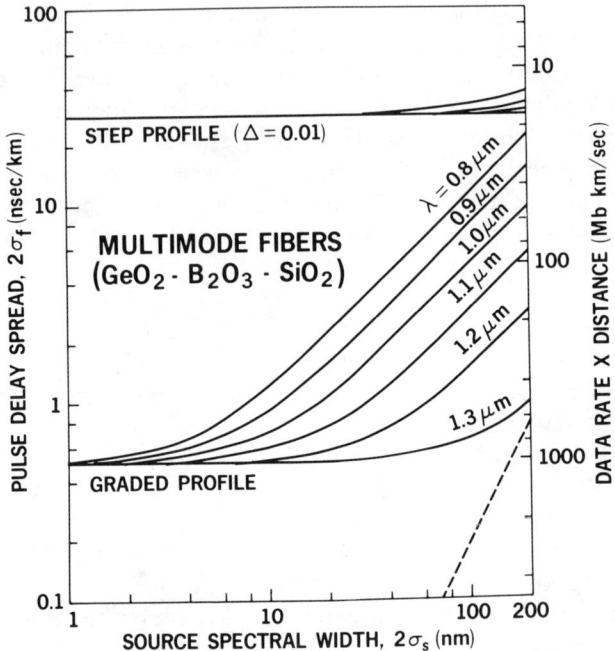

FIG. 2. Pulse-delay spread $2\sigma_f$ and transmission capacity, data-rate × distance, against source spectral width $2\sigma_s$ for germanium borosilicate multimode fibers [From Muska *et al.* (1977). ©1977 IEEE.]

of curves is plotted for various center wavelengths of the source emission spectra. The lower dashed line represents the ultimate limit at λ_0, where material dispersion vanishes, and pulse spreading is proportional to the square of the source spectral width as given by Eq. (5). For InGaAsP surface-emitting LEDs, the rms spectral width of 50 nm sets a maximum transmission bandwidth–distance product of 2.4 Gbit km/sec. However, for an edge-emitting LED with a narrower spectral width ($\sim \sigma_s \simeq 30$ nm), material effects at λ_0 become relatively unimportant, because the limiting bandwidth–distance product in this case is 7 Gbits km sec^{-1}.

III. System Considerations

3. Receiver Sensitivity

When the speed of transmission is a few tens of megabits per second or less, fiber dispersion may not impose a limit on the repeater spacing. Other factors, such as the transmission loss of the fiber cable, the transmitted power launched into the fiber, and the required input power to the receiver (to meet certain performance specifications), then dominate. In an optical receiver the sensitivity or the required input power is ultimately limited by the quantum (or shot) noise associated with the primary-signal photocurrent in the detector. In practice, Johnson (or thermal) noise from the input circuit, shot noise from leakage currents in the detector, and the noise associated with the input transistor amplifier all contribute in varying degrees to degrade the receiver sensitivity. At low bit rates, noise arising from leakage currents is the determining factor, whereas circuit noise and device noise tend to dominate at high bit rates. A comprehensive treatment of the design and performance of optical digital receivers was presented by Personick (1971).

Figure 3 (Li, 1983) summarizes the performance of state-of-the-art optical digital receivers employing currently available devices for operation in the "long-wavelength" region of $1.2-1.6$ μm; i.e., InGaAs $p-i-n$ photodiodes (Lee et al., 1980a, 1981) and GaAs field-effect transistors (FETs) (Gloge et al., 1980b); the ordinate is the (minimum) *average* number of primary-signal photoelectrons required in an interval T, equal to (bit rate)$^{-1}$, to achieve an error probability of 10^{-9} (assuming an equal distribution of zeros and ones); it is proportional to the average optical energy per bit. The abscissa is the bit rate B. Since the optical power P_0 is related to the number of photoelectrons per bit N through the expression

$$P_0 = NBhc/\eta\lambda, \qquad (6)$$

(where h is Planck's constant, η the quantum efficiency of the photodetector, and λ the wavelength of the signal photon), a set of diagonal grid lines could be drawn on the chart to represent receiver sensitivity [decibels above 1 mW (dBm)]. Thus the dashed grid lines in Fig. 3 are for $\lambda = 1.3$ μm and $\eta = 0.7$. The calculated performances of optical receivers consisting of InGaAs $p-i-n$ photodiodes followed by GaAs FET preamplifiers are represented by the solid curves (Gloge et al., 1980a); the two heavy solid lines with slopes proportional to $B^{1/2}$ are for total input capacitance $C_T = 2$ and 1 pf, FET transconductance $g_m = 50$ mA V^{-1}, and total leakage current $I_L = 0$. The limiting factor here is the channel conductance noise of the FET, which is proportional to $C_T^2 B^3/g_m$, so that there is a tradeoff between C_T and g_m. The family of curves with initial slopes proportional to $B^{-1/2}$ for various I_L represents degradation due to leakage current. Since the number of electrons per bit associated with a given leakage current decreases with increasing bit rate, the sensitivities at high bit rates are not affected by I_L, as shown in Fig. 3. The dotted band represents projected performance for future avalanche photodiodes (APDs) with optimal avalanche gains, assuming negligible dark current and a range of carrier ionization-rate ratios k between 0.1 and 0.5 (Smith and Forrest, 1982). The required optimal gains (for $I_L = 0$) are given by $M_{opt} \simeq 0.75(R/k)^{1/2}$, where R is the ratio of the sensitivity of the equivalent $p-i-n$ detector ($M = 1$) to the quantum-noise-limited sensitivity (Smith and Forrest, 1982). For $B = 1$ Gbit sec^{-1} and $k = 0.4$, $M_{opt} \simeq 17$, which is well within the range of values obtained in experimental APDs. The large dots represent the best experimental values for $p-i-n$ photodiode receivers reported in the literature (Smith et al., 1982; Williams, 1982). The two crosses represent the best performance for experimental receivers employing the best InGaAsP/InGaAs APD available to date (Campbell et al., 1983).

4. Degradation in Receiver Sensitivity Due to LED Modulation Bandwidth

The previous considerations assume that the optical signal arriving at the receiver is ideal (rectangular pulses) and that the receiver filter [with transfer function $H_r(r)$] shapes the signal pulses into the desired format for threshold detection. However, the finite-modulation bandwidths of the LED causes the signal power to fall off as modulation frequency is increased and therefore causes distortion of the signal. The falloff can be described by the transfer function (Gloge et al., 1980a)

$$H_s(f) = (1 + if/b)^{-1} \qquad (7)$$

where b is the (3-dB electrical) modulation bandwidth of the LED. In general, the LED output power P_s is proportional to $b^{-\nu}$. For moderately fast surface-emitting LEDs, ν is typically $\frac{2}{3}$ (see Fig. 15 in Chapter 5).

In order to equalize the distortion caused by the finite bandwidth of the LED, the transfer function of the receiver must now be modified to have a response $H'_r(f) = H_r(F)/H_s(f)$. This requirement results in a power penalty and hence in degradation of the receiver sensitivity.

Since the transmitter power P_s is proportional to $b^{-\nu}$ and the required equalized receiver power P'_r is proportional to $[1 + K(B/b)^2]^{1/2}$, where K is a constant associated with the receiver noise, the available optical signal margin between transmitter and receiver (being proportional to P_s/P'_r) shows a maximum as a function of b (Gloge et al., 1980a). This maximum occurs for $b = 0.42B$ (if $\nu = \frac{2}{3}$), and the corresponding degradation of receiver penalty is 2.4 dB over the values given in Fig. 3. A faster LED would reduce this penalty but would provide less output power and hence a smaller overall signal margin.

5. Degradation in Receiver Sensitivity Due to Fiber Chromatic Dispersion

Chromatic dispersion in the fiber introduces pulse spreading or distortion of the signal that must be equalized in the receiver. Again, the required equalization results in a power penalty or reduced receiver sensitivity. At the wavelength of minimum dispersion ($\lambda \simeq 1.3 \ \mu$m), the transfer function attributable to the chromatic dispersion of the fiber is (Gloge *et al.*, 1980a)

$$H_c(f) = (1 + i\pi f z \sigma_s^2/10c\lambda^2)^{-1/2}, \qquad (8)$$

where σ_s is the rms width of the LED emission spectrum. For typical LEDs $\sigma_s = \lambda^2/40 \ \mu$m, where λ is in micrometers.

Figure 4 shows the increase in the required receiver power over the values given in Fig. 3 as a result of fiber chromatic dispersion, plotted against the

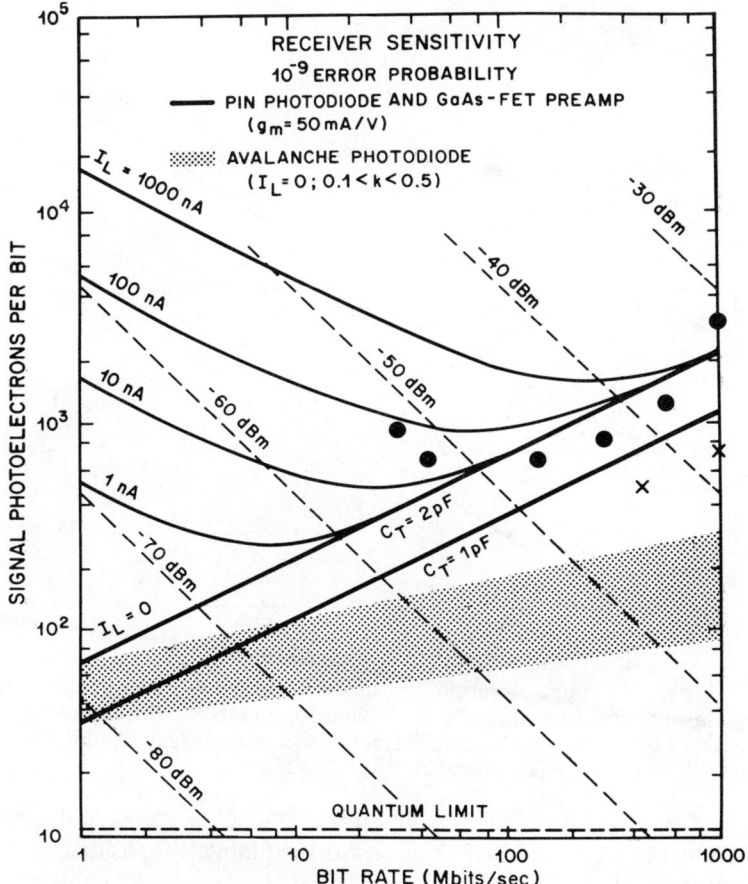

FIG. 3. Receiver sensitivity of state-of-the-art digital lightwave repeaters, expressed in the average number of signal photoelectrons required to achieve an error probability of 10^{-9}, plotted as a function of the bit rate: Solid lines, p-i-n photodiode and GaAs FET preamp ($g_m = 50$ mA V^{-1}); hatched area, APD ($I_L = 0$, $0.1 < k < 0.5$). [From Li (1983). © 1983 IEEE.]

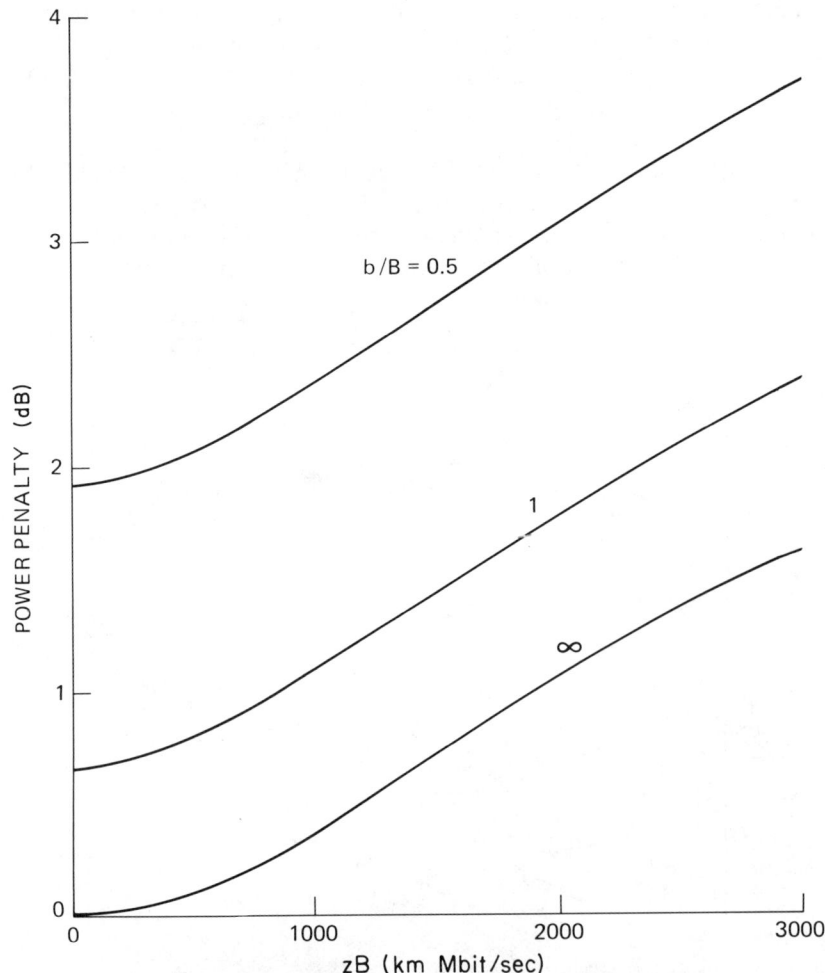

FIG. 4. Increase in optical signal power required at the receiver as a result of material dispersion in germanosilicate multimode fibers plotted against the product of distance z and bit rate B; b is the LED modulation bandwidth [From Gloge *et al.* (1980a). Reprinted by permission from Bell System Technical Journal.]

product of bit rate B times transmission distance z. The calculation also includes the penalty due to LED-bandwidth equalization, as discussed previously in Section 4.

6. Transmission Margin

Figure 5 shows the average optical power margin between the transmitter and the receiver as a function of the bit rate. The band for the LED transmit-

7. LED-BASED MULTIMODE LIGHTWAVE SYSTEMS

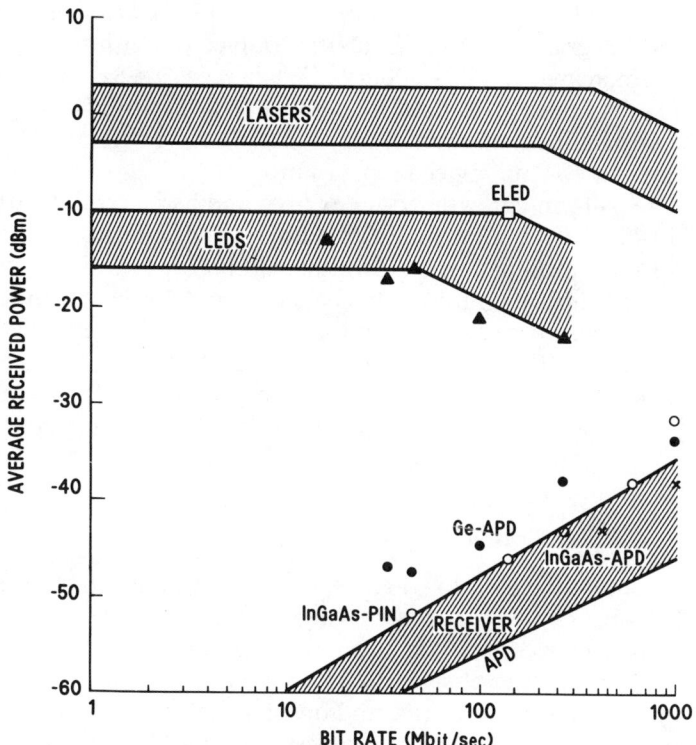

FIG. 5. Average optical power between the transmitter and the receiver as a function of the bit rate.

ter power represents typical values launched into multimode fibers for 1.3-μm operation. The band for the receiver power represents projection of future receiver performance based on results contained in Fig. 3. The lower boundary pertains to future APDs, whereas the upper boundary reflects the best performance obtained with $p-i-n$ photodiodes and FET preamplifiers reported in the literature. The open circles are experimental results obtained with laser transmitters for which chromatic dispersion of the fiber is negligible. The solid dots represent experimental values when surface-emitting LEDs are used, resulting in a degradation of receiver performance by 3–5 dB (due to limited LED bandwidth and fiber chromatic dispersion), compared to those obtained with lasers. The difference between the transmitter power and the receiver power is, of course, the transmission margin.

The bandwidths below 50 MHz, surface-emitting LEDs can launch peak powers in the neighborhood of 50 μW into a 0.23-NA, 50-μm-core graded-index fiber (Gloge et al., 1980a) and more than 100 μW into a 0.2 NA,

62.5-μm-core graded-index fiber (Temkin et al., 1983). Using a monolithically formed integral lens on the LED, the coupled power into a larger core fiber can be increased to 200–300 μW (Wada et al., 1981). The falloff of output power at higher modulation bandwidths is discussed in Chapter 5.

Figure 5 also includes an example of a 1.3-μm edge-emitting LED, shown by an open square. This device launches more than 350 μW (peak power) into a 0.2 NA, 50-μm-core graded-index fiber and had a modulation bandwidth of ~140 MHz (Ure et al., 1982). Although edge-emitting LEDs have larger (~3 times) modulation bandwidths and higher (~2 times) coupling efficiencies than surface-emitting LEDs, their output powers are much more sensitive to temperature changes. The operating lifetime of edge emitters is projected to be ~10^6 hr at room-temperature (A. C. Carter, J. Ure, M. Harding, and R. C. Goodfellow, unpublished work, 1982), compared to >10^9 hr for surface-emitting LEDs (Yeats et al., 1981; Wada et al., 1982; Saul, 1983).

IV. System Experiments and Field Trials

With presently available low-loss (<1 dB km^{-1} at 1.3 μm) multimode fibers an LED-based lightwave system can achieve repeaterless transmission distance up to 20 km and information rates up to a few hundred megabits per second. Applications of such a system include interoffice trunks, subscriber loops, local-area networks, and on-premises data links.

Chromatic-dispersion-limited transmission in multimode fibers was first demonstrated in an experiment on a 137-Mbit/sec data link operating at both 0.9 and 1.23 μm wavelengths using an AlGaAs and an InGaAsP LED, respectively (Muska et al., 1977). In a subsequent experiment (Gloge et al., 1980a), 1.3-μm surface-emitting LEDs, which produced 1-mW output power into the air, were used as the transmitters. The receivers consisted of back-illuminated InGaAs $p-i-n$ photodiodes (Lee et al., 1980a, 1981) and a GaAs FET preamplifiers (Owaga and Chinnock, 1979) and, had measured sensitivities of −46.7 and −36 dBm at 44.7 and 274-Mbit/sec data rates, respectively. Repeaterless transmission distances of 23 km at the low rate and 7 km at the high rate were achieved (Gloge et al., 1980a; Ogawa et al., 1981a). The overall bandwidth of the fiber used was approximately 800 MHz km. These initial experiments demonstrated the feasibility of high bit-rate lightwave systems using 1.3-μm LEDs.

Similar laboratory experiments at both 32 and 100 Mbit/sec were conducted in Japan (Kunita et al., 1979; Shikata et al., 1982). In the 100-Mbit/sec experiment, a small surface-emitting LED with a 23-μm diameter light-emitting area and a 140-nm spectral width (FWHM) centered at the wavelength of 1.25 μm was used. The output power was coupled to a 50-μm core, 0.2-NA graded-index fiber by means of a graded-index-rod lens. A -13-dB coupling efficiency resulted in a launched power of -17.9 dBm into the fiber. The measured fiber bandwidth was 320 MHz km^{-1}. Using a Ge APD for the photodiode, the average received signal power of -44.3 dBm was obtained for 10^{-9} error probability. The power penalty due to fiber dispersion (15 km) was 2.8 dB. Allowing for sufficient system margins, repeater spacings up to 10 km were considered feasible.

The first long-wavelength LED-based system to carry commercial telephone services was installed between two central offices in Sacramento, California, in June 1980. It operated at 12.6 Mbit/sec with interoffice spacing of 6 km. The system utilized existing fiber cables installed for application at $\lambda = 0.85$ μm and therefore was not optimized for 1.3-μm operation. More recently, a 34-Mbit/sec lightwave system was installed for long-haul telephone service (Davis et al., 1982). This system, spanning London and Birmingham in Great Britain, operated with 1.3-μm LEDs over a total distance of 205 km, along which the maximum repeater spacing was 10.5 km. The surface-emitting LEDs, equipped with a truncated spherical lens for improved coupling, produced peak launched powers of -17 dBm into the fiber. InGaAs $p-i-n$ photodiodes and GaAs FETs were used for the front end of the receiver. A sensitivity of -52 dBm has been realized. The mean bandwidth of the fiber was 270 MHz km, and the mean fiber loss was 1.32 dB km^{-1}. Using edge-emitting LEDs, this system can be upgraded to 140 Mbit/sec. (A. C. Carter, J. Ure, M. Harding, and R. C. Goodfellow, unpublished work, 1982; Ure et al., 1982).

Similar application opportunities exist for LED-based lightwave systems in the feeder portion of the subscriber loop where digital systems become increasingly attractive with the introduction of digital end offices. An example is the Fiber SLC™ Carrier system of the Bell System, which first went into commercial service in Chester Heights, Pennsylvania, in November 1982. This was followed by a second installation in early 1983 in Kernersville, North Carolina, with a rapid buildup leading to an anticipated deployment of approximately 500 Fiber SLC™ systems by the end of 1983 (Olsen and Schepis, 1984).

The fibers, specifically designed for LED-based system applications, have a numerical aperture of 0.29, a core diameter of 62.5 μm, and a cladding diameter of 125 μm. The installed fiber-cable loss is less than 1 dB km^{-1} (at $\lambda = 1.3$ μm), and the bandwidth is capable of supporting system bit rates up to 100 Mbit/sec for a maximum feeder length of 20 km. The surface-emitting LEDs with monolithic integral lens are capable of launching more than 100 μW into the fiber. The initial transmission rate in the field trial was at 12.6 Mbit/sec with capabilities of upgrading to 44.5 Mbit/sec and ultimately to 90 Mbit/sec as economic and growth conditions become favorable. The receivers employed similar back-illuminated InGaAs $p-i-n$ photodiodes and GaAs FETs as described in the earlier laboratory experiments (Gloge *et al.*, 1980a).

The use of a digital carrier in the loop plant naturally places many remote feeder terminals in a hostile, unattended environment. Reliability considerations favor the choice of LED/$p-i-n$-photodiode-based systems over laser-based systems. The system margin of 25 dB, achieved at 44.5 Mbit/sec, takes into account a local component temperature range of -40 to $+85\,^\circ$C, degradation of LED power to aging and other system tolerances. A total of 7 dB is allocated to system margin and 18 dB to fiber loss. Device mean lifetimes (to failure) of these LEDs are expected (from the accelerated aging results) to be greater than 10^9 hr at $60\,^\circ$C (Saul, 1983). The best mean life (to failure) for $p-i-n$ photodiodes is estimated at more than 10^9 hr at $60\,^\circ$C, and the worse is about 10^7 hr (Saul *et al.*, 1984).

In addition to using time-division multiplexing (TDM) to increase the system information-carrying capacity by operating at higher bit rates, the wide transmission window of the low-loss fiber in the entire $0.8-1.6$-μm-wavelength region provides opportunities for wavelength-division multiplexing (WDM) as an alternative (Lee, 1982). Besides, in a WDM system, mixed modulation formats for different carrier wavelengths can be transmitted simultaneously without increasing the complexities of digital multiplexing as is necessary in a TDM system. Thus, WDM systems are simpler and may have economic advantages. Possible applications of WDM systems include voice, data, and video services simultaneously transmitted in the subscriber loop.

Nonlinearity in the light-output–current characteristic of an LED can cause degradation of signal-to-noise ratio (SNR) in an AM system. However, FM (modulated on an AM carrier) can be used to advantage, and an improvement of SNR of 20 dB was achieved in an FM system in comparison to an AM system (Straus and Conradi, 1978).

A multichannel FM video optical-fiber transmission experiment (Conradi *et al.*, 1982), which consisted of three FM channels on each of three LED transmitters at 1.1, 1.2, and 1.3 μm was conducted (Fig. 6). The graded-index fiber (66-μm core, 0.23 NA) exhibited an average loss of less than 1.3 dB km^{-1} at 1.3 μm and bandwidths of 203 MHz km^{-1} at 1.3 μm and 132 MHz km^{-1} at 1.1 μm over the 2.54-km transmission length. Three surface-emitting LEDs were employed with output powers of -14.7, -13.9 and -12.6 dBm coupled into the fiber. The three carrier frequencies on each channel were 10, 30, and 50 MHz for the 1.1- and 1.2-μm channels and 30, 50, and 70 MHz for the 1.3-μm channel. The optical multiplexer and demultiplexer ultilized interference filters and Selfoc lenses, with total losses ranging from 3.9 (at 1.3 μm) to 7.9 dB (at 1.2 μm). Signal-to-noise ratios of

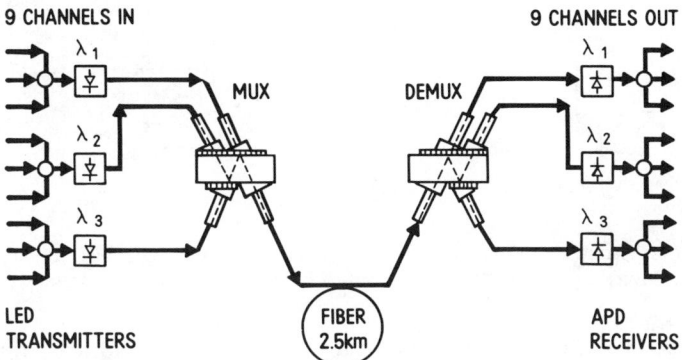

FIG. 6. A simplified schematic of multichannel video/WDM system. [From Conradi et al. (1982). © 1982 Bell-Northern Research LTD.]

49.3 dB for the worst channel and 56.8 dB for the best channel were achieved.

The feasibility of transmitting both video and data services for subscriber-loop applications on a single optical fiber with WDM has been demonstrated (T. P. Benton and L. R. Linnell, unpublished work, 1981). Two LEDs at $\lambda = 1.03$ and 1.3 μm were used as the signal sources for the AM video channel and the 64-kbit/sec data channel, with powers of -12.6 and -15.4 dBm coupled into a 0.29-NA, 62.5-μm-core graded-index fiber. A

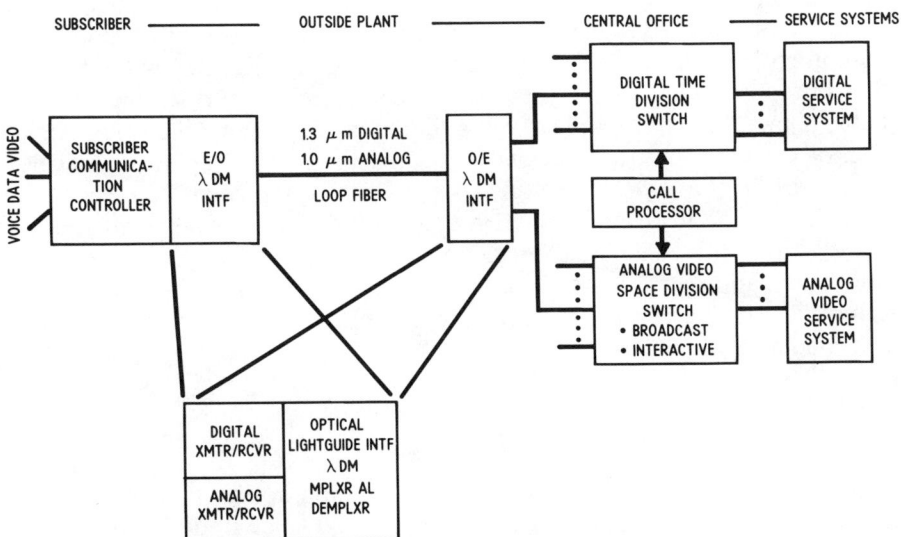

FIG. 7. A schematic of integration of digital and analog wide-band switching system using wavelength-division multiplexing. [From Benton and Linnell (1981).]

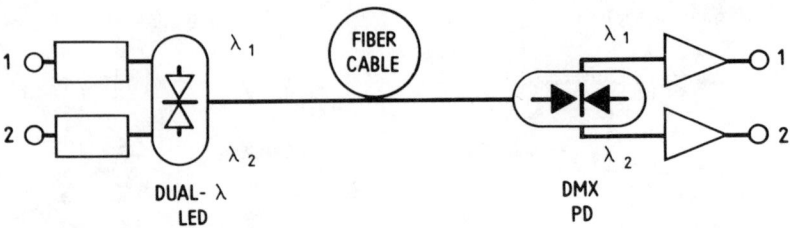

FIG. 8. A schematic of WDM system employing a dual-wavelength LED and a dual-wavelength demultiplexing photodetector. [From Lee (1982).]

graded-index-rod lens interference filter was used for the optical multiplexer, but a dual-wavelength photodetector (Campbell *et al.*, 1980) was employed to serve simultaneously the dual functions of the optical demultiplexer and the photodetector, thus eliminating the need of a separate optical demultiplexer as in a conventional WDM system. An example of a system configuration for a possible future wideband subscriber-loop service using WDM is shown in Fig. 7.

It is also possible to eliminate the use of an optical multiplexer by the use of a dual-wavelength LED (Lee *et al.*, 1980b). Indeed, a digital WDM-system experiment (Ogawa *et al.*, 1981b), operating at 33 Mbit/sec employing both a dual-wavelength LED and a dual-wavelength photodetector has demonstrated reasonably good performance. Figure 8 shows a simplified schematic of the experiment.

The relentless drive to increase the speed of operation of on-premises equipment, such as computers, multiplexers, and digital switches, has served as a stimulus to push higher the transmission speed of short-distance LED-based data links. An example of work addressing this need is the demonstration of a 4.4-km link using a high-speed InGaAsP LED operating at 560 Mbit/sec (Grothe *et al.*, 1983). The specially designed high-speed LED has a modulation bandwidth of 900 MHz and, with a ball-lens, is able to couple 7.3 μW into a 0.2-NA graded-index fiber at a 100-mA drive current. A total link attenuation of 5.3 dB at bit error rate of less than 10^{-9} has been achieved. The highest-speed LED experimental system reported to date operated at 1.6 Gbit/sec over a fiber link of 1 km (Suzuki *et al.*, 1984).

V. Conclusions

It is evident that LED-based lightwave systems operating at the wavelength of minimal dispersion near 1.3 μm of doped-silica graded-index fibers can have many applications that offer advantages of simplicity, reliability, and economy over laser-based systems. Data rates up to a few hundred megabits per second and repeater spans over 10 km are achievable with practical LEDs and $p-i-n$ photodetectors. Using WDM, channel capacity can be increased severalfold and cost savings on the complex electronic multiplexing equipment may be realizable. Present commercial systems are deployed as on-premises data links and for subscriber-loop carrier transmission. Future growth may extend the application of these LED-based systems into the areas of local area networks and wideband services, including voice, data, and video transmission in the subscriber loop.

The ultimate limitation of LED-based multimode-fiber systems are set by the available power of LEDs at high modulation rates and by the chromatic-dispersion-limited bandwidth of multimode fibers. In the area of long-haul, high-capacity transmission, laser-based single-mode lightwave systems are the natural choice. However, the market niche for LED-based fiber systems for short-distance, moderate-bandwidth applications is not at all insignificant, and favorable future growth as new applications evolve may expand further this market sector.

REFERENCES

Benton, T. P., and Linnell, L. R., (1981). Unpublished work.
Campbell, J. C., Dentai, A. G., Lee, T. P., and Burrus, C. A. (1980). *IEEE J. Quantum Electron.* **QE-16,** 601.
Campbell, J. C., Dentai, A. G., Holden, W. S., and Kasper, B. L. (1983). *Tech. Dig.—Int. Electron Devices Meet.,* p. 464.
Carter, A. C., Ure, J., Harding, M., and Goodfellow, R. C., (1982). Unpublished work.
Conradi, J., Kim, H. B., Duck, G., Maciejko, R., Straus, J., and SpringThorpe, A. J. (1982). *Tech. Dig., Top. Meet. Opt. Fiber Commun., 5th, 1982,* Paper TUFF7.
Davis, R., Cleobury, D., and Cronin, D. R. (1982). *Proc. Eur. Conf. Opt. Commun., 8th, 1982,* Paper AXIV-4, p. 460.
French, W. G., MacChesney, J. B., O'Connor, P. B., and Tasker, G. W. (1974). *Bell Syst. Tech. J.* **53,** 951.
Gloge, D., Marcatili, E. A. J., Marcuse, D., and Personick, S. D. (1979). *In* "Optical Fiber Communications" (S. E. Miller and A. G. Chynoweth, eds.), Chapter 4. Academic Press, New York.
Gloge, D., Albanese, A., Burrus, C. A., Chinnock, E. L., Copeland, J. A., Dentai, A. G., Lee, T. P., Li, T., and Ogawa, K. (1980a). *Bell Syst. Tech. J.* **59,** 1365.
Gloge, D., Ogawa, K., and Cohen, L. G. (1980b). *Electron. Lett.* **16,** 366.
Greco, G. J., and Stone, D. E. (1983). *Conf. Rec., Int. Conf. Commun., 1983,* p. 82.
Grothe, H., Muller, G., and Harth, W. (1983). *Electron. Lett.* **19,** 909.
Horiguchi, M., Nakahara, M., Inagaki, N., Kokura, K., and Yoshida, K. (1982). *Proc. Euro. Conf. Opt. Commun., 8th, 1982,* Paper AIII-4.
Jablonowski, D. P., Padgette, D. D., and Merten, J. R. (1982). *Tech. Dig., Top. Meet. Opt. Fiber Commun., 5th, 1982,* Paper TUEE 2.
Kapron, F. P. (1977). *Electron. Lett.* **13,** 96.
Kapron, F. P., Keck, D. B., and Maurer, R. D. (1970). *Appl. Phys. Lett.* **17,** 423.
Keck, D. B., and Bouillie, R. (1978). *Opt. Commun.* **25,** 43.
Keck, D. B., Schultz, P. C., and Zimar, F. (1972). *Appl. Phys. Lett.* **21,** 215.

Kunita, M., Touge, T., and Fujimoto, N. (1979). *Tech. Dig. Top. Meet. Opt. Fiber. Commun., 1979*, Paper TuD1.
Lee, T. P. (1982). *Opt. Laser Technol.* **14**, 15.
Lee, T. P., Burrus, C. A., Dentai, A. G., and Ogawa, K. (1980a). *Electron. Lett.* **16**, 155.
Lee, T. P., Burrus, C. A., and Dentai, A. G. (1980b). *Electron. Lett.* **16**, 845.
Lee, T. P., Burrus, C. A., and Dentai, A. G. (1981). *IEEE J. Quantum Electron.* **QE-17**, 232.
Li, T. (1983). *IEEE J. Selected Area Commun.* **SAC-1**, 356.
Lin, C., Cohen, L. G., French, W. G., and Foertmeyer, V. A. (1978). *Electron. Lett.* **14**, 170.
Lin, C., Liu, P. L., Lee, T. P., Burrus, C. A., Stone, F. T., and Ritger, A. J. (1981). *Electron. Lett.* **17**, 438.
Marcatili, E. A. J. (1977). *Bell Syst. Tech. J.* **56**, 49.
Marcuse, D. (1972). *Bell Syst. Tech. J.* **51**, 1199.
Miller, S. E., Marcatili, E. A. J., and Li, T. (1973). *IEEE Proc.* **61**, 1703.
Muska, W. M., Li, T., Lee, T. P., and Dentai, A. G. (1977). *Electron. Lett.* **13**, 605.
Nagel, S. R., MacChesney, J. B., and Walker, K. L. (1982). *IEEE J. Quantum Electron.* **QE-18**, 459.
Nemchic, J. M., and Buckler, M. J. (1983). *Conf. Rec. Int. Conf. Commun., 1983*, p. 91.
Ogawa, K., and Chinnock, E. L. (1979). *Electron. Lett.* **15**, 650.
Ogawa, K., Chinnock, E. L. Gloge, D., Kaiser, P., Nagel, S. R., and Jang, S. J. (1981a). *Electron. Lett.* **17**, 715.
Ogawa, K., Lee, T. P., Burrus, C. A., Campbell, J. C., and Dentai, A. G. (1981b). *Electron. Lett.* **17**, 857.
Olsen, J. W., and Schepis, A. J. (1984). *Tech. Dig. Opt. Fiber Commun. 1984*, Paper MD1.
Olshansky, R., and Keck, D. B. (1976). *Appl. Opt.* **15**, 483.
Osanai, H., Shioda, T., Moriyama, T., Araki, S., Horiguchi, M., Izawa, T., and Takata, H. (1976). *Electron. Lett.* **12**, 549.
Payne, D. N., and Gambling, W. A. (1975). *Electron. Lett.* **11**, 176.
Personick, S. D. (1971). *Bell Syst. Tech. J.* **52**, 843.
Saul, R. H. (1983). *IEEE Trans. Electron Devices* **ED-30**, 285.
Saul, R. H., Chen, F. S., Shumate, P. W. (1984). *Bell Syst. Tech. J.* (to be published).
Shikata, M., Nomura, H., Suzuki, A., Minemura, K., and Sugimoto, S. (1982). *Tech. Dig., Top. Meet. Opt. Fiber Commun., 5th, 1982*, Paper TUDD3.
Smith, D. R., Hooper, R. C., Smyth, P. P., and Wake, D. (1982). *Electron. Lett.* **18**, 453.
Smith, R. G., and Forrest, S. R. (1982). *Bell Syst. Tech. J.* **61** 2929.
Straus, J., and Conradi, J. (1978). *Conf. Rec. IEEE Natl. Telecommun. Conf., 1978*, Paper 13.4.
Suzuki, A., Inomoto, Y., Hayashi, J., Isoda, Y., Uji, T., Nomura H. (1984). *Electron. Lett.* **20**, 273.
Temkin, H., Zipfel, C. L., DiGiuseppe, M. A., Chin, A. K., Kermidas, V. G., and Saul, R. H. (1983). *Bell Syst. Tech. J.* **62**, 1.
Ure, J., Carter, A. C., Goodfellow, R. C., and Harding, M. (1982). *Conf. Proc. IEEE Spec. Conf. Light Emitting Diodes Photodetectors*, Paper 20. Ottawa-Hull, Ontario, Canada.
Wada, O., Yamakoshi, S., Abe, M., Nishitani, Y., and Sukurai, T. (1981). *IEEE J. Quantum Electron.* **QE-17**, 174.
Wada, O., Yamakoshi, S., Hamaguchi, H., Sanada, T. Nishitani, Y., and Sakurai, T. (1982). *IEEE J. Quantum Electron.* **QE-18**, 368.
Williams, G. F. (1982). *Dig. Tech. Pap., IEEE Int. Solid-State Circuits Conf., 1982*, THPM 13.2.
Yeats, R., Chai, Y. G., Gibbs, T. D., and Antypas, G. A. (1981). *IEEE Electron Device Lett.* **EDL-2**, 234.
Yokota, H., Iwasaki, K., Kumamaru, H., and Takata, H. (1979). *Conf. Proc. Opt. Commun. Conf., 1979*, Paper 19.3.

CHAPTER 8

Semiconductor Laser Noise: Mode Partition Noise

Kinichiro Ogawa

AT&T BELL LABORATORIES
HOLMDEL, NEW JERSEY

I.	Introduction	299
II.	A Simple Model of Mode Partition Noise	300
III.	Analysis of Mode Partition Noise	302
IV.	Mode Partition Coefficient: k Value	311
V.	Theoretical k Value with and without Interaction between Lasing Modes	314
VI.	Statistical Measurement of Mode Partition Noise	322
VII.	Effects of Optical Injection Locking on Mode Partition Noise	324
	Appendix I	328
	Appendix II	329
	References	330

I. Introduction

Semiconductor laser diodes exhibit two fundamental types of noise: (1) the quantum shot noise associated with the total power fluctuation, and (2) mode partition noise carried by each longitudinal mode (Ito *et al.*, 1977). The quantum shot noise should be considered one of the significant noise sources for optical systems requiring high signal-to-noise ratio, such as analog television transmission systems or coherent optical systems that must take into consideration AM or FM noise. The performance of optical fiber digital systems employing direct modulation will not be degraded by this quantum shot noise unless the reflection from the fiber enhances greatly this quantum shot noise.

However, mode partition noise plays a key role in the performance limitation of any optical system (Okano *et al.*, 1980). This chapter discusses mode partition noise, its theoretical background, its effects on system performance, and its distribution function.

II. A Simple Model of Mode Partition Noise

Mode partition noise is caused by the instantaneous fluctuation of the power distribution among the laser longitudinal modes while the total power of all longitudinal modes is kept constant. This partitioning fluctuation among the longitudinal modes is a common problem for multifrequency cavity oscillators, such as gas lasers and early magnetrons (Collins, 1948). This mode-hopping of magnetrons was a serious problem in the radar-development program during World War II.

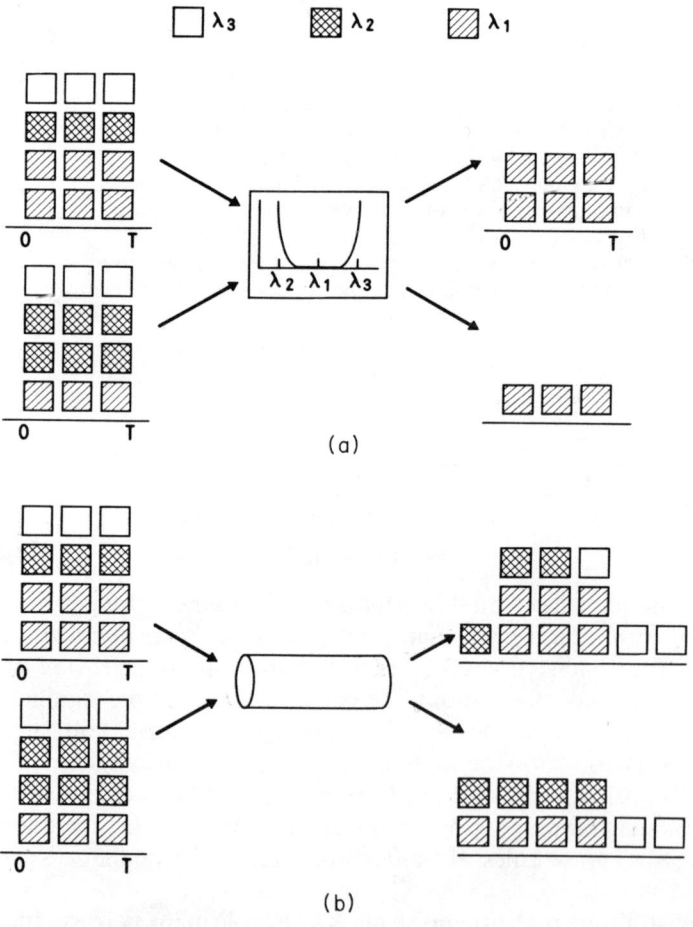

FIG. 1. Simple model of mode partition noise. Each block represents a group of photons that carry the same wavelength. A laser diode emits various combinations of wavelengths (several longitudinal modes) from pulse to pulse: (a) wavelenght-dependent loss; (b) chromatic dispersion.

Figure 1 shows a simple schematic explanation of mode partition noise. When a laser diode is modulated directly with injection current, the power distribution among several longitudinal modes varies from pulse to pulse, even though the total power is kept constant from pulse to pulse. Through a media having some wavelength-dependent dispersion or wavelength-depen-

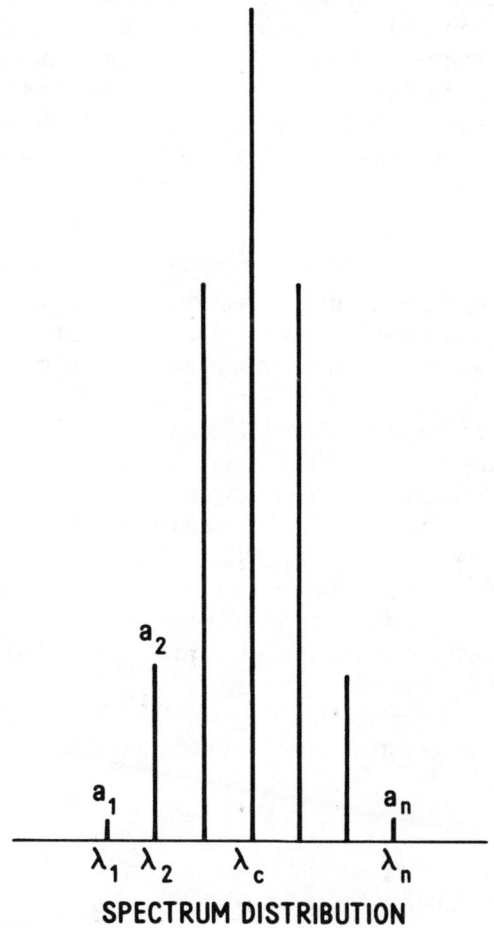

FIG. 2. The time-averaged spectrum distribution of the laser diode. There are N longitudinal modes of which amplitudes are defined by

$$\frac{\exp[-(\lambda_i - \lambda_c)^2/2\sigma^2]}{\sum_{i=1}^{N} \exp[-(\lambda_i - \lambda_c)^2/2\sigma^2]}$$

where λ_c is the center wavelength, λ_i the wavelength of the ith mode, and σ the halfwidth of the spectrum. [From Ogawa (1982). © 1982 IEEE.]

dent loss, these transmitted pulses will deform those pulse shapes, as shown in Fig. 1. These pulse waveform variations from pulse to pulse can be observed as noise at the receiver side, which is called "mode partition noise."

A laser spectral distribution observed experimentally is normally a time-averaged spectrum, not an instantaneous spectrum, which varies from time to time. Since this mode partition noise is a generic function of the laser spectrum fluctuation and the wavelength-dependent dispersion or loss of media, the signal-to-noise ratio due to mode partition noise is independent of the signal power. As shown later (see Fig. 7), mode partition noise can be observed as an error-rate floor (Yamamoto et al., 1980) instead of an improved error rate as the optical power is increased in optical fiber digital systems.

III. Analysis of Mode Partition Noise

The time-averaged spectrum distribution shows N longitudinal modes when a laser diode is directly modulated, as shown in Fig. 2. Those wavelengths are $\lambda_i (i = 1-N)$, and this normalized power carried by each longitudinal mode is a_i ($i = 1-N$). To derive a model of mode partition noise (Ogawa, 1982; Ogawa and Vodhanel, 1982), we used the following assumptions: (1) the total power of all longitudinal modes carried by each pulse is constant; (2) the normalized time-averaged spectrum $p(\lambda_i)$ shows the averaged value of each longitudinal mode determined by an unknown mode partition probability function, $p(a_1, a_2, \ldots a_N)$; (3) within each pulse, the power distribution of the longitudinal mode does not vary, and the received waveform of each longitudinal mode will be distorted through the amplitude and differential delay distribution of media and the waveform-shaping equalizer filter of the receiver.

From assumption (1), the amplitude a_i of the ith longitudinal mode normalized by total power satisfies the following relation:

$$\sum_{i=1}^{N} a_i = 1. \qquad (1)$$

From assumption (2), the observed spectrum can be defined by an unknown mode partition probability function.

$$\overline{a_i} = \int_0^\infty \int_0^\infty \int_0^\infty a_i p(a_1, a_2, \ldots, a_N) \, da_1 \, da_2 \ldots da_N$$
$$= p(\lambda_i) \quad \text{for any } i. \qquad (2)$$

From assumption (3), we define the received pulse waveform of each longitudinal mode by $f(\lambda_i, t)$. Thus, the received signal is expressed by the total summation of each waveform, and we define the received pulse waveform of each longitudinal mode by $f(\lambda_i, t)$ where λ_i is the wavelength of the ith longitudinal mode $f(\lambda_i, t) \leq 1$. Therefore the received signal is expressed by the total summation of each waveform:

$$r(t) = \sum_i f(\lambda_i, t) a_i. \tag{3}$$

The fluctuation at the sampling slot $t = t_0$ is described by the expectation and the second moment of $r(t_0)$ with the mode partition probability function $p(a_1, a_2, \ldots, a_N)$.

$$\sigma^2 = \overline{r^2(t_0)} - (\overline{r(t_0)})^2$$

$$= \sum_i f^2(\lambda_i, t_0) \overline{a_i^2} + 2 \sum_i \sum_{j>i} f(\lambda_i, t_0) f(\lambda_j, t_0) \overline{a_i a_j}$$

$$- \sum_i f^2(\lambda_i, t_0)(\overline{a_i})^2 - 2 \sum_i \sum_{j>1} f(\lambda_i, t_0) f(\lambda_j, t_0) \overline{a_i}\,\overline{a_j}, \tag{4}$$

where

$$\overline{a_i a_j} = \int a_i a_j p(a_1, a_2, \ldots, a_N)\, da_1\, da_2 \ldots da_N,$$

$$\overline{a_i} = \int a_i p(a_1, a_2, \ldots, a_N)\, da_1\, da_2 \ldots da_N.$$

Since a_i is expressed by $(1 - \Sigma_{j \neq i} a_j)$ from Eq. (1), the terms of $\overline{a_i^2}$ can be expressed in terms $\overline{a_i^2} = \overline{a_i(1 - \Sigma_{j \neq i} a_j)} = \overline{a_i} - \Sigma_{j \neq i} \overline{a_i a_j}$. Therefore we obtain the fluctuation σ^2 as follows:

$$\sigma^2 = k^2 \left\{ \sum_i f^2(\lambda_i, t_0) \overline{a_i} - \left[\sum_i f(\lambda_i, t_0) \overline{a_i} \right]^2 \right\}$$

$$= k^2 \left\{ \sum_i [1 - f(\lambda_i, t_0)]^2 \overline{a_i} - \left[\sum_i [1 - f(\lambda_i, t_0)] \overline{a_i} \right]^2 \right\},$$

or

$$\sigma^2 = k^2 \left\{ 2 \sum_i \sum_{j>i} [f(\lambda_i, t_0) - f(\lambda_j, t_0)]^2 \overline{a_i}\,\overline{a_j} \right\},$$

where

$$k^2 = \frac{\sum_i \sum_{j>i} [f(\lambda_i, t_0) - f(\lambda_j, t_0)]^2 (\overline{a_i}\,\overline{a_j} - \overline{a_i a_j})}{\sum_i \sum_{j>i} [f(\lambda_i, t_0) - f(\lambda_j, t_0)]^2 \overline{a_i}\,\overline{a_j}}. \tag{5}$$

To calculate the mode partition noise by this expression, it still requires the mode partition probability function for evaluating k. However, if we assume that every longitudinal mode consists of the stimulated emission, not the spontaneous emission, the relation $\overline{a_i a_j} = \alpha \overline{a_i}\,\overline{a_j}$ (α = const $i \neq j$) will hold for any stimulated emissions. Therefore, the expression of k^2 in Eq. (5) becomes independent from the waveform $f(\lambda_i, t_0)$ and expressed by only the longitudinal mode distribution:

$$k^2 = \frac{\sum_i \sum_{j>i}(\overline{a_i a_j} - \overline{a_i}\,\overline{a_j})}{\sum_i \sum_{j>i}(\overline{a_i a_j})} = \frac{\overline{a_i a_j} - \overline{a_i}\,\overline{a_j}}{\overline{a_i a_j}}, \quad i \neq j. \tag{6}$$

By using the expectation and variance of Eq. (1), the terms of k can be described only by $\overline{a_i}$ and $\overline{a_i^2}$:

$$k^2 = \frac{\sum_i [\overline{a_i^2} - (\overline{a_i})^2]}{1 - \sum_i (\overline{a_i})^2} \quad \text{or} \quad k^2 = \frac{\overline{a_i^2} - (\overline{a_i})^2}{a_i - (\overline{a_i})^2} \quad \text{for any } i. \tag{7}$$

Since $\overline{a_i^2}$ and $\overline{a_i}$ can be measured by an rms voltmeter and a dc meter, the k value can be estimated easily without knowing the cross-correlation term. As can be seen in Eq. (6), we can state that k becomes zero if and only if each longitudinal mode is independent: $\overline{a_i a_j} = \overline{a_i}\,\overline{a_j}$, where $k = 0$. On the other hand, if each longitudinal mode is a mutually exclusive event, we obtain $\overline{a_i a_j} = 0$, where $i \neq j$ and $k = 1$. This indicates that in the case where each pulse contains only one longitudinal mode and its dominant wavelength in each pulse varies randomly, we observe strong mode partition noise. In general, k lies between 0 and 1.

Therefore the mode partition noise σ^2 can be expressed by a laser parameter k value (mode partition coefficient) and wavelength-dependent waveform. Let us examine a simple case of (1) wavelength-dependent loss in which all wavelengths of λ_i ($n \leq i \leq m$) pass through media, and some wavelengths λ_i ($1 \leq i < n, m < i \leq n$) are blocked:

$$\sigma^2 = k^2 \sum_{i=n}^{m} \overline{a_i}\left(1 - \sum_{i=n}^{m} \overline{a_i}\right) \tag{8}$$

where $f(\lambda_i, t_0) = 1$, for $n \leq i \leq m$, and 0, for $0 \leq i < n, n \geq i > m$. Equation (8) indicates that there is no mode partition noise when all longitudinal modes pass through, because the summation of all longitudinal mode power becomes unity ($\sum_{i=1}^{N} a_i = 1$).

Mode partition noise due to the chromatic dispersion of silica fiber is more complicated. Therefore we will simplify Eq. (5) further to obtain a clear physical picture. The received signal at the decision circuit of a two-level digital transmission can be assumed by $\cos(\pi Bt)$, where B is the data rate (bits/sec), and the decision circuit samples the received signal at times of $1/B + n/B$ ($n = 0, \ldots, \infty$), as shown in Fig. 3. Let the differential delay between the ith mode λ_i and the center mode λ_c be Δt_i:

$$f(\lambda_i, t_0) = \cos(\pi B \, \Delta t_i) \simeq \tfrac{1}{2}(\pi B \, \Delta t_i)^2, \tag{9}$$

where $\pi B \, \Delta t_i \ll 1$.

The mode partition noise σ^2 can be expressed by Eq. (5):

$$\begin{aligned}\sigma^2 &= \frac{1}{4} k^2 (\pi B)^4 \left\{ \sum_{i=1}^{N} \overline{(\Delta t_i)^4 a_i} - \left[\sum_{i=1}^{N} \overline{(\Delta t_i)^2 a_i}\right]^2 \right\} \\ &= \frac{1}{4} k^2 (\pi B)^4 \left\{ \sum_{i=1}^{N} (\Delta t_i)^4 p(\lambda_i) - \left[\sum_{i=1}^{N} (\Delta t_i)^2 p(\lambda_i)\right]^2 \right\}. \end{aligned} \tag{10}$$

To simplify the calculation, we describe the spectrum distribution $p(\lambda_i)$ by the continuous Gaussian rather than by the discrete Gaussian distribution:

$$p(\lambda_i) = \frac{1}{(2\pi\sigma^2)^{-1/2}} \exp[-(\lambda_i - \lambda_c)^2/2\sigma^2].$$

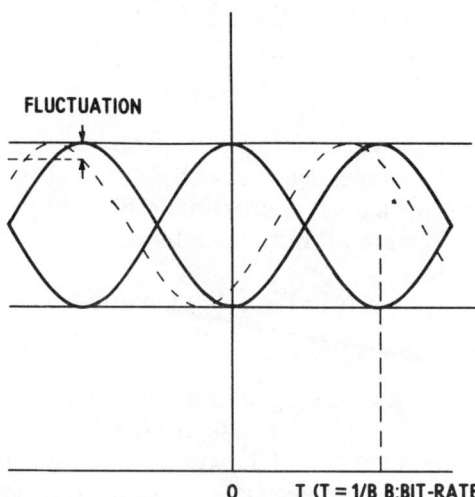

FIG. 3. Eye pattern at the decision circuit. We assume that the received signal after equalization is expressed by (1) the signal of the center wavelength (dominant mode), which varies from mark to space and vice versa and is expressed by $-\cos(\pi Bt)$ or $\cos(\pi Bt)$; (2) a long sequence of marks, which is expressed by 1; and (3) the signal of a long sequence of space, which is expressed by -1. The dashed line indicates the timing point for regeneration; dashed curve represents the signal from the ith mode, which delays τ_i through the fiber; fluctuation occurs at the timing point.

We characterize the optical fiber by its length z and a propagation constant $\beta(\Omega)$, where Ω is the radian light frequency. Then we can calculate the differential delay between Ω_i ($=2\pi c/\lambda_i$) and Ω_c ($=2\pi c/\lambda_c$):

$$\Delta\tau_i = |\ddot{\beta}(\Omega_i - \Omega_c) + \tfrac{1}{2}\dddot{\beta}(\Omega_i - \Omega_c)^2|z, \tag{11}$$

where $\ddot{\beta}$ and $\dddot{\beta}$ are derivatives of β with respect to Ω_i and Ω_c. Therefore by using Eqs. (6), (9), and (11), Eq. (5) becomes

$$\sigma_{pc}^2 = \tfrac{1}{2} k^2(\pi B)^4 [A_1^4 \sigma^4 + 48 A_2^4 \sigma^8 + 42 A_1^2 A_2^2 \sigma^6], \tag{12}$$

where $A_1 = (2\pi c/\lambda_c^2)\ddot{\beta} z$ and $A_2 = \tfrac{1}{2}[2\pi c/\lambda_c^2]^2 \dddot{\beta} z$. Since the average signal power is 1, the noise-to-signal ratio due to the calculated partition noise can be described by

$$N/S = \sigma_{pc}^2 = \tfrac{1}{2} k^2(\pi B)^4 [A_1^4 \sigma^4 + 48 A_2^4 \sigma^8 + 42 A_1^2 A_2^2 \sigma^6]. \tag{13}$$

Equation (13) indicates that the ratio of signal to partition noise depends on the halfwidth of the laser-diode spectrum and the chromatic dispersion of the fiber; it is independent of the signal power. Therefore, even with unlimited large signal power, the overall system signal-to-noise ratio cannot be improved beyond the limit imposed by partition noise. This results in the error rate asymptote in the error rate versus input optical power, which is determined by the signal-to-mode-partition noise. For a given error-rate asymptote, the mode partition noise must satisfy

$$(\sigma_{pc}^2)^{-1} = Q^2, \tag{14}$$

where Q is defined by the given error rate, as shown in Fig. 4.

For silica fiber in the wavelength range between 1.0 and 1.6 μm, $d\tau/d\lambda = s(\lambda - \lambda_0)/c\lambda^2$ represents a good approximation for the group-delay change, where λ_0 is the wavelength of minimum disperison, s a material constant, and c the speed of light.

The coefficients A_1 and A_2 in Eq. (12) therefore become

$$A_1 = [s(\lambda_c - \lambda_0)z]/c\lambda_c^2 \quad \text{and} \quad A_2 = sz/2c\lambda_c^2. \tag{15}$$

Figure 5 shows the bit-rate–distance product, $B \cdot z$ versus the wavelength between 1.1 and 1.6 μm for the case of $\lambda_0 = 1.3\ \mu$m, $s = 0.05$, $Q = 6$ (asymptotic error rate is 10^{-9}), and the worst mode partition noise, $k = 1$. In Fig. 5, one line represents the case where the half-width of the laser spectrum is 1 nm and another represents the case of 2 nm. For other values of the asymptotic error rate between 10^{-6} and 10^{-15}, the limitation of the bit-rate–distance product will vary. Figure 6 shows the correction coefficient η derived by normalized bit-rate–distance product of arbitrary asymptote error rate of 10^{-9}. For an example, when the asymptote error rate is chosen to be 10^{-15} at the wavelength of 1.25 μm, the correction coefficient η is 0.85 from

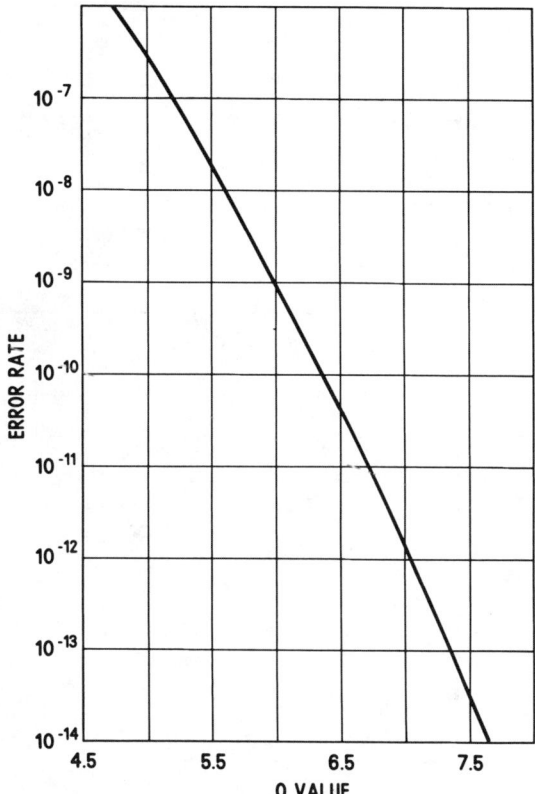

FIG. 4. Q value versus the error-rate. The value of Q is determined by the error rate through the equation $P_0(\text{error rate}) = \sqrt{2}\pi^{-1/2} \int_Q^\infty e^{-x^2/2} dx$. [From Ogawa (1982). © 1982 IEEE.]

Fig. 6. From Fig. 5, the bit-rate–distance product of the asymptote error rate of 10^{-9} is 14 Gbit/sec · km with 2 nm of the halfwidth of the laser spectrum. Thus, for the case of the asymptote error rate of 10^{-15}, the bit-rate–distance product becomes 12 km Gbit/sec ($=0.85 \times 14$ km Gbit/sec).

As shown in Fig. 7, the theory shows an excellent agreement to the experimental results. By using Eqs. (13)–(15), the bit-rate–distance product is obtained for any desired asymptotic error-rate limit. However, it is convenient to know the degradation in terms of the form of the power penalty — the degradation of required input optical power from the case of no partition noise at the error rate of 10^{-9}. The overall signal-to-noise ratio including the receiver noise is described by

$$N/S = \sigma_{\text{pc}}^2 + (N_{\text{rec}}/S) \leq Q^{-2}, \tag{16}$$

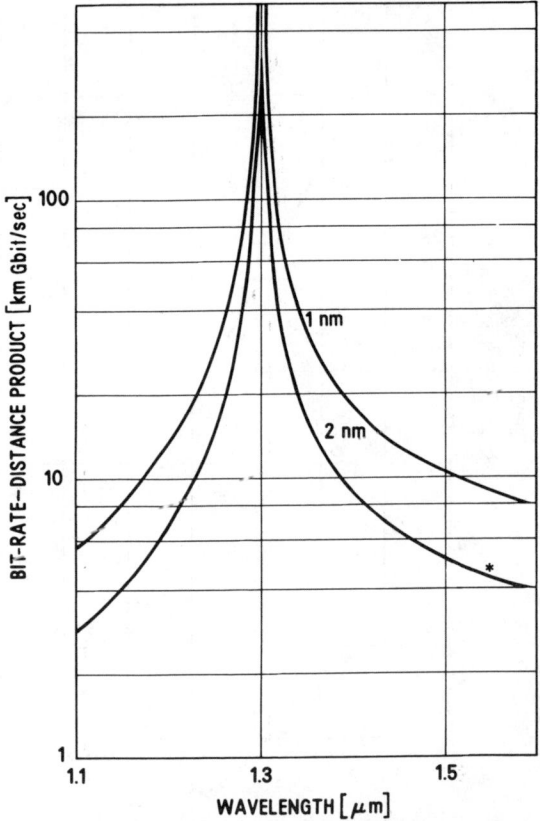

FIG. 5. Bit-rate–distance product [km Gbit/sec] versus wavelength [μm]. The error-rate asymptote is 10^{-9}, beyond which the error-rate will not be improved, even with unlimited input optical signal. There are two curves for two different half-widths of the laser spectrum, 2 and 1 nm. The chromatic dispersion is simulated by $d\tau/d\lambda = 0.05(\lambda - \lambda_c)/c\lambda^2$. [From Ogawa (1982). © 1982 IEEE.]

where N_{rec} is the noise due to the receiver. In the case of no mode partition noise, the signal-to-noise ratio satisfies the following relation:

$$S/N_{\text{rec}} = (6)^2. \tag{17}$$

The power penalty α dB due to the degradation caused by mode partition noise can be defined by Eqs. (16) and (17) as

$$\frac{1}{\sigma_{\text{pc}}^2} (6)^2 \left[\frac{10^2 \alpha/10}{10^2 \alpha/10 - 1} \right], \tag{18}$$

where α dB is the power penalty at the error rate of 10^{-9}. For the systems

8. MODE PARTITION NOISE

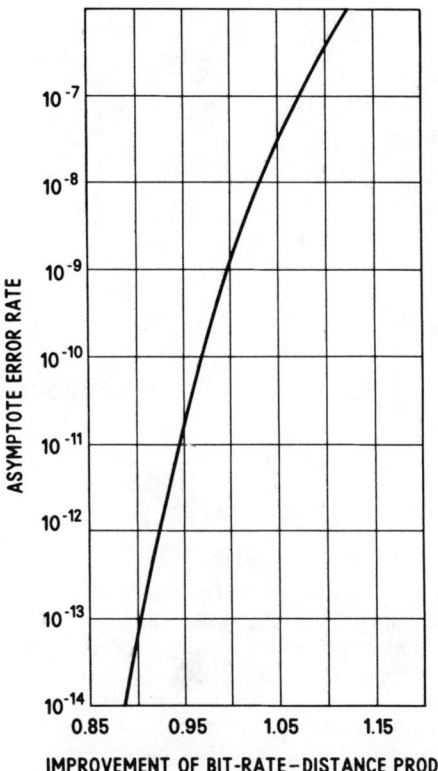

FIG. 6. The correction factor η of the bit-rate–distance produce versus the asymptotic error rate. [From Ogawa (1982). © 1982 IEEE.]

designer, it would also be convenient to know the relationship between the asymptotic error rate and the power penalty α dB at an error rate 10^{-9}, defined by Eq. (19):

$$\alpha = 5 \log[Q^2/(Q^2 - (6)^2)], \tag{19}$$

where $Q^2 = (\sigma_{pc}^2)^{-1}$.

We can simplify Eq. (13) further at the wavelength of the minimum dispersion λ_0 and also at the wavelength far from λ_0. At the wavelength of λ_0, the coefficient A_1 in Eq. (13) becomes zero. Therefore the bit-rate–distance product in the case of the asymptote error rate of 10^{-9} is given by

$$B \cdot z = \frac{2c\lambda_0^2}{\pi\sqrt{6}(24)^{1/4}S\sigma^2} = \frac{1180 \; [\text{nm}^2]}{\sigma^2 \; [\text{nm}^2]} \; \text{km} \frac{\text{Gbit}}{\text{sec}}. \tag{20}$$

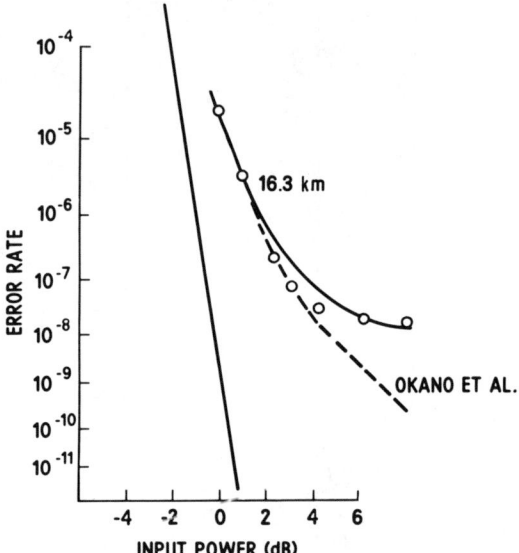

FIG. 7. Comparison to experimental result. The open circles represent the experimental results of error-rate measurements for 280 Mbit/sec. The horizontal axis is the input optical power normalized by the input optical power of 10^{-9} error rate without the partition noise. Experiment is the case of the fiber length of 16.3 km, the spectrum width of 2 λm, and the center wavelength of 1.54 μm taken from the results of S. Yamamoto et al. (1980). The solid line shows the theoretical results by the analytic method proposed here. The dash curve shows the theoretical results by Okano et al. (1980); $\Delta\tau = 17$ psec km^{-1} nm^{-1}.

At the minimum dispersion, the bit-rate–distance product is dependent upon the square of the half-width of laser spectrum. At the wavelength far from the minimum dispersion wavelength λ_0, the coefficient A_2 in Eq. (13) can be neglected. Therefore we obtain

$$B \cdot z = \frac{c\lambda_c^2 (2)^{1/4}}{\pi\sqrt{6} S |\lambda_c - \lambda_0|} \frac{1}{\sigma} = \frac{0.93 \lambda_c^2}{\lambda_c - \lambda_0 |\sigma|} \frac{[\mu m^2]}{[\mu m^2]} \quad \text{(km Gbit/sec)}$$

$$= \frac{153.3}{[d\tau/d\lambda]_{\lambda_c} \sigma} \quad \text{(km Gbit/sec)}, \tag{21}$$

where $|d\tau/d\lambda_c|$ is expressed in picoseconds per nanometer per kilometer at the wavelength of λ_c.

The single-mode fiber bandwidth is also a generic function of the source spectrum and the media chromatic dispersion. Thus, one can expect a certain relationship between the 6-dB bandwidth (optically, 3-dB bandwidth) and the bit-rate–distance product due to the mode partition noise. We consider cases only at the wavelength of minimum dispersion and also far from the wavelength of minimum dispersion. From Yamamoto et al. (1980) the 6-dB bandwidth is given as

$$(f_{6dB})z = \frac{\sqrt{15}}{2\pi} \frac{c\lambda^2}{s\sigma^2} \quad \text{at the wavelength of } \lambda_0,$$

$$(f_{6dB})z = \frac{\sqrt{1.386}}{2\pi} \frac{c\lambda_c^2}{s|\lambda_c - \lambda_0|} \frac{1}{\sigma} \quad \text{at the wavelength far from } \lambda_0. \tag{22}$$

Combining Eqs. (20)–(22), we obtain

$f(6 \text{ dB}) \cdot z/B \cdot z = 5.7$ at the wavelength of minimum dispersion

$f(6 \text{ dB}) \cdot z/B \cdot z = 1.22$ at wavelengths far from (23)
the wavelength of minimum dispersion.

This indicates that the performance limitation due to the mode partition noise is dominant over the limitation due to the bandwidth for any wavelength in the case of a single-mode-fiber system.

IV. Mode Partition Coefficient: k Value

In Part III, we discussed the mode partition noise for the case of $k = 1$. What k value can be exhibited by a semiconductor laser?

From Eq. (7), k can be determined by combining two kinds of spectral measurements, one using a dc voltmeter to measure the expectation of the ith mode power $\overline{a_i}$ and the other using an rms voltmeter to measure the variance of the ith mode power $[\overline{a_i^2} - (\overline{a_i})^2]$. Since laser diodes are modulated by pulsed current, it is necessary to eliminate the modulation signal power to obtain the fluctuation of each mode. There are two methods: the low-pass filter (LPF) method and the sampling method.

As shown in Fig. 8a, the pulse train of the ith mode is described by a pulse amplitude-modulated (PAM) signal. Let τ be the pulse width and T the repetition period. The time-averaged value of each mode, $\overline{a_i}$, is proportional to the dc component $[\overline{a_i}/(\tau/T)]$. From the well-known sampling theorem, rms voltage through a low-pass filter having the cut-off frequency of $[1/2T]$ is proportional to the square root of the variance of a_i, $\{[\overline{a_i^2} - (\overline{a_i})^2] [\tau/T]^2\}^{1/2}$. By choosing the appropriate normalization, k can be measured by using two measured spectral distributions: one is the average spectrum measured with a dc voltmeter, and the other is the fluctuation spectrum measured using an rms voltmeter through a low-pass filter.

Because of sensitivity limitations, it is necessary to amplify the output voltage of the $p-i-n$ detector. Since an ac-coupled amplifier is used, it is not possible to determine directly the dc component $\overline{a_i}$ from the dc voltmeter.

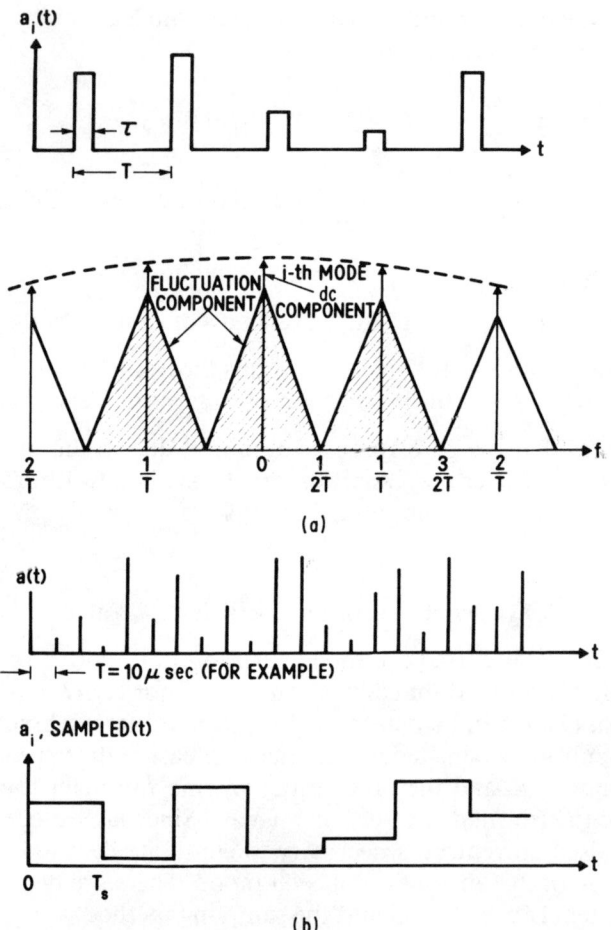

FIG. 8. Measurement principles of the mode fluctuation coefficient. (a) Low-pass filter method: (top) ith-mode pulse amplitude-modulated signal; (bottom) its associated power spectrum, from which one can extract the average and variance of the ith mode by using low-pass filters. (b) Sampling method: (top) ith-mode power spectrum; (bottom) ith-mode pulse train. Output from the sample-and-holder circuit, which has a much slower rate than the original signal. Since the output signal has a 100% duty cycle, one can obtain the average and variance of the ith mode.

However, the dc component ($\overline{a_i}$) is also carried by the fundamental sampling frequency $1/T$, and the quantities $[\overline{a_i^2} - (\overline{a_i})^2]$ and $\overline{a_i}$ can be determined separately from rms voltage measurements using two low-pass filters with cut-off frequencies below and above the sampling frequency of $1/T$. When τ/T is kept smaller than unity and we used the cut-off frequency of f_A where

f_A is between $1/T$ and $2/T$, the rms voltage V_A is

$$V_z = \{2(\overline{a_i})^2(\tau^2/T^2) + 2f_A T[\overline{a_i^2} - (\overline{a_i})^2](\tau^2/T^2)\}^{1/2}.$$

Since the second term, $[\overline{a_i^2} - (\overline{a_i})^2](\tau^2/T^2)$ is measured by a low-pass filter with a cut-off frequency of $1/2T$, we can obtain $\overline{a_i}$ easily.

As shown in Fig. 8b, the pulse train of the ith mode is sampled at a certain time slot and its value held for the next sampling time. The output of the sample-and-hold circuit is now a 100% duty-cycle pulse train with a much slower repetition rate.

Since the pulse train becomes 100% duty cycle through the sample-and-hold circuit, the average $\overline{a_i}$ and the variance $[\overline{a_i^2} - (\overline{a_i})^2]^{1/2}$ can be measured directly with a dc meter and an rms voltmeter, respectively.

Generally, the LPF method will measure k averaged over the width of the pulse. The sampling method will measure k at a specific time within the pulse and, optionally, the k value variation along the pulse. If k is the same throughout the pulse, both methods will give the same value.

Figure 9 shows the pulse-width dependence of k for three types of lasers; an InGaAsP/InP proton-bombarded strip (PBS) laser ($I_{th} = 115$ mA), an InGaAsP/InP buried heterojunction (BH) laser ($I_{th} = 20$ mA), and a buried-crescent heterojunction (BCH) laser ($I_{th} = 20$ mA). For the BH laser, k decreased from 0.65 to 0.14 as the pulse width was increased from 1.4 to

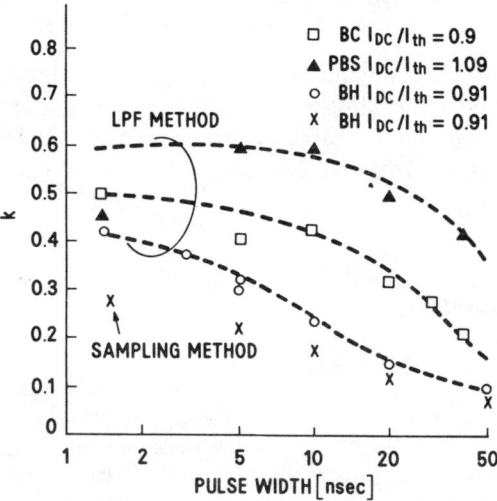

FIG. 9. Pulse-width dependence of k values for three types of laser diodes. InGaAsP/InP BH and BC lasers show that k is decreased by increasing the pulse width from 1.4 to 50 nsec. An InGaAsP/InP PBS laser exhibits almost no pulse-width dependence because of relaxation oscillation. Also, the results from the LPF and sampling methods for the BH laser are shown. Both methods agree fairly well.

50 nsec. The BCH laser exhibits a similar dependence. The PBS laser had generally large k values (0.65–0.36), peaking for pulse widths around 5 nsec. This particular PBS laser has strong relaxation oscillations, which seems to be associated with the large k values.

Figure 10 shows the dc bias dependence of k at the pulse widths of 1.4 nsec for the BH laser and 5 nsec for PBS laser. For the BH laser, k depended on the dc bias current. When the dc bias was lower than the threshold current, k became slightly larger. Presumably, the transition from spontaneous emission to stimulated emission causes larger fluctuations when compared to the transition from stimulated emission to stimulated emission. On the other hand, PBS lasers exhibit the weak dependence of k on the bias current. This can be explained by the large relaxation oscillation.

As indicated in Figs. 9 and 10, the sampling and LPF results agree fairly well. However, when the driving pulse becomes short, the k value from the sampling method becomes smaller, especially for the BH laser. This is because k varied during pulses; k was the largest at the beginning of the pulse and decreased at later times within pulse. As shown in Fig. 11, the k value is almost unity at the beginning of pulse. This indicates that laser diodes start only one mode.

V. Theoretical k Value with and without Interaction between Lasing Modes

The mode partition coefficient k is a useful, universal index to characterize mode fluctuations of laser diodes. Since the k value is an important parameter for system design, it is desirable to know the theoretical expectation of the k value and how to control k values. This section introduces a simple model

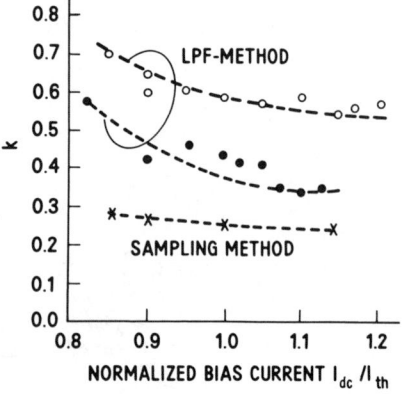

FIG. 10. dc Bias dependence of k. The BH ($\times$, ●) and BC lasers show a strong dc bias dependence of k at pulse width 1.4 nsec. For the PBS (○) laser, k was almost independent of the dc bias (5-nsec pulse width).

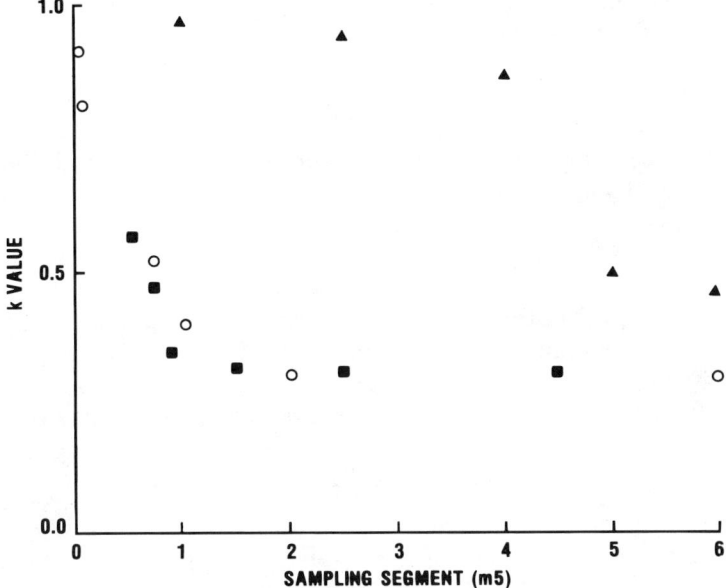

FIG. 11. k-Value variation from beginning to end of pulse. ○, BC InGaAsP/InP, ■, InGaAsP/InP (short cavity), ▲, InGaAsP/InP (short cavity) coated.

for the relationship between the lasing mechanism and the k value and discusses theoretical values of k.

During the transient process due to the pulse modulation, electrons in the conduction band will be distributed according to a probability function that is the product of the density function of the states in the conduction band and the Fermi–Dirac distribution function determined by the injected-electron population. If the entire distribution process of electrons follows this homogeneous distribution process, there is no mode power fluctuation. However, the redistribution process form nonlasing states (spontaneous emission levels) to lasing states (stimulated emission) will cause power fluctuation. According to the following simple model of the redistribution process, laser diodes that have only one lasing mode exhibit a k value of zero. Once laser diodes have more than one lasing mode, the k value is between 0.56 and 1.0 if the spectrum width is very narrow.

Before going into a detailed analysis, let us clarify the notation and discuss the assumptions j for any nonlasing states and i, i', and k for any lasing states.

(1) Although semiconductor laser diodes have no discrete energy levels as do gas lasers, we assume for convenience discrete energy levels that correspond to the discrete resonant wavelengths of the optical cavity. However,

we convert from discrete summation to integration of a continuous function at the end of the calculation.

(2) The probability function for emitting additional photons ΔN from each discrete emission level is assumed to be determined by the photon population density functions $P(N)$ of each level.

(3) Physically, once lasing modes occur, electrons tend to fill the vacant levels caused by lasing. Also, electrons in the nonlasing levels adjacent to a lasing level diffuse into the lasing levels by losing or gaining a small amount of energy from phonon scatterings. In this model, we simplify these processes such that electrons will at first be captured by every level according to the Fermi–Dirac probability but are then redistributed to lasing levels in a way that tends to maintain a thermal distribution. The transition from nonlasing to lasing levels obeys the following rule: All additional electrons injected into the jth nonlasing level will transfer to the ith lasing level with a probability of q_{ji}.

A total of N photons per second are initially emitted according to the probability functions $P_i(N)$ and $P_j(N)$ from the ith lasing level and jth non-lasing level. Now additional photons ΔN are emitted, and the new probability functions $P_i(N + \Delta N)$ and $P_j(N + \Delta N)$ can be derived by using first-order perturbation:

$$(N + \Delta N)P_i(N + \Delta N) = NP_i(N) + \Delta N P_i(N) + \sum_j \Delta N P_j(N) C_{ji},$$

$$(N + \Delta N)P_j(N + \Delta N) = NP_j(N) + \Delta N P_j(N) - \sum_i \Delta N P_j(N) C_{ji}, \qquad (24)$$

where C_{ji} is the transfer coefficient from the jth nonlasing level to the ith lasing level. By Eq. (3), C_{ji} will be one or zero, with probabilities of q_{ji} and $1 - q_{ji}$, respectively. For the discrete-level model, it is hard to justify all additional electrons $\Delta N P_j(N)$ transferring to only one lasing level. In the final calculation, we use a continuous distribution so that the electrons to be transferred are not treated as a group of electrons; however, it is still the quantization of charge that produces the fluctuations of interest.

Equation (24) produces the following linear differential equation for non-lasing and lasing levels.

$$N(\partial P_j/\partial N) = P_j \sum_i C_{ji} \quad \text{for the nonlasing level,}$$

$$N(\partial P_i/\partial N) = \sum_j P_i C_{ji} \quad \text{for the lasing level.} \qquad (25)$$

where N is the photon density and P_j and P_i are the probability density functions for the jth nonlasing level and ith lasing levels, respectively. Also,

P_j, P_i, and C_{ji} have the following relation:

$$\sum_j P_j + \sum_i P_i = 1, \qquad \sum_i C_{ji} = 1. \tag{26}$$

Linear differential equations can be solved completely with Eq. (26) to obtain

$$P_j = P_{j0}\alpha, \qquad \text{for spontaneous emission,}$$

$$P_i = \sum_j P_{j0} C_{ji}(1-\alpha) + P_{i0}, \qquad \text{for stimulated emission,} \tag{27}$$

where $\alpha = N_0/(N + N_0)$ and P_{j0} and P_{i0} are the initial probability densities just at the dc bias level, where there are only nonlasing modes, and N_0 is the initial photon population at the dc bias.

Since we are interested in the mode partition coefficient k, we need to calculate the expectations and second moments of P_j and P_i with Eq. (26) and the probability coefficients q_{ji} of C_{ji}:

$$\overline{P_j} = P_{j0}\alpha, \qquad \overline{P_j^2} = P_{j0}^2 \alpha^2, \qquad \overline{P_i} = \sum_j P_{j0}(1-\alpha)q_{ij} + P_{i0},$$

$$\overline{P_i^2} = \sum_j P_{j0}^2 (1-\alpha)^2 q_{ji} + P_{i0}^2 + 2\sum_j \sum_{j'>j} P_{j0} P_{j'0}(1-\alpha)^2 q_{ji} q_{j'i}$$

$$+ 2P_{i0} \sum_j P_{j0}(1-\alpha) q_{ji}, \tag{28}$$

where $q_{ji} = \overline{C_{ji}}$ and $q_{ji} = \overline{C_{ji}^2}$. From Eq. (28), the fluctuation of both types of emission can be derived:

$$\overline{\sigma_j^2} = 0 \quad \text{for nonlasing modes,}$$

$$\overline{\sigma_i^2} = (q_{ji} - q_{ji}^2)(1-\alpha)^2 \sum_j P_{j0}^2 \quad \text{for lasing modes.} \tag{29}$$

Since we have obtained the average value and fluctuation terms of every mode, we can define k as follows:

$$k^2 = \frac{\sum \sigma_i^2 + \sum \sigma_j^2}{1 - \sum (\overline{P_i})^2 - \sum (\overline{P_j})^2}$$

$$= \frac{(1-\alpha)^2 \sum_j P_{j0}^2 \left[\sum_i (q_{ji} - q_{ji}^2)\right]}{1 - \sum (\overline{P_i})^2 - \sum (\overline{P_j})^2}. \tag{30}$$

Since we define P_i for the lasing-mode and P_j for nonlasing mode, the following assumption can be introduced:

$$\alpha \ll 1, \quad \bar{P}_i \gg \bar{P}_j', \quad \text{for any } i \text{ and } j. \tag{31}$$

The probability function q_{ji} is assumed to be $\bar{P}_i$ for any j; then Eq. (30) can be simplified:

$$k^2 \simeq \sum_j P_{j0}^2, \tag{32a}$$

for the case of more than one lasing emission, and

$$k^2 = 0, \tag{32b}$$

when there is only one lasing emission; Now we need to know the initial spontaneous spectrum and how many modes become lasing modes N. The spontaneous emission spectrum for InGaAsP is expressed as

$$P_{j0} \text{ or } P_{i0} = (2\pi\sigma^2)^{-1/2} \exp(-(\lambda - \lambda_c)^2/2\sigma_s^2), \tag{33}$$

where $\sqrt{2}\sigma_s = 9$ nm for InGaAsP laser diodes. If the wavelength separation between modes is $\Delta\lambda$ and the number of lasing modes is N, k^2 will be defined by simply integrating Eq. (33) from $\Delta\lambda N/\sigma$ to infinity (see Appendix I):

$$k^2 \simeq \pi^{-1} \exp[-(\Delta\lambda N/2\sigma_s)^2]. \tag{34}$$

Figure 12 shows the calculation results of Eq. (36) by $\eta = 0$ (no mode interaction). The horizontal axis is the rms half-spectrum with σ_i determined by

$$\sigma_i = \Delta\lambda N/4\sqrt{2}, \tag{35}$$

where we assume that modes are lasing if their power is 2% of that of primary lasing mode. For the noninteraction case, k has a maximum value of $1/\sqrt{\pi}(0.56)$ and decreases gradually by increasing spectrum width.

Let us consider the case of a laser diode with lasing-mode interactions. In this case, the electron-redistribution process is introduced between lasing levels as well as from lasing level to lasing level. Linear differential equations similar to those of Eqs. (26) and (27) can be obtained. To simplify the solution, the following rules for electron transfer between any two or more lasing levels are considered:

1. All electrons captured by the ith lasing level will transfer to i'th lasing level with probability $q_{ii'}$ given by $\bar{P}_{i'}/(\bar{P}_i + \bar{P}_i')$.

2. When two lasing levels have mode interaction, all carriers in the ith lasing level will transfer to one of the interacting lasing levels i' with a probability of $q_{ii'} = \bar{P}_{i'}/\Sigma_k P_k$.

By using the results of Eq. (25), we obtain the averages and variance of nonlasing and lasing modes. (See Appendix II.) The value of k therefore is expressed by

$$k^2 = \frac{(1-\alpha)^2 \sum_j P_{j0}^2 \left(\sum_i (q_{ji} - q_{ji}^2) + \sum_{i'} (q_{ji'} - q_{ji'}^2) \right)}{1 - \sum (\bar{P}_i)^2 - \sum (\bar{P}_{i'})^2 - \sum (\bar{P}_j)^2}$$

$$+ \frac{2 \sum_{k>k'} \sum_k \left(2 \sum_j \sum_{j'>j} P_{j'0} P_{j0} (1-\alpha)^2 q_{ji} q_{j'k'} \right)}{1 - \sum (\bar{P}_i)^2 - \sum (\bar{P}_{i'})^2 - \sum (\bar{P}_j)^2}$$

$$+ \frac{2 \sum_j P_{j0}(1-\alpha) \sum_k P_{k0} \left(\left(\sum_{i'} q_{ji'} \right)^2 - \sum_{i'} q_{ji'}^2 \right)}{\left(1 - \sum (\bar{P}_i)^2 - \sum (\bar{P}_{i'})^2 - \sum (\bar{P}_j)^2 \right) \left(\sum_{i'} P_{i'} \right)}$$

$$+ \frac{\left(\sum_k P_{k0} \right)^2 \left(\left(\sum_{i'} q_{ji'} \right)^2 - \sum_{i'} q_{ji'}^2 \right)}{\left(1 - \sum (\bar{P}_i)^2 - \sum (\bar{P}_{i'})^2 - \sum (\bar{P}_j)^2 \right) \left(\sum_k q_{jk} \right)^2}. \qquad (36)$$

By assuming a Gaussian spectrum for spontaneous and stimulated emissions, $A \ll 1$ and $\Sigma(\bar{P}_j)^2 \ll \Sigma(\bar{P}_i)^2 + \Sigma(\bar{P}_{i'})^2$, we obtain a simplified expression for k value:

$$k^2 = \frac{1}{\pi} \exp\left[-\left(\frac{\Delta\lambda N}{2\sigma_s} \right)^2 \right]$$

$$+ \left\{ \left[\text{erf}\left(\frac{\Delta\lambda N}{2\sqrt{2}\sigma_s} \right)^2 \right] - \frac{1}{\pi} \exp\left[-\left(\frac{\Delta\lambda N}{2\sigma_s} \right)^2 \right] \right\}$$

$$\times \frac{1 - \pi^{-1}}{\{1 - \text{erf}[2(\Delta/N)] - \pi^{-1} \{1 - \exp[-8(\Delta/N)^2]\}}$$

$$+ 2\left[\text{erf}\left(\frac{\Delta\lambda N}{2\sqrt{2}\sigma_s} \right) \right] \left[1 - \text{erf}\left(\frac{\Delta\lambda L}{2\sqrt{2}\sigma_s} \right) \right]$$

$$\times \left\{ 1 - \frac{1}{1-\pi^{-1}} \left[\text{erf}\left(2\frac{L}{N} \right) - \frac{1}{\pi} \exp\left[-8\left(\frac{L}{N} \right)^2 \right] \right] \right\}, \qquad (37)$$

where $q_{ji} = \bar{P}_i$ and $\text{erf}(x) = (2/\sqrt{\pi}) \int_x^\infty e^{-\alpha^2} d\alpha$, σ_s is the rms half-width of the spontaneous emission spectrum, and N is the number of lasing emission of which the spectrum separation is $\Delta\lambda$, NL, where L is less than or equal to N, is the number of mode interactions, and the stimulated emission spectrum is $\sigma_l = \Delta\lambda N/4\sqrt{2}$.

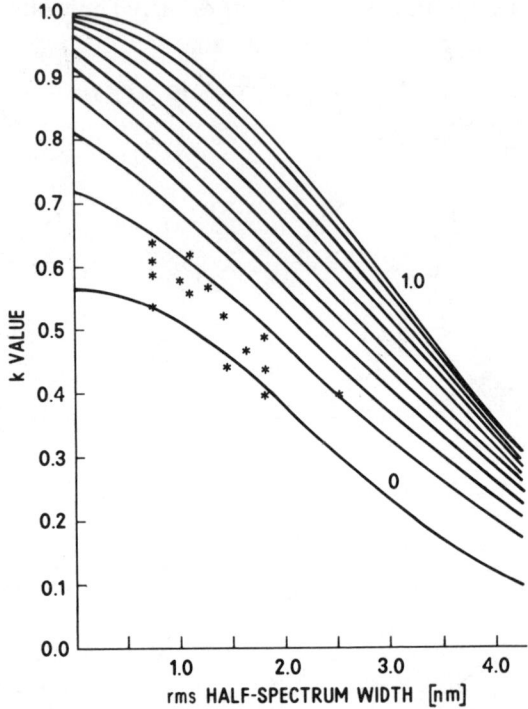

FIG. 12. Dependency of theoretical k value on spectrum width. The parameter η (0–1.0, step = 0.1) is defined such that L modes from total stimulated emission N strongly suppress the other modes N, and by the ratio of interaction, η becomes $L(N-1)/(N-1)N = L/N$. When $\eta = 0$, there is no mode interaction among stimulated emissions; k shows the maximum value of $\pi^{-1/2}$ when σ (rms half-width of the spectrum) becomes small. When $\eta = 1.0$, all modes interact with each other, and k will reach the maximum of 1 and decrease gradually by larger σ. The dots are the experimental results from three kinds of lasers measured at the condition of the pulse width 1.5–5 nsec. For a short pulse, measured k values indicate strong dependency on spectrum width, which is in good agreement with theoretical results.

Figure 12 shows calculation results of Eq. (37), with parameter η from 0.1 to 1.0. Since N is the total number of modes, and L modes interact with N modes, the total number of interactions between modes is $(N-1)L$, and the possible number of mode interaction is $(N-1)N$. Therefore a parameter $\eta(=L/N)$ indicates the fraction of lasing modes with mode interactions among all other lasing modes. When all lasing modes interact with each other ($\eta = 1.0$), the k value has a maximum value of 1 by decreasing the spectrum width.

Figure 12 shows also several measured k values obtained with pulse widths from 1.0 to 5 nsec. The measured points are distributed among the curves for

$\eta = 0$ and 0.2 (10%). Even though this analysis is simple, the theoretical k values are in good agreement with the measured results. The measured results were taken from several different types of lasers: BH, proton-bombarded stripe laser, and BCH InGaAsP/InP lasers.

Since the theoretical k value during high-speed modulation will not be zero except for the case of pure single-mode laser diodes, what k value is best for the system? Figure 13 shows of the improvement of the bit-rate–distance product normalized to that of an rms half-spectrum width of 1 nm and a k value of unity. When σ_l increases, the k value decreases; the bit-rate–distance product decreases due to the direct effect of σ_l even though k is smaller.

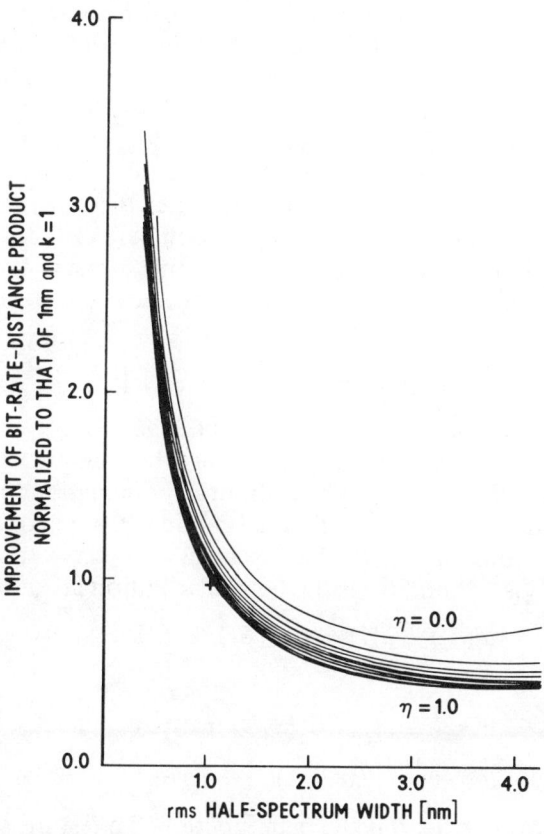

FIG. 13. Figure of merit for the bit-rate–distance product. The bit-rate–distance product improvement over that defined by the worst partition noise ($k = 1$) and $\sigma = 1$ nm (reference point, +) is plotted for various amounts of lasing-mode interactions. Small σ is preferred for the system even though the k value becomes large. Improvement of bit-rate–distance product over the case of worst mode partition is small when direct modulation of laser diodes is involved.

There are two ways to decrease the k values: (1) construct the laser diodes with a pure single lasing-mode emission, or (2) increase σ_l, which will introduce large degradation for the overall system. These theoretical results indicate that a large k value (0.56–0.7) is inevitable for a multimode laser with small spectral width but that the best laser diodes for system design should have narrow spectra to improve the overall bit-rate–distance product.

VI. Statistical Measurement of Mode Partition Noise

Another way of characterizing noise is to study its statistical distribution function instead of describing fluctuation by a mean square value. Pao-Lo Liu et al. (1984) have demonstrated a statistical measurement technique (Liu et al., 1984) by using a sampling method and a multichannel analyzer. Figure 14a shows mode partition in a multimode laser. There were three main modes with average power levels of 62, 27 and 8%. As can be seen in Fig. 14b, all three modes had distributions extending from 0 to 100% of total output power. In addition to mode partition, fluctuation of the total output and the measurement-setup noise had also affected these distributions. Their main effect is to soften and to extend the shoulders of these distributions beyond zero and unity work. Let us examine this measured result with a k value of a laser diode to describe mode partition and the average power of the ith mode $\overline{a_i}$:

$$k = [\overline{a_i^2} - (\overline{a_i})^2]/[\overline{a_i} - (\overline{a_i})^2], \tag{38}$$

where $\overline{a_2} = \int_0^1 a_i p(a_i)\, da_i$ and $\overline{a_i^2} = \int_0^1 a_i^2 p(a_i)\, da_i$ and a_i is the instantaneous power in the ith mode, and p is the distribution function. If the total output power is assumed to be unity without any fluctuations, the maximum span of the distribution of an individual mode is from zero to unity. One well-known distribution function having a real value between zero and unity and satisfying Eqs. (5) and (6) is the beta distribution given by

$$p(a_i) = a_i^x (1 - a_i)^y / B(1 + x, 1 + y), \tag{39}$$

where

$$x = (1/k^2 - 1)\overline{a_i} - 1, \tag{40}$$

$$y = (1/k^2 - 1)(1 - \overline{a_i}) - 1. \tag{41}$$

The normalization factor B is the beta function. To test the validity of this model, we have used the $\overline{a_i}$ values of Fig. 14a and suitably selected k values and obtained the curves shown in Fig. 15. These curves match in a qualitative sense the experimental distributions shown in Fig. 14b. As shown in Fig. 15, the k values are slightly different for each mode. This is because the

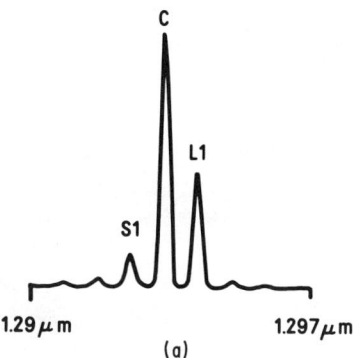

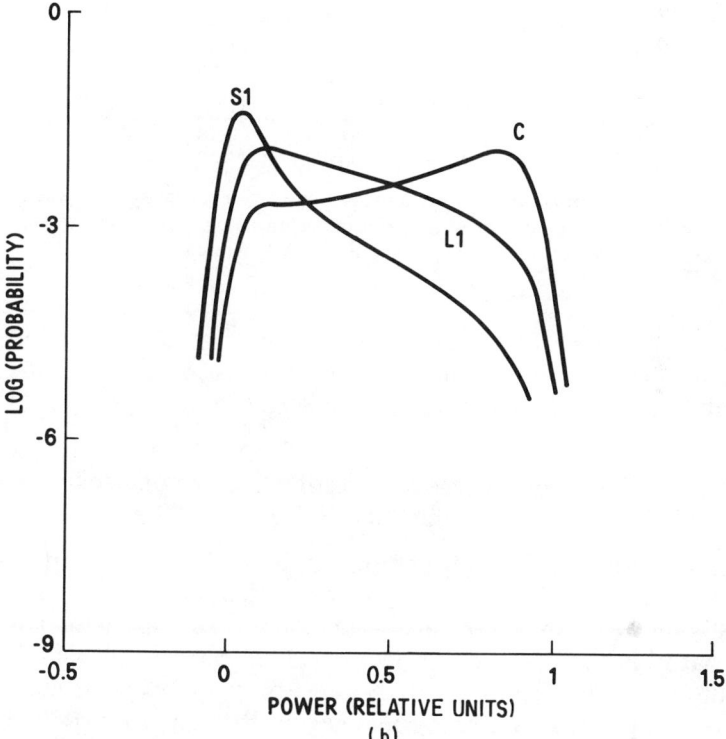

FIG. 14. (a) Spectrum and (b) mode-resolved distributions for a multimode ridge waveguide InGaAsP laser. There were three major modes with power levels of 62, 27 and 8%. Distributions for these modes had a span from zero to unity. In other words, one mode could be zero or could take the whole output power.

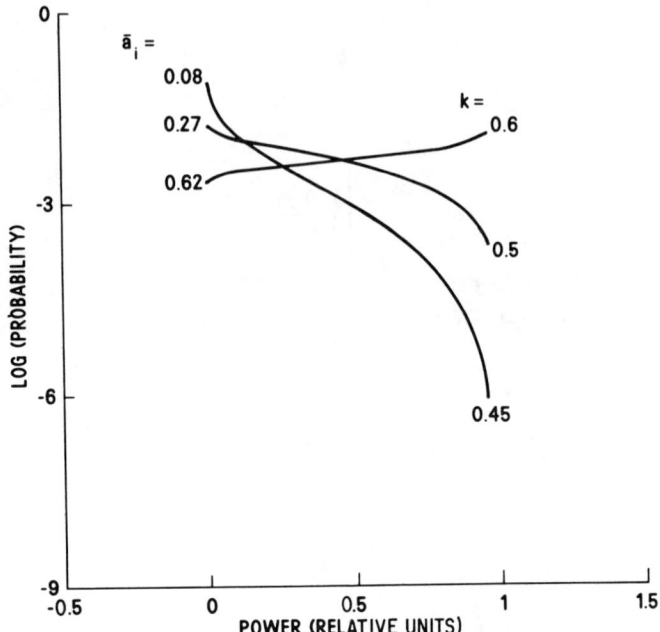

FIG. 15. Calculated probability density functions. We used Eq. with $\bar{a}_i$ values measured from the spectrum of the multimode ridge-waveguide laser shown in Fig. 14; k values were chosen to give the best fit between calculated curves and experimental data.

theoretical model for the k value is based on the stimulated emission. Every longitudinal mode carries some spontaneous emission as well as the stimulated emission (Ogawa and Vodhanel, 1982).

VII. Effects of Optical Injection Locking on Mode Partition Noise

Even a laser diode in a steady-state operation shows mode partitioning at a surprisingly large scale, but its degree of fluctuation in terms of k is still normally smaller than that of pulsed-laser diodes. The key factor in reducing mode partition noise is use of a laser diode that has small k and narrow spectrum width. Optical power injection from a dc-operated master laser into a pulse-operated slave laser could narrow the spectrum width of a slave laser (Yamada *et al.,* 1980; Kobayashi and Kimura, 1980).

The effect of optical injection on mode partitioning was examined, and the rather unexpected result obtained, which was that optical injection in some cases will enhance the mode partitioning.

Let us examine the mode partition coefficient k_{in} expected with light injection from a master laser into a slave laser. We assume that a master laser exhibits a nearly single-mode spectrum, of which the dominant wavelength matches the ith-mode wavelength of the slave laser.

From the definition of k, we have the following relation between k_p and the power amplitude $\overline{a_p}$ of the ith mode of the slave laser in pulse operation without light injection, and between k_{dc} and a_{dc} from the master laser, the mode partition coefficient, and the power amplitude of the cw master laser:

$$\overline{k_p^2} = \frac{[\overline{a_p^2} - (\overline{a_p})^2]}{[\overline{a_p} - (\overline{a_p})^2]}, \qquad k_{\text{dc}}^2 = \frac{[\overline{a_{\text{dc}}^2} - (\overline{a_{\text{dc}}})^2]}{[\overline{a_{\text{dc}}} - (\overline{a_{\text{dc}}})^2]}, \tag{42}$$

where $\overline{a_p^2}$ and $\overline{a_{\text{dc}}^2}$ are the normalized second moments of a_p and a_{dc} and of $\overline{a_p}$ and $\overline{a_{\text{dc}}}$, as well as the average values of a_p and a_{dc} normalized by total mode power, respectively.

Since the locking phenomena depend upon the injected power a_{dc} fluctuates between 0 and 1 with the unknown distribution fuction $p(a_{\text{dc}})$. The slave laser becomes free-running when a_{dc} is less than a certain value α ($\alpha < 1$):

$$\overline{a}_{\text{dc}} = \int_0^1 a_{\text{dc}} P(a_{\text{dc}}) \, da_{\text{dc}}, \tag{43a}$$

$$\overline{a_{\text{dc}}^2} = \int_0^1 a_{\text{dc}}^2 P(a_{\text{dc}}) \, da_{\text{dc}}, \tag{43b}$$

$$q = \int_\alpha^1 P(a_{\text{dc}}) \, da_{\text{dc}}, \tag{43c}$$

where q is the probability that the slave laser is locked by injection. If we assume that the mode intensity of the slave laser fluctuates according to the mode fluctuation of the master laser during the locking, we will obtain the mode amplitude a_{in} of a slave laser with a light injection as follows:

$$\overline{a}_{\text{in}} = \overline{a}_0 q + \overline{a}_p (1-q), \qquad \overline{a_{\text{in}}^2} = \overline{a_0^2} q + \overline{a_p^2}(1-q), \tag{44}$$

where the first terms in both equations are due to the injection locking and the second terms are due to the free-running condition.

By using Eq. (44), k_{in}, the mode partition coefficient of the slave laser with light injection, is defined:

$$k_{\text{in}}^2 = \frac{\overline{a_{\text{in}}^2} - (\overline{a}_{\text{in}})^2}{\overline{a}_{\text{in}} - (\overline{a}_{\text{in}})^2}$$

$$= \frac{k_{\text{dc}}^2 \beta q + k_p^2 \gamma(1-q) + q(1-q)(\overline{a}_0 - \overline{a}_p)^2}{\beta q + \gamma(1-q) + q(1-q)(\overline{a}_0 - \overline{a}_p)^2}, \tag{45}$$

where $\beta = \bar{a}_0 - (\bar{a}_0)^2$, $\gamma = \bar{a}_p - (\bar{a}_p)^2$, and $k_{dc}^2 = k_0^2 = [\overline{a_0^2} - (\bar{a}_0)^2]/[\bar{a}_0 - (\bar{a}_0)^2]$.

As defined in Eq. (43c), the probability q of locking is unknown. However, by using the assumption that the master laser is nearly single mode, we can approximate q:

$$q \approx \bar{a}_{dc} \tag{46}$$

In addition to this assumption, we assume that $\bar{a}_{dc} = \bar{a}_0$ when the slave laser is locked. Therefore we can simplify the expression for the k_{in} injection:

$$k_{in}^2 = \frac{k_{dc}^2 + \bar{a}_{dc}(1 - \bar{a}_p/\bar{a}_{dc})^2 + k_p^2(\bar{a}_p - (\bar{a}_p)^2)/(\bar{a}_{dc})^2}{1 + \bar{a}_{dc}(1 - \bar{a}_p/\bar{a}_{dc})^2 + (\bar{a}_p - (\bar{a}_p)^2)/(\bar{a}_{dc})^2}$$

$$\simeq \frac{k_{dc}^2 + k_p^2(\bar{a}_p - (\bar{a}_p)^2) + (1 - \bar{a}_p)}{2 - (\bar{a}_p)^2} \tag{47}$$

where $\bar{a}_{dc} \simeq 1$. Therefore k_{in} decreases monotonically as $\bar{a}_p$ increases from zero to unity; k_{in} has the minimum value of k_{dc} and the maximum value of $\sqrt{(k_{dc}^2 + 1)/2}$.

When light is injected into one of quenched modes of a slave laser, $\bar{a}_p$ is nearly zero and k_{in}^2 becomes $(k_{dc}^2 + 1)/2$, which is the maximum value. Generally, k_{dc}^2 is much smaller than 1, and therefore k_{in} becomes 0.7, which is often larger than k_p (0.4–0.7).

When light is injected into one of the strong modes of the slave laser, k_{in} approaches k_{dc} as $\bar{a}_p$ increases, which indicates the stabilization effect of light injection. However, it requires large $\bar{a}_p$ to achieve stabilization. The cw master laser diode is nearly single-longitudinal mode. The wavelength separation is 0.8 nm and the measured values of k_{dc} were between 0.3 and 0.5, depending on the dc bias current. The slave laser exhibits several longitudinal modes in pulse operation without light injection, as shown by the upper spectra in Fig. 16a. The wavelength separation is about 1 nm. The two slave laser diodes were selected so that wavelength matching can be achieved by controlling the junction temperature of the master laser and by changing the dc bias currents of both lasers.

The lower spectra of Fig. 16b are those of the pulsed slave laser with light injection. Figure 16 shows several cases where light from the master laser was injected into one of the strong modes of the slave laser and one case where the light master laser was injected onto one of the quenched modes of the slave laser. For all cases, the spectra with light injection are nearly single mode. The slave laser was pulsed at 15 MHz with a 5-nsec pulse width. The k values with and without a light injection were measured by the low-pass filter methods.

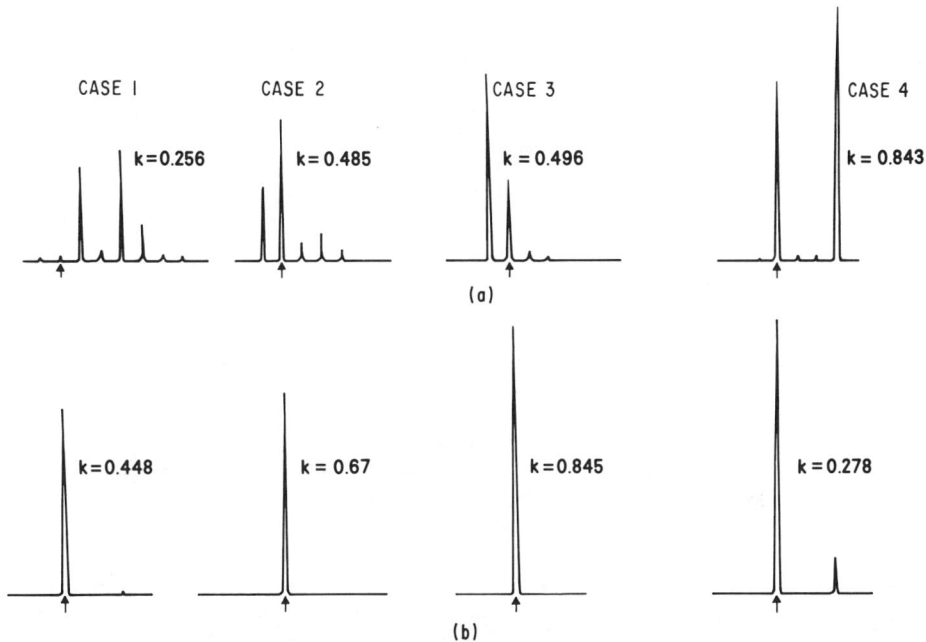

FIG. 16. Spectra (a) without and (b) with injection. The cases 1–3 show the strong injection locking, yielding nearly single-mode operation of the pulsed slave laser. Case 4 shows weak injection locking, which suppresses mode partitioning but does not result in single-mode operation of the slave laser.

Table I shows the measured results. When optical injection is sufficiently effective to achieve the nearly single-mode operation (cases 1–3, Fig. 16), k is much larger than that without injection. Table I also shows the expected k value from k_{dc} and k_{pc} using Eq. (45). The theoretical values of k_{in} and the measured results show good agreement. However, in case 4 in Fig. 16, k is smaller with light injection; however, the spectrum is not single mode.

Optical injection can produce nearly single-mode operation in the slave laser. However, the degree of fluctuation in terms of k value are not improved. This phenomenon can be explained as follows. Since the laser diode in dc operation still fluctuates, there is a probability that $a_{dc} < \alpha$ for some time interval. During that time, the slave laser becomes free-running rather than oscillating in the injected mode. This causes mode jumps, which results in large mode partitioning.

The suppression of mode partitioning was observed, as indicated in Fig. 16, case 4. In this case, there is considerable power in one side mode, which

TABLE I

Effects on Mode Partitioning By Optical Injection Locking

Case	k Value Measured Without injection	k Value Measured With injection	Theoretical with injection
1	0.256	0.448	0.7[a]
2	0.485	0.67	0.7[b]
3	0.496	0.845	0.72[b]
4	0.843	0.278	0.31

[a] $k_{dc} = 0.3$.
[b] $k_{dc} = 0.5$.

indicates weak injection locking. With weak injection locking, the assumption that $\overline{a}_0 = \overline{a}_{dc}$ used in Eq. (47) no longer holds. For this case, Eq. (47) becomes

$$k_{in}^2 = (k_{dc}^2 + k_p^2 \delta + \varepsilon)/(1 + \delta + \varepsilon), \tag{48}$$

where

$$\delta = \frac{(\overline{a}_p - (\overline{a}_p)^2)(1 - \overline{a}_{dc})}{\overline{a}_0 \overline{a}_{dc}(1 - \overline{a}_0)}, \quad \varepsilon = \frac{(1 - \overline{a}_{dc})(\overline{a}_0 - \overline{a}_p)^2}{(\overline{a}_0)(1 - \overline{a}_0)}.$$

Since $\overline{a}_{dc} \simeq 1$, we obtain

$$k_{in}^2 \cong k_{dc}^2. \tag{49}$$

So, under weak injection locking, we can expect reduced mode partitioning but not a narrowing of the spectrum.

The result is very simple. To suppress the mode partitioning, it is necessary to use a quiet master laser (small k). A great deal of study has been done on the distributed feedback (DFB), distributed Bragg reflector (DBR), and multicavity [such as the cleaved coupled cavity (C^3)] lasers to obtain a mode partition noise-free laser.

Appendix I

To calculate the k value without mode interactions, we assume the continuous spectrum given by Eq. (30). Thus, k^2 is expressed by

$$k^2 = \sum P_{j0}^2$$

$$= \frac{2}{\sqrt{2}\pi\sigma_s} \int_{\lambda'=\lambda} \int_{\Delta\lambda N/2}^{\infty} \exp\left[-\frac{\lambda^2}{2\sigma_s^2}\right] \exp\left[-\frac{\lambda'^2}{2\sigma_s^2}\right] d\lambda\, d\lambda'$$

$$= \frac{1}{\sqrt{2}\pi\sigma_s} \int_0^{2\pi} d\theta \left(\delta = \frac{\pi}{4}\right) + \delta\left(\theta - \frac{3}{4}\pi\right) \int_{\Delta\lambda N/\sqrt{2}}^{\infty} \exp\left[-\frac{(\gamma^2)}{2\sigma_s^2}\right] \gamma\, d\gamma$$

$$= \frac{1}{\pi} \exp\left[-\left(\frac{\Delta\lambda N}{2\sigma_s}\right)^2\right], \tag{AI.1}$$

where N is the number of stimulated emissions, and $\Delta\lambda$ is the wavelength separation of two modes.

Appendix II

For the case when lasing modes interact with one another, we consider another redistribution process among lasing levels described by

$$\frac{\partial P_{i'}}{\partial N} = \sum_k C_{ki} P_k, \tag{AII.1}$$

where $C_{ki'} = 1$ or 0, with the probability of $q_{ki'}$ or $1 - q_{ki'}$, respectively. Combining the results of Eqs. (25) and (AII.1), the expectation and the second moments of the lasing modes i, which have no mode interaction, and the lasing modes i, with mode interaction, are obtained. The lasing mode i with no mode interaction has the same expression as Eq. (27).

For the lasing mode with mode interactions,

$$\overline{P_{i'}} = \left\{ \sum_k \left[\sum_j P_{j0}(1-d) q_{jk} + P_{k0} \right] \right\} \frac{q_{ji'}}{\sum_k q_{jk}},$$

$$\overline{P_{i'}^2} = \left\{ \sum_k \left[\sum_j P_{j0}^2 (1-\alpha)^2 q_{j'k} \right] \right\} \frac{q_{ji'}}{\sum_k q_{jk}}$$

$$+ 2 \left\{ \sum_k \left[\sum_j \sum_{j'>j} P_{j0} P_{j'0} (1-\alpha)^2 q_{jk} q_{j'k} \right] \frac{q_{j'k}}{\sum_k q_{jk}} \right\}$$

$$+ 4 \left\{ \sum_k \left[\sum_{k'>k} \sum_j \sum_{j'>k} P_{j0} P_{j'0} (1-\alpha)^2 q_{jk} q_{k'k'} \right] \frac{q_{ji'}}{\sum_k q_{jk}} \right\}$$

$$+ \left\{ \left[\sum_k P_{k0} \right]^2 \frac{q_{ji'}}{\sum_k q_{jk}} \right\} \quad \text{(AII.2)}$$

$$+ 2 \left[\sum_k P_{k0} \right] \left[\sum_k \sum_k P_{j0}(1-\alpha) q_{jk} \right] \frac{q_{ji'}}{\sum_k q_{jk}},$$

$$\overline{\sigma_{i'}^2} = \sum_j P_{j0}^2 (1-\alpha)^2 [q_{ji'} - q_{ji'}^2]$$

$$+ 2 \sum_k \sum_{k' > k} \left[2 \sum_j \sum_{j' > j} P_{j0} P_{j'0} (1-\alpha)^2 q_{jk} q_{j'k'} \right] \frac{q_{jk'}}{\sum_k q_{jk}}$$

$$+ 2 \sum_j P_{j0}(1-\alpha) \sum_k P_{k0} [q_{ji'} - q_{ji'}^2]$$

$$+ \left[\sum P_{j0} \right]^2 \left[\frac{q_{ji'}}{\sum_k q_{jk}} - \frac{(q_{ji'})^2}{\left[\sum_k q_{jk} \right]^2} \right],$$

where we used the assumption $q_{ji} = q_{j'i} = q_{ki}$ for any j, j', or k.

References

Ito, T., Machida, S., Nawata, K., and Ikegami, T. (1977). *IEEE J. Quantum Electron.* **QE-13,** 574–579.

Kobayashi, S., and Kimura, T. (1980). *IEEE J. Quantum Electron.* **QE-16,** 915–917.

Ogawa, K. (1982). *IEEE J. Quantum Electron.* **QE-17,** 849–855.

Ogawa, K., Vodhanel, R. S. (1982). *IEEE J. Quantum Electron.* **QE-18,** 1090–1093.

Okano, Y., Nakagawa, K., and Ito, T. (1980). *IEEE Trans. Commun.* **COM-28,** 238–243.

Liu, Pao-Lo, Fencil, L. E., Ko, J.-S., Kaminow, I. P., Lee, T. P., and Burrus, C. A. (1983). *IEEE J. Quantum Electron.* **QE-19,** 1348–1351.

Yamada, J., Kobayashi, S., Machida, S., and Kimura, T. (1980). *Jpn. J. Appl. Phys.* **19**(11), L689–L692.

Collins, G. B. (1948). "MIT Radiation Laboratory Series 6: Microwave Magnetrons." McGraw-Hill, New York.

Yamamoto, S., Kasaguchi, H., and Seki, N. (1980). *Tech. Meet. Rep. Commun. Sys. IEEE* **CS80-144**(10), 61–66.

Index

A

Absorption, 11
Activation energy, 247, 260, 264, 268, 274
Anti-index guided lasers, 71
Astigmatism, 72
Auger recombination, 14, 16, 23, 51, 94
Avalanche photodiode, 287
Average power drift, 163

B

Band structure, 3, 20
  InGaAsP, 3, 20
Beam-direction shift, 62
Beam waist, 72
Bit-rate–distance product, 308
Bragg wavelength, 83
Buried-heterostructure laser, 35, 244, 248
Burn-in, 241, 253
Burrus structure, 243, 248, 264

C

Carrier density, 7
Carrier injection efficiency, 142
Carrier leakage, 14, 209
  lifetime, 204, 206
Chromatic dispersion, 285, 288, 292
Coherence, 73
Coupled-cavity lasers, 82
Coupling, to fiber, 221
  direct, 221
  with lens, 225
Cumulative failures, 241
Current confinement, 47
Current leakage, 46
  spreading, 26
Cut-off condition, 41

D

Dark-line defect, 241, 249, 258, 266, 272
Dark-spot defect, 268, 269
Density matrix formalism, 76
Diffusive leakage current, 7, 15
Dislocation, 241, 250, 253, 255, 259, 272, 273
  temperature dependence, 252
Distributed feedback (DFB) laser, 83
Dual-wavelength photodetector, 296
DX center, 78

E

Edge emitter, 196, 249, 268, 286, 292
Effective refractive index approximation, 67
Electrical derivative characterization, 49
Electrothermomigration, 270
Error rate, 307
External quantum efficiency, 128
Extinction degradation, 166
  ratio, 163

F

Facet degradation, 69
Far-field patterns, 69
Freak behavior, 265
Free-carrier absorption, 94

G

Gain, 9, 64
  coefficient, 64
Gain guiding, 27
Gaussian distribution, 305
Graded-index laser, 284
Grating feedback, 83

H

Heterodyne communications, 75
Heterojunction laser, 4
 buried-heterostructure, 35, 41, 47, 55, 67, 321
 gain-guided, 25, 71
 index-guided, 33, 67
 InGaAsP, 3
 proton-bombarded stripe, 158, 321
 ridge-waveguide, 36, 67
Heterojunction, 5–9
Homogeneous broadening, 73
Hot-wall vapor deposition, 116, 118

I

Index-change ratio, 65, 74
Index-guided lasers, 33
Infant failure, 240, 266
Interfacial recombination velocity, 8, 128, 199
Intermodal dispersion, 283
Intermodal pulse spreading, 284
Internal quantum efficiency, 142
Intervalence-band absorption, 14, 16
Intraband carrier relaxation time, 73

K

Kink, 31, 43, 62

L

Laser material, for $\lambda > 2~\mu m$
 AlInAsSb, 96
 $Hg_xCd_{1-x}Te$, 102
 $InAs_{1-x-y}P_xSb_y$, 97
 $In_{1-x}Ga_xAs_ySb_{1-y}$, 96
 PbCdS, 98
 PbGeS, 98
 PbEuSeTe, 99
 $Pb_{1-x}Sn_xSe$, 98
 PbSnSeTe, 100
 PbSnSSe, 100
 $Pb_{1-x}Sn_xTe$, 98
 PbSSeTe, 100
Lattice-matched heterostructure, 94

LED, see Light-emitting diode
Light–current characteristics, 69, 160, 213
Light-emitting diode, 194, 239
 failure rate, 240
 power, 198
Linewidth, 73
Liquid-phase epitaxy
 InAsPSb, 107
 InGaAsSb, 104
 PbSnSeTe, 114
 PbSnTe, 112
Log-normal failure distribution, 241

M

Mean time to failure, 241, 242, 260, 264, 267, 268
Metallo-organic compounds, 103
Miscibility gaps, 95, 104
Mode partition coefficient, 311
 partition noise, 300, 322
Mode-competition noise, 81
Mode-hopping noise, 79
Modulation bandwidth, 206
 characteristics, 33, 55
Molecular beam epitaxy, 103, 117, 169
MTTF, see Mean time to failure

N

Nonlinearity, 31

O

Optical feedback, 79
Optical injection locking, 324

P

Parasitics, 207
Phase diagram
 IV–VI compound semiconductor, 110
 InAsPSb, 107
 InGaAsSb, 106
 PbSnTe, 112
Point defect, 262

Power versus bandwidth, 220
Profile dispersion, 284

R

Radiance saturation, 214
Relative intensity noise, 78

S

Self-absorption, 199
Self-focusing effect, 65
Source-dissolution–successive-growth technique, 104
Space communication, 184
Spectral characteristics, 53
Spectral narrowing, 73
Spectral peak shift, 73
Stacking faults, 259
Stimulated emission, 9
Stripe geometry, 62
Superluminescent diode, 196
Surface emitter, 194, 240

T

Tap-controlled transmitter, 179
Temperature dependence, 12
  T_0, 14
  threshold current, 12
Thermal impedance, 243
Threshold, 7, 11, 29
  current density, 11, 29, 43
Transmission bandwidth, 283
Transmission cathodoluminescence, 259
Transmission system, 239
Transmitter functional lifetime, 156, 170
  temperature dependence, 175
Transverse-mode control, 62
Transverse-mode instability, 62
Tunneling current, 94

W

Waveguiding, 37
  confinement factor, 97, 124
  cut-off condition, 41
Wavelength division multiplexing, 294, 296

Contents of Volume 22

Part A

Kazuo Nakajima, The Liquid-Phase Epitaxial Growth of InGaAsP
W. T. Tsang, Molecular Beam Epitaxy for III–V Compound Semiconductors
G. B. Stringfellow, Organometallic Vapor-Phase Epitaxial Growth of III–V Semiconductors
G. Beuchet, Halide and Chloride Transport Vapor-Phase Deposition of InGaAsP and GaAs
Manijeh Razeghi, Low-Pressure Metallo-Organic Chemical Vapor Deposition of $Ga_xIn_{1-x}As_yP_{1-y}$ Alloys
P. M. Petroff, Defects in III–V Compound Semiconductors

Part B

J. P. van der Ziel, Mode Locking of Semiconductor Lasers
Kam Y. Lau and Amnon Yariv, High-Frequency Current Modulation of Semiconductor Injection Lasers
Charles H. Henry, Spectral Properties of Semiconductor Lasers
Yasuharu Suematsu, Katsumi Kishino, Shigehisa Arai, and Fumio Koyama, Dynamic Single-Mode Semiconductor Lasers with a Distributed Reflector
W. T. Tsang, The Cleaved-Coupled-Cavity (C^3) Laser

Part C

R. J. Nelson and N. K. Dutta, Review of InGaAsP/InP Laser Structures and Comparison of Their Performance
N. Chinone and M. Nakamura, Mode-Stabilized Semiconductor Lasers for 0.7–0.8- and 1.1–1.6-μm Regions
Yoshiji Horikoshi, Semiconductor Lasers with Wavelengths Exceeding 2 μm
B. A. Dean and M. Dixon, The Functional Reliability of Semiconductor Lasers as Optical Transmitters
R. H. Saul, T. P. Lee, and C. A. Burrus, Light-Emitting Device Design
C. L. Zipfel, Light-Emitting Diode Reliability
Tien Pei Lee and Tingye Li, LED-Based Multimode Lightwave Systems
Kinichiro Ogawa, Semiconductor Noise-Mode Partition Noise

Part D

Federico Capasso, The Physics of Avalanche Photodiodes
T. P. Pearsall and M. A. Pollack, Compound Semiconductor Photodiodes
Takao Kaneda, Silicon and Germanium Avalanche Photodiodes
S. R. Forrest, Sensitivity of Avalanche Photodetector Receivers for High-Bit-Rate Long-Wavelength Optical Communication Systems
J. C. Campbell, Phototransistors for Lightwave Communications

Part E

Shyh Wang, Principles and Characteristics of Integratable Active and Passive Optical Devices
Shlomo Margalit and Amnon Yariv, Integrated Electronic and Photonic Devices
Takaaki Mukai, Yoshihisa Yamamoto, and Tatsuya Kimura, Optical Amplification by Semiconductor Lasers

Contents of Previous Volumes

Volume 1 Physics of III–V Compounds

C. Hilsum, Some Key Features of III–V Compounds
Franco Bassani, Methods of Band Calculations Applicable to III–V Compounds
E. O. Kane, The $k \cdot p$ Method
V. L. Bonch-Bruevich, Effect of Heavy Doping on the Semiconductor Band Structure
Donald Long, Energy Band Structures of Mixed Crystals of III–V Compounds
Laura M. Roth and Petros N. Argyres, Magnetic Quantum Effects
S. M. Puri and T. H. Geballe, Thermomagnetic Effects in the Quantum Region
W. M. Becker, Band Characteristics near Principal Minima from Magnetoresistance
E. H. Putley, Freeze-Out Effects, Hot Electron Effects, and Submillimeter Photoconductivity in InSb
H. Weiss, Magnetoresistance
Betsy Ancker-Johnson, Plasmas in Semiconductors and Semimetals

Volume 2 Physics of III–V Compounds

M. G. Holland, Thermal Conductivity
S. I. Novkova, Thermal Expansion
U. Piesbergen, Heat Capacity and Debye Temperatures
G. Giesecke, Lattice Constants
J. R. Drabble, Elastic Properties
A. U. Mac Rae and G. W. Gobeli, Low Energy Electron Diffraction Studies
Robert Lee Mieher, Nuclear Magnetic Resonance
Bernard Goldstein, Electron Paramagnetic Resonance
T. S. Moss, Photoconduction in III–V Compounds
E. Antončik and J. Tauc, Quantum Efficiency of the Internal Photoelectric Effect in InSb
G. W. Gobeli and F. G. Allen, Photoelectric Threshold and Work Function
P. S. Pershan, Nonlinear Optics in III–V Compounds
M. Gershenzon, Radiative Recombination in the III–V Compounds
Frank Stern, Stimulated Emission in Semiconductors

Volume 3 Optical of Properties III–V Compounds

Marvin Hass, Lattice Reflection
William G. Spitzer, Multiphonon Lattice Absorption
D. L. Stierwalt and R. F. Potter, Emittance Studies
H. R. Philipp and H. Ehrenreich, Ultraviolet Optical Properties
Manuel Cardona, Optical Absorption above the Fundamental Edge
Earnest J. Johnson, Absorption near the Fundamental Edge
John O. Dimmock, Introduction to the Theory of Exciton States in Semiconductors
B. Lax and J. G. Mavroides, Interband Magnetooptical Effects
H. Y. Fan, Effects of Free Carriers on Optical Properties

Edward D. Palik and George B. Wright, Free-Carrier Magnetooptical Effects
Richard H. Bube, Photoelectronic Analysis
B. O. Seraphin and H. E. Bennett, Optical Constants

Volume 4 Physics of III–V Compounds

N. A. Goryunova, A. S. Borschevskii, and D. N. Tretiakov, Hardness
N. N. Sirota, Heats of Formation and Temperatures and Heats of Fusion of Compounds $A^{III}B^V$
Don L. Kendall, Diffusion
A. G. Chynoweth, Charge Multiplication Phenomena
Robert W. Keyes, The Effects of Hydrostatic Pressure on the Properties of III–V Semiconductors
L. W. Aukerman, Radiation Effects
N. A. Goryunova, F. P. Kesamanly, and D. N. Nasledov, Phenomena in Solid Solutions
R. T. Bate, Electrical Properties of Nonuniform Crystals

Volume 5 Infrared Detectors

Henry Levinstein, Characterization of Infrared Detectors
Paul W. Kruse, Indium Antimonide Photoconductive and Photoelectromagnetic Detectors
M. B. Prince, Narrowband Self-Filtering Detectors
Ivars Melngailis and T. C. Harman, Single-Crystal Lead–Tin Chalcogenides
Donald Long and Joseph L. Schmit, Mercury–Cadmium Telluride and Closely Related Alloys
E. H. Putley, The Pyroelectric Detector
Norman B. Stevens, Radiation Thermopiles
R. J. Keyes and T. M. Quist, Low Level Coherent and Incoherent Detection in the Infrared
M. C. Teich, Coherent Detection in the Infrared
F. R. Arams, E. W. Sard, B. J. Peyton, and F. P. Pace, Infrared Heterodyne Detection with Gigahertz IF Response
H. S. Sommers, Jr., Microwave-Based Photoconductive Detector
Robert Sehr and Rainer Zuleeg, Imaging and Display

Volume 6 Injection Phenomena

Murray A. Lampert and Ronald B. Schilling, Current Injection in Solids: The Regional Approximation Method
Richard Williams, Injection by Internal Photoemission
Allen M. Barnett, Current Filament Formation
R. Baron and J. W. Mayer, Double Injection in Semiconductors
W. Ruppel, The Photoconductor–Metal Contact

Volume 7 Application and Devices: Part A

John A. Copeland and Stephen Knight, Applications Utilizing Bulk Negative Resistance
F. A. Padovani, The Voltage–Current Characteristics of Metal–Semiconductor Contacts
P. L. Hower, W. W. Hooper, B. R. Cairns, R. D. Fairman, and D. A. Tremere, The GaAs Field-Effect Transistor
Marvin H. White, MOS Transistors
G. R. Antell, Gallium Arsenide Transistors
T. L. Tansley, Heterojunction Properties

Volume 7 Application and Devices: Part B

T. Misawa, IMPATT Diodes
H. C. Okean, Tunnel Diodes
Robert B. Campbell and Hung-Chi Chang, Silicon Carbide Junction Devices
R. E. Enstrom, H. Kressel, and L. Krassner, High-Temperature Power Rectifiers of $GaAs_{1-x}P_x$

Volume 8 Transport and Optical Phenomena

Richard J. Stirn, Band Structure and Galvanomagnetic Effects in III–V Compounds with Indirect Band Gaps
Roland W. Ure, Jr., Thermoelectric Effects in III–V Compounds
Herbert Piller, Faraday Rotation
H. Barry Bebb and E. W. Williams, Photoluminescence I: Theory
E. W. Williams and H. Barry Bebb, Photoluminescence II: Gallium Arsenide

Volume 9 Modulation Techniques

B. O. Seraphin, Electroreflectance
R. L. Aggarwal, Modulated Interband Magnetooptics
Daniel F. Blossey and Paul Handler, Electroabsorption
Bruno Batz, Thermal and Wavelength Modulation Spectroscopy
Ivar Balslev, Piezooptical Effects
D. E. Aspnes and N. Bottka, Electric-Field Effects on the Dielectric Function of Semiconductors and Insulators

Volume 10 Transport Phenomena

R. L. Rode, Low-Field Electron Transport
J. D. Wiley, Mobility of Holes in III–V Compounds
C. M. Wolfe and G. E. Stillman, Apparent Mobility Enhancement in Inhomogeneous Crystals
Robert L. Peterson, The Magnetophonon Effect

Volume 11 Solar Cells

Harold J. Hovel, Introduction; Carrier Collection, Spectral Response, and Photocurrent; Solar Cell Electrical Characteristics; Efficiency; Thickness; Other Solar Cell Devices; Radiation Effects; Temperature and Intensity; Solar Cell Technology

Volume 12 Infrared Detectors (II)

W. L. Eiseman, J. D. Merriam, and R. F. Potter, Operational Characteristics of Infrared Photodetectors
Peter R. Bratt, Impurity Germanium and Silicon Infrared Detectors
E. H. Putley, InSb Submillimeter Photoconductive Detectors
G. E. Stillman, C. M. Wolfe, and J. O. Dimmock, Far-Infrared Photoconductivity in High Purity GaAs
G. E. Stillman and C. M. Wolfe, Avalanche Photodiodes
P. L. Richards, The Josephson Junction as a Detector of Microwave and Far-Infrared Radiation
E. H. Putley, The Pyroelectric Detector — An Update

Volume 13 Cadmium Telluride
Kenneth Zanio, Materials Preparation; Physics; Defects; Applications

Volume 14 Lasers, Junctions, Transport
N. Holonyak, Jr. and M. H. Lee, Photopumped III–V Semiconductor Lasers
Henry Kressel and Jerome K. Butler, Heterojunction Laser Diodes
A. Van der Ziel, Space-Charge-Limited Solid-State Diodes
Peter J. Price, Monte Carlo Calculation of Electron Transport in Solids

Volume 15 Contacts, Junctions, Emitters
B. L. Sharma, Ohmic Contacts to III–V Compound Semiconductors
Allen Nussbaum, The Theory of Semiconducting Junctions
John S. Escher, NEA Semiconductor Photoemitters

Volume 16 Defects, (HgCd)Se, (HgCd)Te
Henry Kressel, The Effect of Crystal Defects on Optoelectronic Devices
C. R. Whitsett, J. G. Broerman, and C. J. Summers, Crystal Growth and Properties of $Hg_{1-x}Cd_x$Se Alloys
M. H. Weiler, Magnetooptical Properties of $Hg_{1-x}Cd_x$Te Alloys
Paul W. Kruse and John G. Ready, Nonlinear Optical Effects in $Hg_{1-x}Cd_x$Te

Volume 17 CW Beam Processing of Silicon and Other Semiconductors
J. F. Gibbons, Beam Processing of Silicon
Arto Lietoila, Richard B. Gold, James F. Gibbons, and Lee A. Christel, Temperature Distributions and Solid Phase Reaction Rates Produced by Scanning CW Beams
Arto Lietoila, James F. Gibbons, Applications of CW Beam Processing to Ion Implanted Crystalline Silicon
N. M. Johnson, Electronic Defects in CW Transient Thermal Processed Silicon
K. F. Lee, T. J. Stultz, and J. F. Gibbons, Beam Recrystallized Polycrystalline Silicon: Properties, Applications, and Techniques
T. Shibata, A. Wakita, T. W. Sigmon, and J. F. Gibbons, Metal–Silicon Reactions and Silicide Formation
Yves I. Nissim and James F. Gibbons, CW Beam Processing of Gallium Aresenide

Volume 18 Mercury Cadmium Telluride
Paul W. Kruse, The Emergence of $(Hg_{1-x}Cd_x)$Te as a Modern Infrared Sensitive Material
H. E. Hirsch, S. C. Liang, and A. G. White, Preparation of High-Purity Cadmium, Mercury, and Tellurium
W. F. H. Micklethwaite, The Crystal Growth of Cadmium Mercury Telluride
Paul E. Petersen, Auger Recombination in Mercury Cadmium Telluride
R. M. Broudy and V. J. Mazurczyck, (HgCd)Te Photoconductive Detectors
M. B. Reine, A. K. Sood, and T. J. Tredwell, Photovoltaic Infrared Detectors
M. A. Kinch, Metal-Insulator-Semiconductor Infrared Detectors

Volume 19 Deep Levels, GaAs, Alloys, Photochemistry

G. F. Neumark and K. Kosai, Deep Levels in Wide Band-Gap III–V Semiconductors
David C. Look, The Electrical and Photoelectronic Properties of Semi-Insulating GaAs
R. F. Brebrick, Ching-Hua Su, and Pok-Kai Liao, Associated Solution Model for Ga–In–Sb and Hg–Cd–Te
Yu. Ya. Gurevich and Yu. V. Pleskov, Photoelectrochemistry of Semiconductors

Volume 20 Semi-Insulating GaAs

R. N. Thomas, H. M. Hobgood, G. W. Eldridge, D. L. Barrett, T. T. Braggins, L. B. Ta, and S. K. Wang, High-Purity LEC Growth and Direct Implantation of GaAs for Monolithic Microwave circuits
C. A. Stolte, Ion Implantation and Materials for GaAs Integrated Circuits
C. G. Kirkpatrick, R. T. Chen, D. E. Holmes, P. M. Asbeck, K. R. Elliott, R. D. Fairman, and J. R. Oliver, LEC GaAs for Integrated Circuit Applications
J. S. Blakemore and S. Rahimi, Models for Mid-Gap Centers in Gallium Arsenide

Volume 21 Hydrogenated Amorphous Silicon: Part A

Jacques I. Pankove, Introduction
Masataka Hirose, Glow Discharge; Chemical Vapor Deposition
Yoshiyuki Uchida, dc Glow Discharge
T. D. Moustakas, Sputtering
Isao Yamada, Ionized-Cluster Beam Deposition
Bruce A. Scott, Homogeneous Chemical Vapor Deposition
Frank J. Kampas, Chemical Reactions in Plasma Deposition
Paul A. Longeway, Plasma Kinetics
Herbert A. Weakliem, Diagnostics of Silane Glow Discharges Using Probes and Mass Spectroscopy
Lester Guttman, Relation between the Atomic and the Electronic Structures
A. Chenevas-Paule, Experimental Determination of Structure
S. Minomura, Pressure Effects on the Local Atomic Structure
David Adler, Defects and Density of Localized States

Volume 21 Hydrogenated Amorphous Silicon: Part B

Jacques I. Pankove, Introduction
G. D. Cody, The Optical Absorption Edge of a-Si:H
Nabil M. Amer and Warren B. Jackson, Optical Properties of Defect States in a-Si:H
P. J. Zanzucchi, The Vibrational Spectra of a-Si:H
Yoshihiro Hamakawa, Electroreflectance and Electroabsorption
Jeffrey S. Lannin, Raman Scattering of Amorphous Si, Ge, and Their Alloys
R. A. Street, Luminescence in a-Si:H
Richard S. Crandall, Photoconductivity
J. Tauc, Time-Resolved Spectroscopy of Electronic Relaxation Processes
P. E. Vanier, IR-Induced Quenching and Enhancement of Photoconductivity and Photoluminescence
H. Schade, Irradiation-Induced Metastable Effects
L. Ley, Photelectron Emission Studies

Volume 21   Hydrogenated Amorphous Silicon: Part C

Jacques I. Pankove, Introduction
J. David Cohen, Density of States from Junction Measurements in Hydrogenated Amorphous Silicon
P. C. Taylor, Magnetic Resonance Measurements in a-Si:H
K. Morigaki, Optically Detected Magnetic Resonance
J. Dresner, Carrier Mobility in a-S:H
T. Tiedje, Information about Band-Tail States from Time-of-Flight Experiments
Arnold R. Moore, Diffusion Length in Undoped a-S:H
W. Beyer and H. Overhof, Doping Effects in a-Si:H
H. Fritzsche, Electronic Properties of Surfaces in a-Si:H
C. R. Wronski, The Staebler–Wronski Effect
R. J. Nemanich, Schottky Barriers on a-Si:H
B. Abeles and T. Tiedje, Amorphous Semiconductor Superlattices

Volume 21   Hydrogenated Amorphous Silicon: Part D

Jacques I. Pankove, Introduction, Hybrid Structures
D. E. Carlson, Solar Cells
G. A. Swartz, Closed-Form Solution of $I-V$ Characteristic for a-Si:H Solar Cells
Isamu Shimizu, Electrophotography
Sachio Ishioka, Image Pickup Tubes
P. G. LeComber and W. E. Spear, The Development of the a-Si:H Field-Effect Transistor and Its Possible Applications
D. G. Ast, a-Si:H FET-Addressed LCD Panel
S. Kaneko, Solid-State Image Sensor
Masakiyo Matsumura, Charge-Coupled Devices
M. A. Bosch, Optical Recording
A. D'Amico and G. Fortunato, Ambient Sensors
Hiroshi Kukimoto, Amorphous Light-Emitting Devices
Robert J. Phelan, Jr., Fast Detectors and Modulators
P. G. LeComber, A. E. Owen, W. E. Spear, J. Hajto, and W. K. Choi, Electronic Switching in Amorphous Silicon Junction Devices

AUG 0 1 1985